WASTEWATER SOLIDS INCINERATION SYSTEMS

Prepared by the **Incineration Task Force** of the **Water Environment Federation**

James E. Welp, P.E., *Chair*
Peter Brady, B.E., *Vice-Chair*

Antonio V. Almeida
Allen Baturay, P.E.
Timothy L. Bauer
Bob Bower
N. Kelly Brown
Peter Burrowes
Daniel L. Busch
S. Rao Chitikela, Ph.D., P.E., BCEE
David W. Cooley, P.E.
Brian Copeland, P.E.
Frank A. Dachille, P.E.
Ky Dangtran, Ph.D.
Gregory Brian David
Robert P. Dominak, P.E.
Stephen Greenwood, P.E.
David Gregory
Scott E. Harder, P.E.
Michael W. Heitz, P.E.
Webster F. Hoener, P.E.
Greg G. Homoki, P.E.
Robert L. Lamal
Robert M. Lantz, P.E., DEE
Craig Lawniczak
F. Michael Lewis
Lee A. Lundberg, P.E.
Michael Moore
Tom Nitka
Eugenio Perez
Raymond C. Porter
Lee Potts, Ph.D.
James M. Rowan, P.E.
Alan B. Rubin, Ph.D.
Frank C. Sapienza, P.E.
Ken Stevens
James E. Welp
Ben C. Wester
Robert E. Williamson

Under the Direction of the **Residuals Subcommittee** of the **Technical Practice Committee**

2009

Water Environment Federation
601 Wythe Street
Alexandria, VA 22314-1994 USA
http://www.wef.org

WASTEWATER SOLIDS INCINERATION SYSTEMS

WEF Manual of Practice No. 30

Prepared by the Incineration Task Force of the Water Environment Federation

Wastewater Solids Incineration Systems, MOP 30

ISBN: 978-1-57278-389-8

IMPORTANT NOTICE

The material presented in this publication has been prepared in accordance with generally recognized engineering principles and practices and is for general information only. This information should not be used without first securing competent advice with respect to its suitability for any general or specific application.

The contents of this publication are not intended to be a standard of the Water Environment Federation (WEF) and are not intended for use as a reference in purchase specifications, contracts, regulations, statutes, or any other legal document.

No reference made in this publication to any specific method, product, process, or service constitutes or implies an endorsement, recommendation, or warranty thereof by WEF.

WEF makes no representation or warranty of any kind, whether expressed or implied, concerning the accuracy, product, or process discussed in this publication and assumes no liability.

Anyone using this information assumes all liability arising from such use, including but not limited to infringement of any patent or patents.

About WEF

Formed in 1928, the Water Environment Federation (WEF) is a not-for-profit technical and educational organization with 35,000 individual members and 81 affiliated member associations representing an additional 50,000 water quality professionals throughout the world. WEF and its member associations proudly work to achieve our mission of preserving and enhancing the global water environment.

For information on membership, publications, and conferences, contact

Water Environment Federation
601 Wythe Street
Alexandria, VA 22314-1994 USA
(703) 684-2400
http://www.wef.org

Manuals of Practice of the Water Environment Federation

The WEF Technical Practice Committee (formerly the Committee on Sewage and Industrial Wastes Practice of the Federation of Sewage and Industrial Wastes Associations) was created by the Federation Board of Control on October 11, 1941. The primary function of the committee is to originate and produce, through appropriate subcommittees, special publications dealing with technical aspects of the broad interests of the federation. These publications are intended to provide background information through a review of technical practices and detailed procedures that research and experience have shown to be functional and practical.

Contents

Chapter 5 Combustion Technology

Chapter 6 Heat Recovery and Reuse

Chapter 7 Emissions Control and Monitoring

Chapter 8 Ash Handling and Recycling

Chapter 9 Instrumentation and Control

Chapter 10 Incinerator Operations

Chapter 11 Incinerator Maintenance

Appendix A Combustion Fundamentals

Appendix B Incineration Subsystems

Appendix C Determination of Wastewater Solids Incineration Related Costs

Appendix D Case Studies

Preface

Information from *Incineration* (MOP OM-11), published in 1988, and *Sludge Incineration: Thermal Destruction of Residues* (MOP FD-19), published in 1992, was revised and combined to create *Wastewater Solids Incineration Systems* (MOP 30). This manual is a comprehensive guide to the safety, permitting, design, operation, and maintenance of the incineration, or thermal oxidation, process and includes new sections about safety, permitting, and instrument and control systems. The combustion theory and technology sections have been updated and case studies of recent new and updated facilities are presented. Sections on heat recovery and reuse, emission control and monitoring, and ash handling and recycling have been expanded and upgraded.

This Manual of Practice was produced under the direction of James E. Welp, P.E., *Chair*. The principal authors of this Manual of Practice are as follows:

Chapter 1	James E. Welp, P.E.
Chapter 2	James M. Rowan, P.E.
Chapter 3	Raymond C. Porter
Chapter 4	F. Michael Lewis
Chapter 5	Peter Burrowes
Chapter 6	Lee A. Lundberg, P.E.
Chapter 7	Frank C. Sapienza, P.E.
Chapter 8	David W. Cooley, P.E.
Chapter 9	Michael W. Heitz, P.E.
Chapter 10	Robert P. Dominak, P.E.
Chapter 11	Robert P. Dominak, P.E.
Appendix A	F. Michael Lewis
Appendix B	James E. Welp, P.E.
Appendix C	Robert P. Dominak, P.E.
Appendix D	Peter Brady, B.E.
Appendix E	Peter Brady, B.E.
Appendix F	Ky Dangtran, Ph.D.

Contributing authors are as follows: Timothy L. Bauer, P.E. (Chapter 8); John Borghesi, P.E. (Chapter 8); Ky Dangtran, Ph.D. (Chapter 5); Scott E. Harder, P.E. (Appendix C); Webster Hoener, P.E. (Chapter 2); Greg G. Homoki, P.E. (Chapter 6); Lee A. Lundberg, P.E. (Chapter 4); and Robert L. Paulson (Chapter 5).

Authors' and reviewers' efforts were supported by the following organizations:

Alpine Technology, Inc., Austin, Texas
Alstom Power, Inc., Energy Recovery Systems, Wexford, Pennsylvania
BDP Industries, Greenwich, New York
Black & Veatch, Chicago, Illinois; Cincinnati, Ohio; Kansas City, Missouri; Nashville, Tennessee
Carlson Associates, Catharpin, Virginia
CDM, Inc., Cambridge, Massachusetts
CH2M Hill, Inc., Corvallis, Oregon; Boston, Massachusetts; Kitchener, Ontario, Canada; Mendota Heights, Minnesota
City of Kansas City Water Services Department, Kansas City, Missouri
Degremont Technologies Infilco, Richmond, Virginia
Earth Tech U.K.
Environmental Financial Group, Inc., Minneapolis, Minnesota
ENVIROSTRATEGIES, LLC, Oakton, Virginia
ERM, Exton, Pennsylvania
F. Michael Lewis, Inc., El Segundo, California
Green Bay Metropolitan Sewerage District, Green Bay, Wisconsin
Hampton Roads Sanitation District, Newport News, Virginia
HDR Engineering, Seattle, Washington
Malcolm Pirnie, Cleveland, Ohio
Metropolitan Sewer District of Greater Cincinnati, Ohio
Minergy, Neenah, Wisconsin
Northeast Ohio Regional Sewer District, Cleveland, Ohio
Veolia Water North America, Houston, Texas
Von Roll, Inc., Norcross, Georgia

List of Figures

List of Tables

Table **Page**

WASTEWATER SOLIDS INCINERATION SYSTEMS

Chapter 1

Introduction

Interest in incineration as a reliable and cost-effective method of solids handling has grown for several reasons, including

- Approximately 95% reduction in volume and 70% reduction in weight of solids, which greatly reduces transportation requirements.
- Complete destruction of pathogens, viruses, and organic compounds in solids.
- Potential for heat energy recovery for autogenous combustion and process use, building heating, or power generation.
- Decreasing public acceptance of biosolids land application in some locations.
- Increasing complexity of treatment processes to produce Class A biosolids for land application.

Although facilities have practiced incineration since the early 1900s, in the last few decades they have focused on land application and reuse of biosolids to take advantage of its nutrient value. New regulations were also discussed that might have precluded incineration as a management option, including

- Federal regulations (Title 40 Part 503 of the Code of Federal Regulations) cover operating and monitoring of the biosolids program. The regulations provide risk-based guidelines and establish limits for total hydrocarbons or carbon monoxide, concentration, and overall process removal efficiencies of various metals.

- Review of toxic substances such as dioxins and furans in incinerator emissions. As a result of the review, the U.S. Environmental Protection Agency (U.S. EPA) said that the amount of dioxins and furans in emissions from incinerators were so small they posed no threat to public health and need not be regulated (U.S. EPA, 2000).
- Review of solid waste regulations. The U.S. EPA decided not to include incinerators under the solid waste regulations (U.S. EPA, 2007).

Great strides have been made in dewatering and cake conveyance technologies. The ability to produce and convey cake with a solids concentration of 27 to 30% has made it possible to achieve autogenous combustion without thermal pretreatment, which has greatly simplified solids processing and increased reliability or the incineration system.

1.0 THE CARBON CYCLE

The carbon cycle, an important but often misunderstood process, is the natural pathway of carbon through our ecosystem. The carbon cycle includes the following basic steps:

- People eat carbon-containing food.
- People excrete waste carbon as carbon dioxide through exhalation and as body wastes.
- Body wastes are collected and treated in wastewater treatment plants where microorganisms consume carbon-bearing wastes in the aeration and digestion processes and release carbon dioxide and waste.
- Carbon-bearing solids are generated. During incineration, the high heat, oxygen- rich environment of an incinerator allows for oxidation of carbon in the solids to carbon dioxide.
- Plants convert carbon dioxide from air and sunlight (photosynthesis) and the cycle begins again.

Figure 1.1 illustrates the carbon cycle and demonstrates that it is not related to any wastewater processing or solids management process.

Most public discussion about the carbon cycle deals with greenhouse gas emissions and carbon produced from fossil fuels. The carbon produced by burning fossil

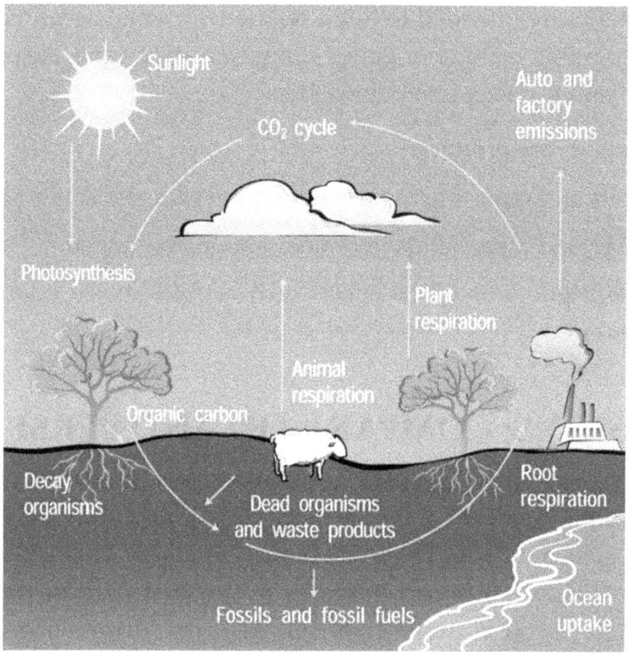

FIGURE 1.1 Illustration of the carbon cycle (courtesy of *Windows to the Universe*, http://www.windows.ucar.edu/, at the University Corporation for Atmospheric Research (UCAR)©. The Regents of the University of Michigan; All Rights Reserved.).

fuels is not part of the natural carbon cycle and may contribute to climate change. Wastewater solids, however, are part of the natural carbon cycle and do not contribute to carbon content in the atmosphere.

Incineration offers opportunities to reduce use of fossil fuels and to offset fuel consumption either through heat recovery and energy production or by greatly reducing the volume of material or ash to be hauled.

2.0 PUBLIC RELATIONS

A public relations program can provide a basis for communication and education, helping to ensure a successful incineration program. Utilities have used several tools to communicate with the public, including meetings, working groups, and advisory committees.

These programs are intended to build relationships and accountability with the public. Most programs include ongoing communications and cover the following topics related to public perception of biosolids:

- How biosolids are produced.
- Regulations and reporting requirements governing biosolids processing.
- Incineration and other solids treatment options evaluated.
- Project descriptions and local effects such as traffic, noise, odors, and construction.

3.0 EMISSIONS: COMPARISON AND PERSPECTIVE

Emissions need to be compared on an "apples-to-apples" basis. Many programs that involve long hauling distances generate more emissions during transportation than is generated by incineration and hauling a much smaller volume of ash. Trucks produce non-point emissions that are sometimes overlooked, because they are not regulated as wastewater or air-permitted sources. Also, truck emissions are regulated as "best case" emissions within a controlled environment at constant speed, which is significantly lower than actual on-the-road emissions.

In the past 10 years, both incinerator emissions and truck emissions have been reduced significantly. The previous limit for particulate emissions for incinerators was 0.65 g/kg (1.3 lb/dry ton), whereas the newer fluid bed facilities or upgraded multiple-hearth facilities with enhanced scrubbers are reporting particulates in the range of 0.05 to 0.25 g/kg (0.1 to 0.5 lb/dry ton). As for hydrocarbons, there are very few fluid bed facilities that are producing even 10% of the maximum allowable 100 ppm total hydrocarbons (THC) limit prescribed in the 40 CFR Part 503. Multiple-hearth facilities have also been able to comply with the THC limits. Nitrogen oxide (NOx) emissions for fluid bed incinerators have also declined by more than 50%, from more than 2.5 to 1 g/kg (5 lb/dry ton to 2 lb/dry ton) or less.

For most people, emissions are what they see coming out of the stack. Most new facilities and upgraded older facilities produce no visible emissions. Even vapor plumes can be avoided with reheating or subcooling of exhaust gases.

4.0 ECONOMIC ISSUES

The economics of incineration have improved in the past few decades primarily due to advances in dewatering technology. Dewatering can produce cake with 27 to 30% total solids concentration, which supports autogenous combustion, meaning that no auxiliary fuel is required. Several facilities were shut down in the early 1970s because of the high costs of auxiliary fuel.

Typical incineration facilities have experienced annualized operating costs of $155 to $310 per dry ton with a capital cost component of $40 to $60 per dry ton. These costs include operation and maintenance, electricity, auxiliary fuel, and reuse or disposal of ash (WEF, 2002). If additional energy is recovered and used to generate electricity, the costs may be offset by $30 to $50 per dry ton.

When comparing alternatives, it is important to include the costs of preprocessing (anaerobic digestion, dewatering, etc.) and post-processing (transportation, tipping fees, etc.). An approach to an economic evaluation is presented in Appendix C.

5.0 USE OF THE MANUAL

Chapter 2, Safety, presents an overview of safety and personnel issues associated with the incineration process.

Chapter 3, Permitting and Emission Regulations, provides an overview of the current regulations and the incinerator permitting process.

Chapter 4, Combustion Theory, provides basic background information on combustion.

Chapter 5, Combustion Technology, provides an overview of current combustion technologies in use, including both fluid bed incinerators and multiple-hearth incinerators, as well as introducing other emerging technologies such as gasification, plasma arc, and the SlurryCarb™ process.

Chapters 6, 7, and 8 provide information on heat recovery and reuse, emission control and monitoring, and ash handling and recycling.

Chapter 9, Instrumentation and Control, provides an overview of process control and monitoring requirements.

Chapters 10 and 11 provide an overview of operation and maintenance of fluid bed and multiple-hearth incinerators.

6.0 REFERENCES

U.S. Environmental Protection Agency (2000) Unified Agenda, Agenda of Regulatory and Deregulatory Actions, 65 Fed. Reg. 23430, 23460.

U.S. Environmental Protection Agency (2007) *Standards for the Use or Disposal of Sewage Sludge.* (40 CFR Part 503).

http://yosemite.epa.gov/r10/water.nsf/NPDES+Permits/Sewage+S825 (accessed March 2008).

Water Environment Federation, Residuals and Biosolids Committee, Bioenergy Technology Subcommittee (2002) *Thermal Oxidation of Sewage Solids.* http://www.wef.org/NR/rdonlyres/979954FA-CAA0-45C2-91F9-4D07107F7A4B/0/ThermalOxidation.pdf (accessed March 2008).

7.0 SUGGESTED READING

U.S. Environmental Protection Agency (2001) Standards for the Use or Disposal of Sewage Sluge: Final Notice, 66 Fed. Reg. 66028.

Chapter 2

Safety

1.0 GENERAL SAFETY

1.1 Purpose

This section presents basic guidance for maintaining a safe incineration facility. The first priority for safety is to protect plant personnel, contractors, and visitors at the incineration facility. The second priority is to protect the physical plant of the incineration facility, including equipment and structures.

1.2 Scope

This section serves as an introduction to safety practices at an incineration facility. It is not intended to be all-inclusive, but to identify the basic needs and sources for further investigation as applicable. Referenced codes, industry standards, and other sources listed here should be consulted for details that may apply to a specific case. Protection of the environment, including public safety, is not part of this section. Chapter 3, Permitting and Emissions Regulations, provides guidance on environmental protection for incineration facilities.

1.3 Plant Safety

Although this section addresses safety issues particular to incineration, this system cannot be considered separately from overall plant safety. Adequate incineration safety procedures are dependent on a strong overall plant safety program, which creates a culture that emphasizes consideration of safety in all decisions and procedures.

2.0 REGULATIONS, CODES, AND STANDARDS

The design and operation of incineration facilities is governed by numerous federal, state, and local regulations, codes, and standards. Some regulations and codes are enforceable, based on the applicability for a particular facility considering the operating entity and location, while others serve as guidelines. In all cases the underlying safety principles should be considered and incorporated into the design and operation of incineration facilities. Some of the regulations, codes, and standards that have a particular application to incineration facilities are listed below.

2.1 Occupational Safety and Health Standards

The Occupational Safety and Health Act established general national Occupational Safety and Health Standards (29 CFR 1910) (2006). These standards apply directly to privately owned treatment facilities and it is common practice for publicly owned treatment facilities to meet these standards also. Some of the items of particular application to incineration facilities include the following:

- Walking-working surfaces,
- Means of egress,
- Occupational health and environmental control (ventilation and noise exposure),

- Personal protective equipment,
- Permit for entry to confined spaces,
- Control of hazardous energy (lockout/tagout),
- Fire protection,
- Machinery and machine guarding,
- Electrical safety, and
- Fall protection.

2.2 Building, Fire, and Mechanical Codes

Local codes typically include safety requirements that are applicable to incineration facilities. Code requirements are location-specific and are based on the particular codes adopted for the area. Some of the International Code Council's model code regulations are identified below as an example of the codes applicable to incineration facilities.

- International Building Code: egress requirements for incinerator rooms.
- International Fire Code: incinerator requirements related to fire code authority, means of egress for incinerator rooms.
- International Fuel Gas Code: commercial-industrial type incinerators constructed and installed according to NFPA 82, gas piping installation requirements, including sizing, materials, support, shutoff valves, and flow controls.

2.3 National Fire Protection Association

The National Fire Protection Association (NFPA) maintains codes and standards that can apply to different aspects of fire and explosion safety for municipal wastewater incineration equipment. Several of these codes and their applicability are discussed here.

- NFPA 30, *Flammable and Combustible Liquids Code:* Applies to the storage, handling, and use of liquid fuels, including fuel oil used as auxiliary fuel for incinerators (2008a).
- NFPA 31, *Standard for the Installation of Oil-Burning Equipment:* Applies to the installation of stationary oil-burning equipment and appliances (2006a).

- NFPA 54, *National Fuel Gas Code:* Applies as a safety code for the installation of fuel gas systems, equipment, and related accessories (2006b). It governs the design, sizing, and installation of gaseous fuel systems for incinerators that use natural gas, propane, or other similar fuel. Items covered include piping materials, operating pressures, over- and under-pressure protection, need for shutoff valves, and air for combustion.
- NFPA 82, *Standard on Incinerators and Waste and Linen Handling Systems and Equipment:* Applies to installation and use of waste storage rooms, containers, handling systems, incinerators, compactors, and linen and laundry handling systems (2004). Much of this standard appears directed at incinerators burning solid waste. The explanatory material recognizes that there are many different incineration technologies and designs for burning a wide range of wastes, from solids to liquids, sludge, and fumes. The standard is not intended to cover or include all design details for each incineration technology. However, it includes many requirements that appear applicable to wastewater sludge incinerators, such as the requirements for auxiliary fuel, air for combustion and ventilation, incinerator design, placement, and clearances.
- NFPA 86, *Standard for Ovens and Furnaces:* Applies to ovens, dryers, or furnaces used for industrial processing of materials and operating at approximately atmospheric pressure (2007). Of particular interest are the provisions for safety equipment and application for fuel-firing burners. Although it contains no direct reference to combustion of wastewater solids, this standard is typically applied to municipal incinerators. It is used to specify safety equipment and practices associated with the use of burners, particularly purging requirements.
- NFPA 820, *Fire Protection in Wastewater Treatment and Collection Facilities:* Establishes minimum requirements for prevention and protection against fire and explosions in wastewater treatment facilities (2008b). Addresses hazard classifications for specific areas and processes, including incinerators. This standard applies to incineration, but generally addresses the requirements for buildings where incinerators are located as follows:
  - No requirement for ventilation in the incinerator area. Ventilation is typically provided for other purposes such as heat removal.
  - Unclassified area for electrical equipment.
  - Requires limited-combustion, low-flame spread, or noncombustible building materials.
  - Requires a fire suppression system such as sprinklers.

2.4 Insurance and Other Industry Standards

In addition to the regulations, codes, and standards listed above, insurance and other industry standards can apply to incineration systems, such as burner safety systems. It is important to identify the requirements of a facility's insurance carrier. In many cases, insurance standards can be more stringent than local codes or NFPA regulations and should be considered when selecting and designing safety provisions for incineration systems, especially where particular hazards are present. Standards that should be considered for fuel valve train and burner safety systems include the following:

- Insurance standards, Industrial Risk Insurers (IRI) or Factory Mutual (FM).
- Industry standards, including Underwriters Laboratories (UL).

3.0 INCINERATOR SAFETY CONSIDERATIONS

The design and operation of an incinerator need to incorporate adequate safety features and procedures to address the hazards of high surface temperatures, the handling of fuel, and the combustion of the solids unique to incineration. Safety provisions and operating considerations are described below.

3.1 Hot Equipment Surfaces and Personnel Protection

Provisions to protect personnel against burns from hot surfaces should include the following:

- Where possible, ducts and equipment surfaces should be insulated to keep surface temperatures at 60°C (140°F) or lower while operating at an ambient temperature of 32°C (90°F) or higher.
- Incinerators and some of their exhaust breeching are refractory-lined (insulated) but must operate at surface temperatures above 60°C (140°F). The reactor walls of fluid bed incinerators are often designed to keep shell temperatures higher than 100°C (212°F) to prevent condensation from exhaust leaks to the inside wall of the metal, requiring protection of personnel against the hot surfaces.
- Barriers such as expanded metal shields or barrier fences should be used to protect personnel from contact with hot ducts and equipment. Where barrier fences are used, locked gates should be installed to prevent unauthorized access.
- Facility personnel should be provided with protective gear such as gloves, clothing, and eye shields while operating or servicing a hot incinerator.

In addition, multiple-hearth incinerators pose unique personnel safety problems because they include observation ports, which are used to check combustion conditions within the incinerator. Many units are equipped with sliding or hinged covers that can expose the observer to hot incinerator gases when open, particularly if the incinerator should lose negative pressure. Personnel working near or on those incinerators should wear proper eye protection, clothing, and other equipment. Modifications to provide better protection include glass covers for observation ports to allow visual checking without direct exposure to incinerator gases.

Multiple-hearth incinerators are equipped with hearth doors to provide access for maintenance of the plows and arms and removal of clinkers. Such maintenance is normally scheduled for periods when the incinerator is out of service. In the event that emergency "hot" maintenance is required, it must be performed under conditions that pose risk to personnel. Special safety precautions, including protective clothing, gloves, and hoods must be observed.

3.2 Fuel Safety Provisions

Either oil, natural gas, or both, can be used to initially heat the incinerator and as supplemental fuel when needed. Fuel supply and combustion systems should include the following safety features:

- Design of supplemental fuel supply piping, with consideration for sizing, materials, configuration, support, shutoff valves, and pressure and flow controls in accordance with applicable standards.
- Design of supplemental fuel safety systems, with consideration for gas conditioning, burner equipment standards, pilot type, ignition type, flame monitoring, combustion air pressure monitoring, fuel pressure control and monitoring, emergency fuel shutoff, venting, and purging, in accordance with applicable standards.

3.3 Fire and Explosion Protection

Modern incinerators, both fluid bed and multiple hearth, have a good safety record in prevention of fire and explosions. Here are some of the provisions that should be included in the design of either fluid bed or multiple-hearth incinerator systems:

- Solids handling systems configured to reduce the risk of spills and accumulation of solids that could dry and produce combustible dust, which could lead to explosion.

- Purging of the combustion area before startup.
- Adequate instrumentation, including temperature, pressure, and feed monitors.
- Adequate equipment sizing and control provisions to ensure proper combustion conditions.
- Flue gas ductwork and equipment designed to prevent leakage of exhaust gas into the building.

Special design considerations for a fluid bed incinerator include the following:

- The combustion of the dewatered cake takes place in the reactor vessel, generally under positive pressure. The reactor vessel should be designed according to the applicable structural and welding standards for the pressures to be encountered.
- The reactor uses lances for auxiliary fuel when the incinerator is at operating temperature. A preheat burner is used for the initial warm up of the incinerator. Operating practices, equipment, and interlocks must be selected to prevent accumulation of natural gas, fuel oil, or other fuels in the reactor, particularly in a cooled reactor. These provisions include use of block and bleed valves on the fuel supply lines and removable oil lances to prevent leakage to the incinerator bed when the incinerator is not operating.
- The reactor refractory should be installed in a manner that prevents development of pockets between the shell and refractory, which could allow combustible gases to accumulate and cause minor explosions that cause damage. Pockets would also permit condensate to collect on the interior of the reactor shell which could lead to corrosion.

Special design considerations for a multiple-hearth incinerator include the following:

- An emergency bypass damper and ductwork should be provided for venting of combustion gases in the event of power failure.
- Purging of the incinerator before starting burners.

4.0 OPERATING SAFETY PROVISIONS

Experienced operators with proper training and well-defined operating procedures have an important role in the safe operation of an incineration system. Well-trained and experienced operators recognize unusual and unsafe conditions and act to reduce the risk of injury to personnel, fires, and explosions.

4.1 Provide Well-Trained Operators

Safety issues as they relate to operators and plant operation are listed below.

- Provide extended, typically 60 to 90 days, hands-on training to personnel who will be operating a new incineration system.
- Require the manufacturer of an incinerator system to train operators to recognize abnormal conditions and instruct them how to restore normal operation or to shut down the system, and to understand the instrumentation used for monitoring, alarms, and emergency shutdowns.
- Prepare standard operating procedures (SOPs) for startup, normal operation and shutdown, and emergency shutdown. Additional SOPs may be developed for "how to get out of trouble" under other operating conditions.

4.2 Use Effective Operating Procedures

Correct operating procedures are equally as important and include the following:

- Optimize incinerator operation to establish parameters for low feed, high feed, and other operating scenarios related to feed and dewatered cake solids content.
- Monitor moisture content of solids at least once per shift to confirm cake feed conditions.
- Operate the incinerators continuously to reduce the potential for problems associated with startup and shutdown.
- Ensure that safety procedures are enforced during shutdowns and startups when the potential for problems is highest.
- Provide tamperproof control and monitoring systems to ensure that safe shutdown and startup procedures are implemented under emergency conditions.

5.0 HAZARD AND OPERABILITY REVIEWS

Some industries in North America and in Europe use hazard and operability (HAZOP) reviews to identify significant hazards to health, safety, and the environment, as well as significant operability problems associated with process systems. In North America, HAZOPs are beginning to be used for solids processing systems, including incinerators. Once potential problems have been identified, they must be resolved, mitigated, or eliminated through design changes to improve the safety and operability of the facility under design.

For incineration systems, HAZOP reviews are a systematic and structured method to identify potential safety problems and the appropriate mitigation measures to reduce the risks. HAZOP reviews can be performed during all stages of design of an incineration facility, from concept through various levels of completion. HAZOP may also be used for existing facilities.

HAZOP reviews generally include the following:

- Description of the facility being reviewed. The review may include an entire system, a part of it, or specific items of equipment. Drawings, particularly process and instrumentation diagrams, are used to identify nodes for study and to chart progress of the review.
- Nodes or parts of the facility are selected for individual review using parameters such as flow, temperature, pressure, and level and deviation guidewords such as more, less, obstructed, or reverse flow, or higher or lower temperatures, to discuss causes and effects of deviations and to determine consequences.
- For each cause, the consequences are discussed, and safeguards (anything that could prevent or alleviate consequences) are identified. Recommendations for modifications such as design changes or drawing changes, and requests for additional information are also recorded. Such recommendations require follow-up to ensure mitigation of concerns.
- When all nodes have been examined, the HAZOP study is completed. However, additional review may be needed based on responses to the recommendations.

After completing the review, the design and operations team evaluates the safeguards and makes recommendations for incorporation into the design. If changes are recommended, an additional HAZOP review should be considered.

HAZOP reviews increase awareness by designers, owners, equipment suppliers, and operators of potential risks associated with operation of the process. Reviews should be timed to allow design changes to be made before construction begins. Reviews also force a more thorough discussion of the functions of facilities, which benefits all parties by helping them understand operation before construction is completed.

As outlined above, a HAZOP review can provide safety benefits by ensuring that the system has been thoroughly reviewed for safety concerns and providing an opportunity to educate plant personnel of some of the safety issues involved in operating an incineration system.

6.0 REFERENCES

National Fire Protection Association (2004) *Standard on Incinerators and Waste and Linen Handling Systems and Equipment.* Document no. 82, Quincy, Massachusetts.

National Fire Protection Association (2006a) *Standard for the Installation of Oil-Burning Equipment.* Document no. 31, Quincy, Massachusetts.

National Fire Protection Association (2006b) *National Fuel Gas Code.* Document no. 54, Quincy, Massachusetts.

National Fire Protection Association (2007) *Standard for Ovens and Furnaces.* Document no. 86, Quincy, Massachusetts.

National Fire Protection Association (2008a) *Flammable and Combustible Liquids Code.* Document no. 30, Quincy, Massachusetts.

National Fire Protection Association (2008b) *Standards for Fire Protection in Wastewater Treatment and Collection Facilities.* Document no. 820, Quincy, Massachusetts.

Occupational Safety & Health Administration (2006) *Occupational Safety and Health Standards* (29 CFR 1910). U.S. Department of Labor, Washington, D.C.

Chapter 3

Permitting and Emissions Regulations

1.0 PERMITTING AND EMISSIONS REGULATIONS

Since the early 1970s, the Clean Water Act (CWA) and the Clean Air Act (CAA) have served as the foundation of the U.S. Environmental Protection Agency's (U.S. EPA's) regulatory authority. Both programs have been amended over the years to continue to improve the environment and define the nation's regulatory agenda.

The CWA has resulted in tremendous improvements in the quality of the wastewater effluent from treatment plants and in the quality of the water in receiving waterways. The CWA Amendments of 1987 required the U.S. EPA to establish standards for use or disposal of "sewage sludge," including incineration, and were incorporated into Title 40 Part 503 of the Code of Federal Regulations (40 CFR 503). The CAA includes requirements for the combustion of wastewater treatment plant (WWTP) residuals, which is recognized in the Part 503 regulations. Wastewater treatment plants that are thermally treating residuals must meet the requirements of both the CWA and CAA.

The CAA has greatly reduced emissions to the atmosphere and improved air quality in urban areas. The CAA not only established emission limits for individual emission units, it also considered other sources at the facility and the attainment status of the ambient air in the region in which the facility is located. Construction of a new facility cannot commence until all of these considerations have been addressed. With the passage of the CAA Amendments of 1990, many new programs were implemented, including emission limits for hazardous air pollutants from specific source groups and the Title V Operating Permit program to document and track emissions on a continuing basis for the entire plant.

Thermal treatment processes for WWTP residuals must be reviewed before construction of the facility or completion of planned modifications. The CAA requires that a facility undergo a two-step air permit review process and receive approval from the reviewing agency (U.S. EPA regional office or delegated state or local regulated authority) before construction can begin. The attainment status of the facility site, local meteorology or terrain features, and state regulatory requirements will affect the specific emission limits that may apply to a facility. Once the facility is completed, a performance test is conducted to demonstrate that the pre-construction limits have been met. An operating permit application then needs to be submitted and approval issued for the facility to continue operating. Careful consideration to the air permitting requirements for a thermal treatment system must be made at the early planning and design phases of a project.

This chapter reviews air quality permitting requirements and emission limitations that apply to thermal treatment technologies processing WWTP residuals. The goal is to provide WWTP managers, operators, engineers, and planners with an understanding of air quality permitting requirements for this source group. The chapter is organized by project phase, including planning, implementation, construction, startup, and operation.

Online resources for environmental regulations include the U.S. EPA Home Page (http://www.epa.gov). The Government Printing Office (GPO) also provides online access to the Code of Federal Regulations (http://www.gpoaccess.gov/cfr/index.html). State environmental regulations can be found online for most state and local agencies. Some states caution that the online regulations are provided for informational purposes only, the published version is the official document.

2.0 PROJECT PLANNING: PRE-PERMITTING

The nature of the regulatory review and emission limitations depend on the type of facility, the magnitude of the potential emissions, and attainment status for the area where the plant is located. If the facility is located in an area that is not in attainment of the National Ambient Air Quality Standards (NAAQS), stringent air pollution control requirements and emission limits will apply (CAA, Title I, Part D, Plan Requirements for Nonattainment Areas, as amended). A facility located in an area meeting the NAAQS may still be subject to a comprehensive permit review (CAA, Title I, Part C, Prevention of Significant Deterioration of Air Quality, as amended), but the resulting emission limits may be less stringent than a facility located in a nonattainment area.

To anticipate the nature of the air permitting review and the applicable emission limitations, air permitting should begin in the project planning stage when detailed information about the proposed facility is still being developed. Information that can be obtained at this planning stage includes:

- Attainment status of criteria pollutants for the proposed facility's location.
- Classification of the facility as to its status as a major or minor air emissions source considering ownership of the adjacent facilities and industrial classification.
- Potential emission rates of the new facility or net change in emissions from an existing facility.

2.1 Attainment Status

Air quality standards in the United States are mandated by the CAA and its amendments. The U.S. EPA Office of Air Quality Planning and Standards has set NAAQS for six principal pollutants, called "criteria" pollutants. The six pollutants defined in 40 CFR 50 are:

- Carbon monoxide (CO);
- Sulfur dioxide (SO_2);
- Nitrogen dioxide (NO_2);
- Ozone (O_3);
- Various categories of particulate matter, including particulate matter less than 10 μm in size (PM-10) and particulate matter smaller than 2.5 μm (PM-2.5); and
- Lead (Pb).

U.S. EPA has identified two types of standards for these pollutants: (1) primary ambient air quality standards, which define levels of air quality necessary to protect public health with an adequate margin of safety; and (2) secondary standards, which define levels needed to protect the public welfare from any known or anticipated adverse effects of a pollutant. Such standards are subject to revision. Additional primary and secondary standards may be promulgated as the U.S. EPA deems necessary to protect public health and welfare.

Geographic areas in which the NAAQS for all criteria pollutants are met are called "attainment areas"; areas in which one or more standards are violated are called "nonattainment areas." A nonattainment area must develop and implement a plan to meet and maintain CAA standards. When a nonattaining region again meets the standard, the area can be redesignated as a "maintenance area." A maintenance area is a geographic region redesignated by the U.S. EPA from nonattainment to attainment as a result of monitored attainment of the standard and U.S. EPA approval of a plan to maintain air quality standards for at least a 10-year period. This determination is made on a pollutant-specific basis; for example, an area can be in nonattainment for ozone and in attainment for other criteria pollutants. Because emissions of nitrogen oxides (NOx) and volatile organic compounds (VOCs) can lead to the formation of ozone, regions designated as being in nonattainment for ozone also have more restrictive limits for NOx and VOCs. For example, an area may be classified as a serious nonattainment area for ozone resulting in lower emission threshold limits for NOx and VOC but be classified as attainment for CO, SO_2, and particulate matter.

The official listing of attainment status designations is in 40 CFR 81 Subpart C-Section 107 Attainment Status Designations. This subpart of the federal regulations lists attainment areas by state and air quality control region. These regulations can be accessed through the online resources described above. The U.S. EPA also operates a Web page called the Green Book that lists nonattainment areas for criteria pollutants (http://www.epa.gov/oar/oaqps/greenbk). This Web site offers a variety of ways (by state, county, or pollutant) to search area attainment designations.

2.2 Facility Classification

A facility's classification is determined by three factors: common owner or operator, adjacent facilities, and the same industrial classification. For a municipal facility, the operation of a wastewater treatment plant may be under the direction of a public works director who may also oversee other municipal facilities. A regional authority may have responsibilities only for the wastewater treatment operations even if the treatment facility serves people in multiple communities. An operator of a privately operated plant would be responsible only for the operations at that plant. It is possible for the solids-handling facility to be privatized even if the rest of the WWTP was run by a municipal or regional authority. In this case, the owner/operator of the solids-processing facility would have responsibility only for emissions associated with that process.

If a regional authority has responsibility for more than one WWTP and associated solids processing, the treatment plants would be considered separate facilities as long as they are not adjacent to one another. A public roadway passing between the two facilities is not sufficient to treat the facilities as separate operations. For example, a regional authority may have several regional WWTPs that are separated geographically. The operation of the WWTPs would be considered separate facilities. If, however, the solids-processing facility for all the regional plants was located adjacent to one of the WWTP, then that WWTP and the solids-processing facility would be considered one facility, as they are under common ownership and are adjacent to one another.

The CAA Amendments of 1990 also group facilities by major industrial classification. Sanitary Services are part of the U.S. Department of Labor Standard Industrial Classification (SIC) Major Group 4900 (the DOL main page for accessing the SIC is http://www.osha.gov/pls/imis/sic_manual.html). The subgroup (Industrial Group 4950) includes collection systems and refuse systems. Thus, a municipality or regional authority that operates a wastewater treatment plant and a solid waste management facility on adjacent properties must consider both operations as part of the same facility.

2.3 Potential Emissions

Potential emissions are the emissions that would occur on an annual basis if the facility was operated at its design capacity continuously. Potential emissions can be reduced if there is a physical limitation that constrains the process from operating at its design capacity on a continuous basis or a federally enforceable operational limitation was adopted which limits operation (for example, the total quantity of solids to be processed on an annual basis). If the potential emissions from a new source are greater than the emission thresholds for federal permit review, as presented in Table 3.1, then the facility is a new major emission source. A facility with emissions less than major emission source thresholds are considered a minor source and are subject to state permit review requirements.

If the proposed action is a modification to an existing facility, then emissions from the existing facility need to be quantified to determine whether the existing source is a major or minor air emissions source and whether the proposed change will result in a net increase or decrease in air emissions. If the net change in emissions is greater than the emission thresholds for a modification, then the proposed action is a major modification. The process of determining what emission credits apply when calculating a net change can be rather involved and may require emission offset credits. Consultation with the governing regulatory authority may be needed to be sure that changes in net emissions are being calculated appropriately.

In the planning phase, the details associated with the proposed action may not be fully developed. Yet estimating potential emissions is a key to determining future regulatory review. A preliminary estimate of emissions for a multiple-hearth or fluid bed incinerator can be made using emission factors. Emission rates based on generic emission factors should not be the sole method for determining emission limitations. Source specific emission testing and vendor performance guarantees are preferred methods for setting emission limits.

State and local regulatory agencies also have established emission thresholds that determine whether a facility is subject to regulatory review. These thresholds are lower than the major source and major modification thresholds defined in Table 3.1. Facilities should consult with the appropriate regulatory authority to determine what the emission thresholds are and what information is needed to meet the state and local permit review requirements.

Emission factors that can be used to estimate emission rates from a proposed incinerator or to identify applicable emission limits can be found in the U.S. EPA's

TABLE 3.1 Emission threshold levels for major sources and major modifications.

Pollutant	Pollutant attainment status	Major source threshold[a] (ton/yr[b])	Major modification threshold[c] (ton/yr[b])
CO	CO attainment	100	100
	CO serious nonattainment	50	50
SO_2	SO_2 attainment or nonattainment	100	40
PM-10	PM-10 attainment	100	15
	PM-10 nonattainment	70	15
PM-2.5	PM-2.5 attainment or nonattainment	100	15
NO_2	NO_2 or ozone attainment	100	40
	Ozone attainment or marginal or moderate nonattainment or ozone transport region	100	25
	Serious ozone nonattainment	50	25
	Severe ozone nonattainment	25	25
	Extreme ozone nonattainment	10	Any increase in actual emissions
VOC	Ozone attainment	100	40
	Serious ozone nonattainment or ozone transport region	50	25
	Severe ozone nonattainment	25	25
	Extreme ozone nonattainment	10	Any increase in actual emissions
Lead	Pb attainment or nonattainment	100	0.6

[a] 40 CFR 51.165 (a) (1) (iv) major stationary source.
[b] ton/yr × 0.9072 = Mg/a.
[c] 40 CFR 51.165 (a) (1) (v) major modification.

Technology Transfer Network (http://www.epa.gov/ttn/chief/ap42/index.html). Chapter 2, Solid Waste Disposal, contains a section on combustion of WWTP solids. Emission factors for both multiple-hearth and fluid bed incinerators are provided. The RACT/BACT/LAER Clearinghouse (RBLC) is a database that contains emission limitations from across the country (http://cfpub.epa.gov/rblc/htm/bl02.cfm). The database can be searched for emission limitations that have been adopted for WWTPs.

3.0 PROJECT IMPLEMENTATION: PERMIT TO CONSTRUCT

Emission limits for specific source categories are established under the new source performance standards (NSPS) (CAA Sec. 111, Standards of Performance for New Stationary Sources) and National Emission Standards for Hazardous Air Pollutant (NESHAP) regulations (CAA Sec. 112). These limits are minimum requirements. The prevention of significant deterioration (PSD) provisions (CAA Sec. 165, Preconstruction Requirements) can result in more stringent limits as a result of the best available control technology (BACT) review process. The new source review (NSR) (CAA Sec. 173, Permit Requirements) process includes an emission control evaluation that results in the lowest achievable emission rate (LAER).

To begin the air permit review process, a sufficient amount of information is needed to define the process requirements so that emission rates can be calculated and control strategies can be evaluated. However, the permit process is an iterative process between the applicant and the reviewing authority where emission limits and performance criteria are evaluated. Thus, it is best to begin the permit process before final process design decisions have been made. In a traditional design-bid-build construction process, the 30% design point is a good time to prepare and submit the air permit application. Enough technical information has been prepared to define the process and it is early enough in the design process to make modification to process equipment or air pollution control devices.

3.1 Permit Application Requirements

The type of review and threshold levels are based on the attainment status of the region in which the facility is located. Review requirements depend on the total facility annual potential emission rate or the annual potential emission rate for the proposed modification. A facility is defined as a contiguous facility, within the same industrial classification code and under common ownership. It is easy to understand that the incinerator for processing wastewater residuals would be considered part of the WWTP. However, adjacent landfills or solid waste processing facilities are also within the same industrial classification. If the regional authority or municipality also owns these facilities, the emissions from these facilities may be counted when determining the review status of a proposed solids thermal treatment facility.

A typical permit application would contain the following elements, although the format and presentation would vary by the reviewing agency.

- Permit forms—each agency has a standard set of forms they require permit applicants to use. Many agencies have the forms available electronically as downloadable documents that can be obtained from the agency Web site or interactive online forms that feed directly into a database.
- Process description—a detailed process description that describes the operation of the facility in sufficient detail to support and confirm emission calculations and control strategies. Often process flow diagrams and process equipment data sheets are provided.
- Emission estimates—the basis for each pollutant emission rate should be presented with supporting information such as emission test results from similar units, vendor performance guarantees, or mass balance calculations.
- Control technology assessment—this may be a determination of the BACT or LAER, depending on the attainment status of the region.
- Emission limitations—an assessment to be sure that statutory emission limitations are achieved. Both federal and state emission limits should be identified.
- Air quality compliance—a dispersion modeling analysis may be necessary to determine whether the NAAQS or the PSD increments are exceeded.
- Special issues—state and local agencies may request additional demonstrations to show that hazardous air pollutants, noise, and odor are within acceptable limits.

Permit review times will vary depending on the complexity of the permit application. To ensure that the permit is reviewed in the shortest amount of time, it is best to meet with the regulatory agency in advance, understand what information they need to complete their review, and provide as complete a permit application as possible. A typical permit timeline is as follows:

- Determination of permit completeness—2 to 4 weeks,
- Technical review—4 to 12 weeks,
- Response to technical comments—2 to 8 weeks,
- Public comment period—4 to 8 weeks, and
- Issuance of draft permit conditions—2 to 4 weeks.

The above timeline does not include time spent preparing the initial permit application.

3.2 Federal Regulatory Requirements

The federal regulatory permit review requirements are defined in 40 CFR 51 and 52. The U.S. EPA has delegated review authority to many state and local regulatory agencies so the state and local agencies can review the application and assess compliance with both the federal, state, and local permitting requirements. For state and local agencies where the U.S. EPA has retained review authority, the regional U.S. EPA office will serve as the reviewing authority for compliance with the federal regulations. The state or local agencies will still review and comment on state-specific requirements.

3.2.1 *Nonattainment New Source Review*

New source review is conducted for facilities located in areas where pollutant concentrations are greater than the NAAQS. The degree to which the standards are not being met increases the requirements to provide emission reductions and emission offsets. Emission control strategies must demonstrate the LAER.

3.2.2 *Prevention of Significant Deterioration*

For pollutants that are in attainment with the NAAQS, a PSD review (40 CFR 52.21) is required. This review seeks to maintain ambient air concentrations below the NAAQS by limiting air quality impact to incremental limits above the baseline concentrations. Emission control strategies must demonstrate that they represent BACT considering environmental, energy, and economic effects. A demonstration of potential effects to protection of wildlife areas and national parks may also be required.

3.2.3 *New Source Performance Standards*

For combustion technologies, the NSPS for WWTPs applies. This standard is defined in 40 CFR 60, Subpart O and establishes a particulate emission limit of 0.65 g/kg dry solids input (1.3 lb/ton). Most existing units are capable of meeting this emission limit. The rule also establishes opacity limits and associated monitoring, testing, record keeping, and reporting.

3.2.4 *National Emission Standards for Hazardous Air Pollutants*

The NESHAP are defined in 40 CFR 61. The limitations for beryllium (Subpart C) and mercury (Subpart E) are applicable to units that incinerate or dry WWTP solids. The emission limit for beryllium to the atmosphere from stationary sources shall not exceed 10 g (0.022 lb) of beryllium over a 24-hour period. The emission limit for mercury to the atmosphere from incineration plants, drying plants, or a combination of these that process WWTP residuals shall not exceed 3.2 kg (7.1 lb) of mercury per 24-hour period.

In addition to the criteria pollutants regulated by the NAAQS, there is another set of federally regulated air pollutants known as hazardous air pollutants (HAPs). Hazardous air pollutants are a set of 188 chemicals specifically regulated by the U.S. EPA that are known or believed to cause human health effects in excess of levels specified by the agency.

Sources that emit more that 9 Mg/a (10 ton/yr) of an individual HAP or more than 23 Mg/a (25 ton/yr) of multiple HAPs are considered a significant source of HAPs. Significant sources of HAPs may be subject to the maximum achievable control technology (MACT) regulations (40 CFR 63). Subpart VVV defines MACT standards for WWTPs. However, the standards in this subpart not address solids-processing or thermal treatment systems.

If a WWTP treats industrial wastes and is considered a part of a miscellaneous organic chemical manufacturing process unit (MCPU), then emission control requirements may apply. These additional requirements are defined in 40 CFR 63 Subpart FFFF National Emission Standards for Hazardous Air Pollutants: Miscellaneous Organic Chemical Manufacturing.

3.3 State and Local Regulatory Requirements

Even if a thermal treatment facility is relatively small, state pre-construction permitting requirements may apply. Although the review requirements may not be as stringent, the same emission limits may apply. Most states require a BACT analysis to determine the appropriate air pollution control equipment and emission limits.

Many states also have special regulations governing toxic air pollutant emissions. Some state rules define a control technology assessment for toxic air emissions (T-BACT). Other states require a dispersion modeling assessment to show toxic air pollutant emissions are in compliance with allowable ambient levels. State air toxics programs also may establish emission limits of some hazardous air pollutants.

Additional regulations may also apply with respect to noise and odors. Special compliance demonstrations or pre-construction assessments of potential noise and odor effects may be required. Applicable standards may be developed on a case-by-case basis and include communication with the public and involvement of other interested parties.

Visual plume emissions are regulated by opacity limits as defined in many state regulations. Meeting these opacity limits may not prevent the formation of a yellow plume. Because of the negative response from the public that may result from the formation of a visible yellow plume, mitigation of the visual impact from the plume may be necessary.

4.0 CONSTRUCT: COMMENCE CONSTRUCTION

To accelerate the time it takes to bring a project from concept to operation and save money, a design-build approach has been proposed as an alternative to the traditional design-bid-build approach. Although there is nothing in the CAA that would preclude a design-build approach, some cautions are warranted. The design-build approach seeks to streamline the design and procurement process so that the proposed facility can be constructed sooner. As noted in previous sections, construction cannot commence until the air permit has been issued. To commence construction means that no person may construct a new source or alter an existing source; certain site preparation work may begin, but no permanent facilities may be constructed.

The restriction on construction may also extend to binding agreements or obligations that make commitments to process equipment or construction services that can not be canceled or modified without substantial loss to the owner or operator. The air permit process may result in more stringent emission limitations that could affect process equipment or control technology selection. Thus, it is incumbent on both the facility owner/operator and provider of design-build services to share the risks associated with delays in receiving an air permit or facility design changes due to increased emissions.

4.1 Facility Operation: Permit to Operate

Once the proposed facility is constructed, a process of performance testing begins. The pollutant emissions to be tested are defined in the permit to construct. Testing procedures are defined in 40 CFR 60 Appendix A or methods agreed to by the reviewing authority. A testing protocol is developed that describes how the testing will be conducted and the emissions reported. If the measured emission rates are greater than emission limits established in the permit to construct, then immediate mitigation measures are implemented. Measures may include modifications to air pollution control equipment or re-permitting of the facility.

4.2 Federal Title V Operating Permit Program

Shortly after the facility begins operation, the operating permit program assures continued compliance and reporting of actual emissions from the major facilities.

The operating permit program (Title V of the CAA) consolidates the emission limits established for the entire facility and defines a means to monitor compliance with the limits.

Once the facility is constructed, a performance test may be required by the pre-construction permit that seeks to demonstrate that the emission limits established in the permit have been met by the operating facility. Satisfactory completion of the emission testing and submission of an operation permit application may be needed for continued operating of the facility.

If the facility is a major source, a Title V Operating Permit (40 CFR 70, State Operating Permit Programs) may be needed. The operating permit identifies all air emission sources present at the facility, summarizes the emission limitations and special conditions that have been established by the pre-construction permit, and outlines a process by which continued compliance with emission limitations can be demonstrated.

An accidental release prevention program may be required if hazardous material is used or stored onsite above the threshold limits (40 CFR 68). Often this requirement may apply to the disinfection process of the wastewater treatment plant and not the solids-processing area.

Sampling and testing is also required as part of the wastewater residuals management program (40 CFR 503). Solids testing and other reporting requirements are defined for land application and thermal treatment systems.

4.3 State Operating Permit Program

The state operating permit program applies to major emission sources that have accepted federally enforceable operating limits to restrict annual emissions to levels less than the major source thresholds or large minor emission sources whose potential emissions are less than the major source threshold but greater than the state operating permit program. The state operating permit program is similar to the federal operating permit program except that compliance with the program is administered through the state.

4.4 Accidental Release Prevention Program

The Chemical Accident Prevention Program requirements are defined in 40 CFR 68. Facilities that use or store more than the threshold quantities must prepare and implement a risk management plan (RMP). The operation of an incinerator does not trigger this requirement as the chemicals used to operate the unit are not hazardous or are not stored in quantities greater than the threshold limit.

The general duty clause, like much of the RMP regulation, is performance-based, and the method for compliance is for the most part to be determined by the source. U.S. EPA guidance states that owners that use extremely hazardous substances "must adhere, at a minimum, to industry standards and practices (as well as local, state, and federal laws and regulations) in order to be in compliance with the General Duty Clause." Accordingly, all potentially hazardous substances need to be stored in a building, have separate filling areas and piping and full vessel containment, and be separated in physical distance per applicable codes and standards to ensure no mixing can occur if a vessel loses its entire stored chemical. In addition, the fill pipes, tanks, and loading areas must be clearly marked.

4.5 Wastewater Residuals Management (Part 503)

Performance standards for the treatment and disposal of WWTP residuals required by CWA Amendments of 1987 are contained in Title 40, Part 503 of the Code of Federal Regulations (40 CFR 503), also referred to as Part 503 rules (promulgated 58 FR 9387, Feb. 19, 1993). Subpart B contains the requirements for the placement of biosolids on land application sites. This subpart defines the various classes of biosolids and their suitability for land application. Subpart E defines the requirements for residuals fired in an incinerator.

For multiple-hearth and fluid bed incinerators, pollutant limits are based on emission limits established elsewhere in the regulations, atmospheric dispersion, and control efficiency. Specific limits are derived for seven metals, total hydrocarbons, or carbon monoxide.

- Particulate matter emission limit of 0.18 g/m^3 (0.08 gr/ cu ft) dry gas at standard temperature and pressure corrected to 12% carbon dioxide.
- Beryllium emission of 10 g per 24-hour period.
- Mercury emission limit of 3200 g per 24-hour period.
- Lead, arsenic, cadmium, chromium, and nickel feed cake limits based on ambient air quality and health risk specific concentrations.
- Total hydrocarbon monthly average concentration of 100 ppm by volume, corrected to 0% moisture and 7% oxygen.
- Carbon monoxide monthly average concentration of 100 ppm by volume, corrected to 0% moisture and 7% oxygen.

Continuous emission monitoring for carbon monoxide may be conducted instead of monitoring for total hydrocarbons. A typical combustion unit will have air pollution control devices including wet scrubbers, dry and wet electrostatic precipitators, and fabric filters. Afterburners or regenerative thermal oxidizers may be used to control volatile organic carbon emissions and / or odorous compounds. The efficiency of the air pollution control system is included in the overall control efficiency of the combustion process. Specific performance standards are not established for the control devices.

Management practices are defined that require sufficient monitoring equipment is installed to ensure performance standards are met. These practices include a continuous emission monitor for total hydrocarbons or carbon monoxide and instruments to monitor oxygen and moisture concentrations in the stack. Combustion temperatures must be monitored at least daily to ensure that operating combustion temperatures established during the performance test are not exceeded by more than 20%. Other parameters that monitor the performance of the air pollution control may also be required to ensure their proper operation at a frequency appropriate for that device. Frequency of sampling and analysis for metal concentrations will be determined by concentrations expected in the WWTP residuals.

Records that define performance of the incinerator and actual emissions rates must be reported annually and kept for five years.

5.0 AIR QUALITY COMPLIANCE

The compliance assurance monitoring (CAM) requirements are defined in 40 CFR 64. The CAM rules apply to a pollutant specific emissions unit at a major source that is required to obtain a Title V Permit. The rules apply to an emissions unit that is subject to an emission standard or limitation, uses a control device to achieve compliance, and exceeds the uncontrolled emission criteria. Operation of an incinerator is required to achieve particulate control limits using a particulate control device. A CAM plan demonstrating that the scrubber is operated as intended and achieves the required emission limitations needs to be developed.

Permitting requirements under the CAA have evolved over the past 30 years. As the CAA is amended, new interpretations of existing regulations are made and additional requirements are added. Thus, it is necessary for thermal treatment units to review carefully the air quality permitting requirements as they apply to each location. Satisfactory completion of air emissions testing and preparation of operating permit applications may be needed for continued operation. Periodic monitoring, recordkeeping, and reporting may be a continued condition of plant operation.

Chapter 4

Combustion Theory

1.0 THERMODYNAMIC PROPERTIES OF SOLIDS

Before delving into the combustion of wastewater treatment plant (WWTP) residuals, it is important to develop a basic understanding of their composition. From a thermodynamic perspective, the dewatered feed cake going into the incinerator is composed of three key components: moisture, ash, and volatiles.

In a furnace, the moisture changes from liquid to vapor during the evaporation process; there is no change in chemical composition. The ash is typically chemically inert and does not undergo any chemical reactions in a furnace. It is the volatile (combustible) fraction that reacts with the oxygen in the air and in doing so changes composition and liberates heat.

An effective study of the combustion process depends on a detailed knowledge of specific fuel characteristics. The key tool in developing this information is an analysis by a laboratory that specializes in such moist solids fuels. The American Society of Testing and Materials (ASTM) and other organizations have published standards that govern this kind of testing and which describe in detail the lab procedures to be followed to generate consistent, reproducible test results. Feed cake can benefit from several tests:

- Proximate analysis—moisture, volatile matter, fixed carbon, ash.
- Ultimate analysis—moisture, ash, carbon, hydrogen, nitrogen, oxygen, chlorine (sometimes), sulfur, heating value.

- Ash ultimate analysis—up to 14 significant constituents.
- Ash fusion temperature—performed in oxidizing and reducing atmospheres.

The ASTM proximate and ultimate analysis tests were originally developed for coal.

When applied to fuels such as WWTP residuals or biomass, modifications to these procedures are sometimes necessary. One study resulted in the recommendation to reduce the temperature used when determining the ash ultimate analysis in order to eliminate erroneous results because of the volatilization of some of the alkali fractions (Miles et al., 1996). Unfortunately, no similar study exists for WWTP residuals. Until additional data are developed, the modified biomass analysis procedure is recommended.

Because of the wide variability between incoming streams of WWTPs and differences between wastewater processes, actual equipment designs or modifications based on values published in the literature are generally not worthwhile. In many cases, this practice can lead to inaccurate conclusions. WWTPs should take an adequate number of samples to cover seasonal variations and use a sampling technique that ensures a good composite. These practices are almost as important as obtaining a proper laboratory analysis on feed cake from a particular site. When wet weather flows are known to cause a variation in the ratio of ash to volatiles, fuel analysis during both wet and dry periods is recommended. Because of daily variations due to stratification in tanks or other changes in operation, taking a series of samples throughout the week or on different days over a multi-week period for analysis is recommended. Typically, analysis of the volatiles (moisture, ash-free basis) varies much less than one would expect, and multiple samples can yield a composite number by averaging the results for the volatile fraction of the feed cake.

Most WWTPs routinely perform basic analyses for fixed and volatile solids, often on a daily basis, which can be used to develop a powerful database of historical information. The WWTP can use these data with the more comprehensive analyses performed as a part of the incineration system design or operations process, or both. This invaluable information can help pinpoint the true range of operation that the incineration system will experience and can guide the designer in making provisions for the degree of process flexibility needed in the incineration system. The sampling program should be customized to reflect the characteristics of the specific plant.

TABLE 4.1 Typical proximate analysis of digested biosolids.

Parameter	As received	Dry basis
Moisture (%)	70.00	0.00
Ash (%)	14.16	47.19
Volatiles (%)	14.98	49.94
Fixed carbon (%)	0.86	2.87
HHV		
kJ/kg	3936	13 120
Btu/lb	1692	5 640
HHV (moisture, ash-free basis):		
kJ/kg		24 840
Btu/lb		10 680

Incineration systems may process raw sludge or digested biosolids, or both. Some modern merchant facilities receive a wide variety of feed types. An example of fuel analyses of digested biosolids is shown in Tables 4.1 through 4.3. These data are representative of the information that should be received from a well-qualified laboratory with broad experience in analyzing biomass, municipal solid waste, and WWTP residuals.

Laboratory analysis should always be reviewed to make sure that it is thermodynamically reasonable. The importance of this step cannot be overstressed. The standard ASTM procedures were developed for coal, and occasionally there are chemicals or compounds present that can skew results. Another important fact to bear in

TABLE 4.2 Typical ultimate analysis of digested biosolids.

Parameter	Dry basis (%)	Moisture, ash-free (%)
Carbon	29.64	56.12
Hydrogen	4.29	8.12
Oxygen	13.85	26.22
Nitrogen	3.66	6.94
Sulfur	1.37	2.60
Ash	47.19	0.00

TABLE 4.3 Typical elemental analysis of ash in digested biosolids.

Parameter	Dry basis (%)
SiO_2	44.94
Al_2O_3	14.30
TiO_2	0.80
Fe_2O_3	14.69
CaO	11.63
MgO	2.63
Na_2O	1.21
K_2O	0.54
P_2O_5	5.94
Cl	0.06
SO_3	1.96
CO_2	1.30

mind when reviewing these analyses is that the term "volatiles" in proximate analyses has a different meaning than "volatile solids" testing common in municipal WWTP applications. The volatiles in proximate analysis do not include fixed carbon; volatile solids in WWTP residuals analysis do include the fixed carbon.

There are several ways to check for thermodynamic validity. The best way is to take the weight fraction values of C, H, O, N, and S and insert them to one of several equations that are commonly used to calculate higher heating value (HHV) from an ultimate analysis. One of the most frequently used is the DuLong equation:

$$\text{kJ/kg} = 337\,\text{C} + 1419\,(\text{H} - \text{O}/8) + 143.2\,\text{S}$$
$$\text{Btu/lb} = 144.95\,\text{C} + 610\,(\text{H} - \text{O}/8) + 57.70\,\text{S}$$

The DuLong equation was originally derived for high rank coals, however, and may give high results when used for determining the HHV of WWTP residuals. Another option is to use the Mott-Spooner equation:

$$\text{kJ/kg} = 336.1\,\text{C} + 1419.3\,\text{H} + 94.2\,\text{S} - 153.3\,\text{O} + 0.72\,\text{O}_2$$
$$\text{Btu/lb} = 144.5\,\text{C} + 610.2\,\text{H} + 40.5\,\text{S} - 65.9\,\text{O} + 0.310\,\text{O}_2$$

This equation is sometimes known as the modified DuLong equation and was developed for coals with more than 15% oxygen. Because there is often significant oxygen content in municipal WWTP residuals, the Mott-Spooner equation may give slightly better results.

Channiwala (1992) collected data on more than 200 species of biomass and fitted the following equation to the data:

$$\text{HHV} = 0.3491\ \text{C} + 1.1783\ \text{H} - 0.1034\ \text{O} - 0.0211\ \text{A} + 0.1005\ \text{S} - 0.0151\ \text{N}$$

where HHV is expressed in kilojoules per gram.

Of all of the above equations, Channiwala's equation has the best correlation with primary and waste activated sludges and digested biosolids.

The U.S. EPA "F-Factor," used in making emission rate calculations, is another good tool for verifying that the reported HHV and ultimate analysis are reasonable. This factor is the ratio of the theoretical volume of dry gases at 0% excess air (0% O_2) given off by the complete combustion of a known amount of fuel to the gross caloric value (HHV) of the burned fuel.

The value of the F-Factor (F_d) is computed according to the following equation:

$$F_d = 10^6\ (3.64\ \text{H} + 1.53\ \text{C} + 0.57\ \text{S} + 0.14\ \text{N} - 0.46\ \text{O})\ /\ (\text{HHV})$$

Where

F_d = dry F-Factor at 0% O_2, dry standard cu ft (dscf)/10^6 Btu and

HHV = higher heating value, Btu/lb (cu ft/35.315 = m^3; Btu × 1054.8 = J).

F-factors are reasonably constant for a given fuel category. Average values for some typical fuels are presented in Table 4.4.

TABLE 4.4 Average F-factors (F_d) for selected fuels.

Fuel	F_d, m^3/J (dscf/10^6 Btu)
Bituminous coal	2.63×10^{-4} (9820)
Oil	2.47×10^{-4} (9220)
Gas	2.34×100^{-4} (8740)
Wood bark	2.58×100^{-4} (9640)
Wood chips	2.49×100^{-4} (9280)

Municipal WWTP residuals will vary widely because of origin and methods of treatment; however, the normal range of F_d is approximately 9200 − 10 000 dscf/10^6 Btu. Any analysis yielding F_d values outside this range should be questioned before using the values in a heat and material balance. Running new samples or re-running the analyses on the same sample is generally advisable. As mentioned earlier, even a correctly performed laboratory analysis can sometimes yield skewed results because of interactions by some chemicals present, or because of the lack of homogeneity within a particular sample that leads to the analysis of a nonrepresentative subsample.

When comparing heating values, it is always best to use a moisture ash-free (MAF) basis to avoid skewed results from varying collection and wastewater treatment processes. Using an MAF basis makes it easier to see how a particular feed cake compares to others, because it minimizes differences arising from the location (i.e., different regions of the country) and dewatering techniques.

2.0 ACTUAL FURNACE OPERATING CONDITIONS

All combustion calculations in the previous sections have assumed complete combustion. In actual practice, complete combustion is difficult to achieve, and furnaces always operate with some excess air. The amount of excess air required to reduce CO and total hydrocarbons (THC) to regulatory limits is dependent on furnace design and the combined effects of the three "Ts" of combustion: time, temperature, and turbulence. Based on a feed cake with specific values for total solids, volatile solids, and heating value, the only way to minimize auxiliary fuel use is to reduce furnace exhaust temperature and excess air to the lowest values that allow safe, stable operation while achieving regulatory compliance.

Simply making a heat and material balance for a given excess air value and theoretical temperature of the products of combustion (TTPC) does not mean that the furnace will be able to perform. Often in multiple-hearth furnace operation, excess air must be increased to greater than that necessary to achieve good combustion of hydrocarbons and CO. This excess air is necessary to prevent temperatures on the combustion hearth in the middle of the furnace from getting too high. Some fluid bed furnaces are operated with freeboard water sprays that are used to lower the exit temperature while auxiliary fuel is added to the bed to keep it at the minimum combustion temperature. These practical considerations must be kept in mind when designing new systems or modifying operating parameters of an existing operating system.

3.0 EFFECTS OF OPERATING PARAMETERS

Using heat and material balances to examine the effects of the various parameters is useful. In Figure 4.1, the TTPC or furnace exhaust temperature has been varied for a constant feed cake input at three different air preheat temperatures. The concept of availability is again used and represents the *y* axis. Gross heat input includes not only the heat of combustion contributed by the volatile solids but also the sensible heat in the preheated combustion air. The TTPC is plotted along the *x* axis. The intersection of each curve with the zero availability line represents the temperature that the particular example feed cake would reach if burned in a furnace with zero heat loss under the conditions listed. As temperatures fall below that point, availability is positive and heat would have to be removed. At temperatures greater than that point,

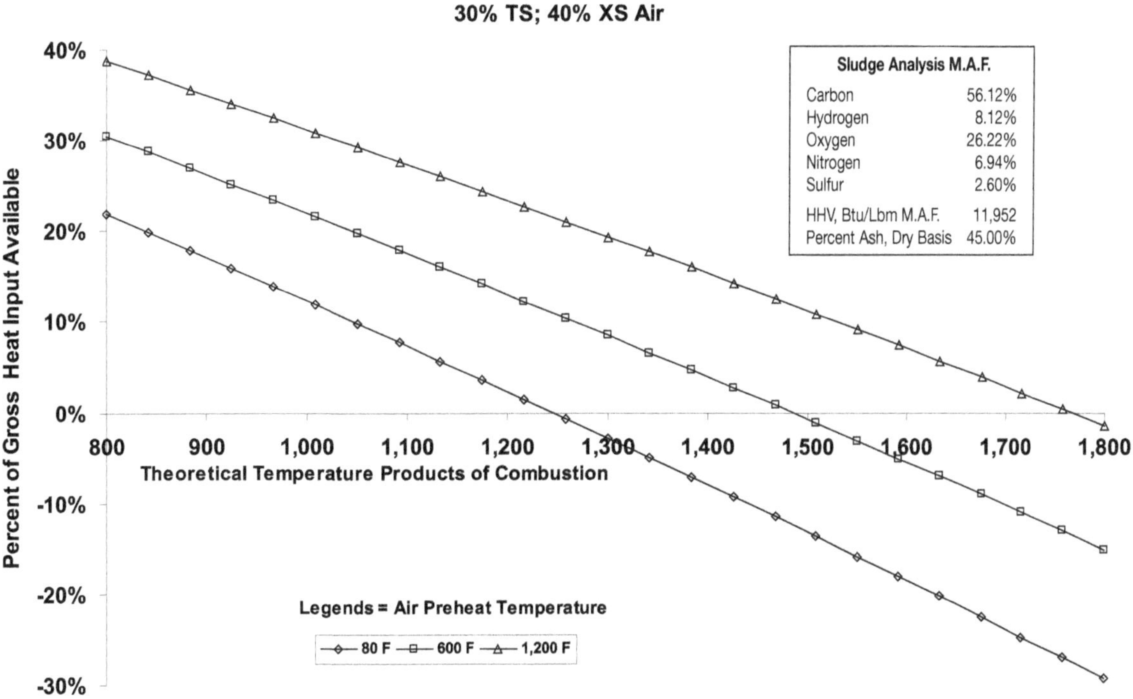

FIGURE 4.1 Available heat versus exhaust temperature (XS = excess; M.A.F. = moisture ash-free; Btu/lb × 2.326 = kJ/kg; 0.555 6[°F] = °C).

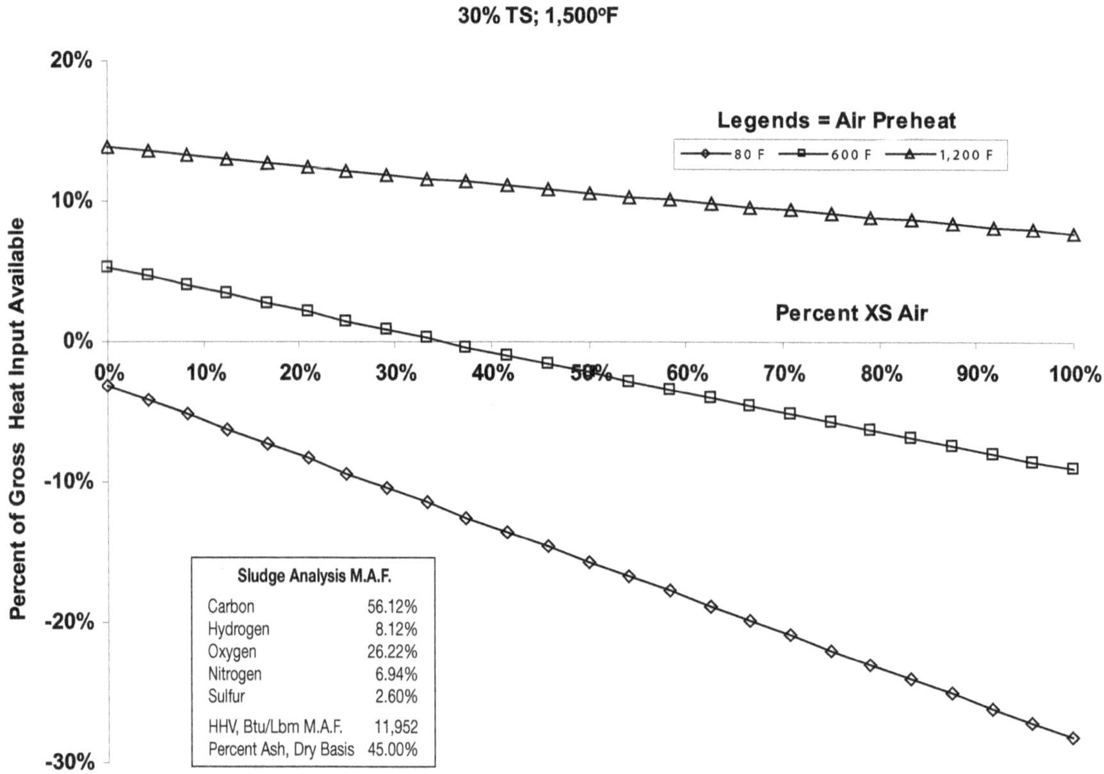

FIGURE 4.2 Available heat from sludge incineration versus percent excess air (XS = excess; M.A.F. = moisture ash-free; Btu/lb × 2.326 = kJ/kg; 0.555 6[°F] = °C).

availability is negative and heat has to be added. Figure 4.2 shows a similar graph, but in this case the x axis represents the percent of excess air used.

In Figure 4.3, the theoretical temperature of TTPC at 40% excess air versus a feed cake heating value (Btu per pound of feed cake moisture [or kJ/kg in SI units]) at three different air preheat temperatures is shown. Two additional lines on Figure 4.3 represent 75% excess air and inlet air temperatures of 80°F and 1200°F (27°C and 649°C) for reference. The lower inlet air temperature case at 75% excess air would correspond to the operating conditions of a typical multiple-hearth furnace. The higher inlet air temperatures at 40% excess air, on the other hand, typify a modern fluid bed system.

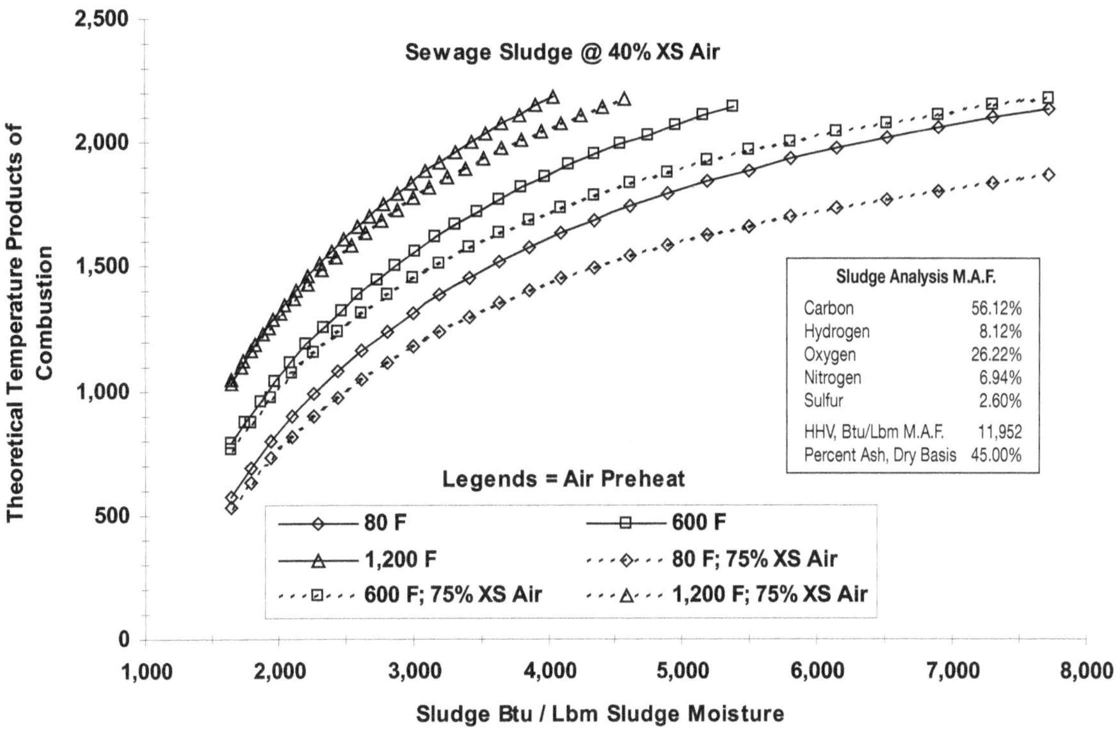

FIGURE 4.3 Theoretical temperature of the products of combustion versus the ratio of feed cake energy input to feed cake moisture (XS = excess; M.A.F. = moisture ash-free; Btu/lb × 2.326 = kJ/kg; 0.555 6[°F] = °C).

4.0 REFERENCES

Channiwala, S. A. (1992). On Biomass Gasification Process and Technology Development. Ph.D. Thesis, Indian Institute of Technology, Bombay.

Miles, T. R.; Miles, T. R., Jr.; Baxter, L. L.; Bryers, R. W. (1996) *Alkali Deposits Found in Biomass Power Plants: A Preliminary Investigation of Their Extent and Nature*; Report no. TP-433-8142; National Renewable Energy Laboratory: Golden, Colorado.

5.0 SUGGESTED READINGS

ASME Research Committee on Industrial and Municipal Wastes (1974) *Combustion Fundamentals for Waste Incineration*; American Society of Mechanical Engineers: New York.

Baukal, C. E.; Schwartz, R. E., Eds. (2001) *The John Zink Combustion Handbook*; CRC Press: Boca Raton, Florida.

Domalski, E. S.; Jobe, T. L., Jr.; Milne, T. A., Eds. (1987) *Thermodynamic Data for Biomass Materials and Waste Components*; American Society of Mechanical Engineers: New York.

Greb, F. W.; Lewis, F. M. (1986) Thermodynamics of Fluidized Bed Sludge Incineration. *Proceedings of the 16th National Conference on Municipal Treatment Plant Sludge Management*; Orlando, Florida, May 28–30; Hazardous Materials Control Research Institute: Silver Spring, Maryland.

Lewis, F. M. (1975) Fundamentals of Pyrolysis Processes for Resource Recovery and Pollution Control. *Proceedings of the 68th Annual Air Pollution Control Association Annual Meeting and Exhibition*; Boston, Massachusetts, June; Air Pollution Control Association: Pittsburgh, Pennsylvania.

Lewis, F. M. (1975) Heat and Material Balances for Non-Autogeneous Wastes. In *Incinerator and Solid Waste Technology*; Stephenson, J. W., Ed.; American Society of Mechanical Engineers: New York; pp 103–111.

Lewis, F. M.; Haug, R. T. (1985) Thermodynamic Optimization of Sludge Combustion Systems. *Paper Presented at the Power Magazine and Synfuels Conference on Energy from Municipal Wastes*; Washington, D.C., Oct.; Power Magazine: Houston, Texas.

Lewis, F. M.; Lundberg, L. A. (1990) Design, Upgrading and Operation of Multiple Hearth and Fluidized Bed Sludge Incinerators to Meet New Emission Regulations. *Proceedings of the 83rd Annual Air Pollution Control Association Annual Meeting and Exhibition*; Pittsburgh, Pennsylvania, June 24–29; Air Pollution Control Association: Pittsburgh, Pennsylvania.

Meyer, C. A.; McClintock, R. B.; Silvestri, G. J.; Spencer, R. C., Jr., Eds. (1983) *Thermodynamic and Transport Properties of Steam*, 5th ed.; American Society of Mechanical Engineers: New York.

Reed, R. J., Ed. (1978) *North American Combustion Handbook*, 2nd ed.; North American Manufacturing Company: Cleveland, Ohio.

Stultz, S. C.; Kitto, J. B., Eds. (1992) *Steam: Its Generation and Use*, 40th ed.; The Babcock & Wilcox Company, a McDermott Company: Barberton, Ohio.

Yaws, C. L., Ed. (1977) *Physical Properties: A Guide to the Physical, Thermodynamic and Transport Property Data of Industrially Important Chemical Compounds*; A Chemical Engineering Publication; McGraw-Hill: New York.

Water Environment Federation (1992) *Sludge Incineration: Thermal Destruction of Residues*; Manual of Practice no. FD-19; Water Environment Federation: Alexandria, Virginia.

Chapter 5

Combustion Technology

(continued)

1.0 INTRODUCTION

Incineration is the most commonly used thermal oxidation process for treatment of wastewater residuals. Other processes—such as plasma-assisted oxidation, pyrolysis, and vitrification—are new processes that are not yet fully developed for this application.

Wastewater treatment plants (WWTPs) have used incineration for solids disposal for more than 70 years. The first multiple-hearth furnace (MHF) was built in 1935 in Dearborn, Michigan. Until the late 1960s, incineration using MHF was the thermal technique of choice. Today there are 250 to 260 aging MHFs in operation in North America.

In the 1970s, fluid bed incineration became the preferred thermal technique, primarily because of tighter emission regulations and the increasing cost of auxiliary fuel.

The Lynwood, Washington, WWTP installed the first municipal fluid bed in 1962. It was a small unit of 1.2 m (4 ft) internal freeboard diameter. Lynnwood was originally sized for 91 kg/h (200 lb/hr) of dry solids. To accommodate plant expansion, the original unit was replaced in 1989 with a 2.9 m (9.5 ft) diameter unit sized for 390 kg/h (860 lb/hr). Today, fluid bed technology has matured. Many larger fluid bed units can be found, such as the St Paul, Minnesota facility. More recently, Lakeview, Ontario, installed two units with dry feed capacities of approximately 3967 kg/h (8750 lb/hr) at 30% total solids and 4167kg/h (9187 lb/hr) at 27% total solids, respectively. With four units of 4167 kg/h each under construction, the Lakeview plant of Region of Peel in Ontario, Canada will become one of the largest fluid bed incineration plants in the world (total installed dry-solids capacity of 400 metric tonne/d (441 ton/d). Since 1962, more than 130 fluid bed incinerators have been installed in North America alone.

Since 1988, 53 new municipal fluid bed systems and one new multiple-hearth system have been installed in North America. Of the fluid bed installations, 18 replaced existing MHFs.

This chapter focuses on fluid bed rather than multiple-hearth technology because of the industry trend of installing fluid bed incinerators in new or replacement installations. Readers are encouraged to refer to the many publications available for information on MHFs.

2.0 FLUID BED INCINERATION SYSTEM

This section addresses six specific fluid bed incinerator topics:

- Principles of fluidization, as applicable to incinerators.
- Principles of fluidization, specific to fluid bed incinerators.

- Fluid bed incinerator design considerations.
- A description of major fluid bed incinerator system components.
- Features of the furnace and subsystems.
- Advantages of the technology.

2.1 Principles of Fluidization

Correlation and theoretical formulas on fluidization technology are well developed in the literature. Thus, these formulas are not presented in detail in this chapter. Abstracts that are important to understanding the technology, however, are included.

2.2 Definition of Fluidization

Fluidization is defined by Kunii and Levenspiel (1969) as an operation by which fine solids are transformed into a fluidlike state through contact with an upflowing gas or liquid. This method of contacting has a number of unusual characteristics. Fluidization engineering attempts to take advantage of this behavior. In applying fluid bed incineration, air is used instead of a liquid to supply oxygen to the combustion and fluidize fine solids. Because the focus is on the application of fluid bed to incineration, the following will deal primarily with gas-fluidized systems.

Depending on airflow rates, various kinds of contacting of a batch of solids by air are illustrated in Figure 5.1. In this figure, superficial air velocity increases from left to right. As air passes upward at a low flow rate through a bed of fine particles, the air merely percolates through the void spaces between stationary particles because the velocity is not high enough to displace the sand particles. This configuration is considered a *fixed bed* because there is no particle movement.

With an increase in airflow rate, particles move apart and a few will vibrate and move in restricted regions. This situation is an *expanded bed* that is approaching fluidization and is characterized by a higher pressure drop than required for fluidization.

At higher air velocity, a point is reached when all particles are suspended in the upward flowing gas. At this point, the frictional force between particle and air counterbalances the weight of the particle. As a result, the vertical component of compressive force between adjacent particles disappears. Pressure drops in sections of the bed approximately equal the weight of fluid and particles in that section. The bed is considered to be barely fluidized and is referred to as a bed at *minimum or*

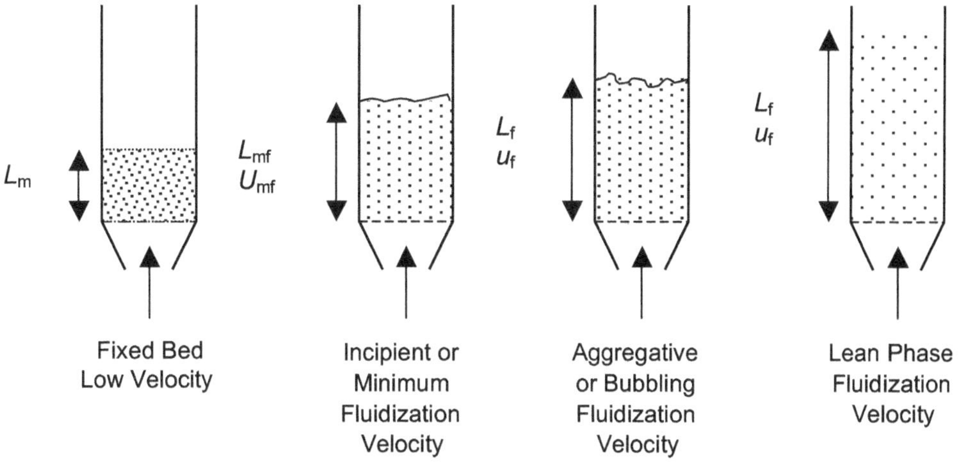

FIGURE 5.1 Various kinds of contacting of a batch of solids by air (airflow rate increases from left to right).

incipient fluidization, but at a lower pressure drop than before fluidization. The air velocity at incipient fluidization is called minimum fluidization air velocity (U_{mf}).

Large instabilities with bubbling and channeling of gas are observed when airflow rate increases beyond minimum fluidization. At higher flow rates, agitation becomes more violent, and the movement of solids becomes more vigorous. In addition, the bed does not expand much beyond its volume at minimum fluidization. Such a bed is called an aggregative fluidized bed, a *bubbling fluidized bed*, or simply a fluidized bed.

Gas-fluidized beds are considered dense-phase fluidized beds as long as there is a clearly defined upper limit or surface to the bed. However, at a sufficiently high fluid flow rate, the terminal velocity of the solids is exceeded, the upper surface of the bed disappears, entrainment becomes appreciable, and solids are carried out of the bed with the gas. This state is called a dispersed, *dilute*, or *lean-phase fluidized bed* with pneumatic transport of solids.

2.3 Liquid-like Behavior of a Fluidized Bed

Liquid-like behavior occurs in a fluidized bed. A dense-phase, gas-fluidized bed looks like a boiling liquid and, in many ways, exhibits liquid-like behavior (Figure 5.2). For example, a large, light object will float and a heavy object will sink. The pressure now

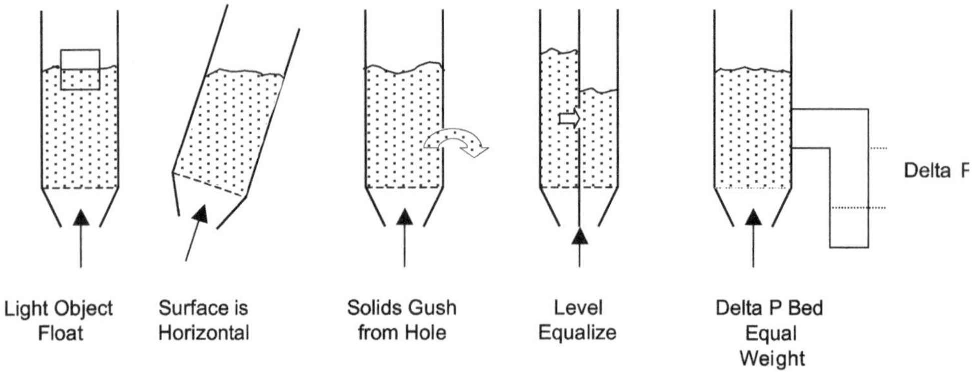

FIGURE 5.2 Liquid behavior of gas fluidized beds.

varies proportionally to height, like a liquid column. The pressure difference between any two points in a bed is approximately equal to the static head of the bed between the two points. The bed also has liquid-like flow properties. Solids will gush in a jet from a hole in the side of the container and can be made to flow like a liquid from vessel to vessel.

When applied to incineration, the liquid-like behavior of the fluidized bed offers the following advantages:

- Smooth, liquid-like flow of particles allows continuous, automatically controlled operations with ease of handling.
- Rapid mixing of solids leads to nearly isothermal conditions throughout the reactor; hence, the operation can be controlled simply and reliably.
- Circulation of solids between two fluidized beds makes it possible to transport vast quantities of heat produced or needed in large reactors.
- Fluidized beds are suited to large-scale operations.
- Heat- and mass-transfer rates between gas and particles are high when compared with other modes of contacting.
- Rate of heat transfer between a fluidized bed and an immersed object is high; hence, heat exchangers within fluidized beds require relatively small surface areas.

The advantages and economy of fluidized contacting have led to its successful use at WWTPs and in industrial operations such as calcining, ore roasting, drying (indirect-type fluid bed dryer by in-bed tube heat exchanger has been used largely in the powder industries), and oil cracking.

2.4 Fluidization Gas Velocity

The term fluidization has been used in the literature to refer to dense-phase fluidization and lean-phase fluidization and circulation systems involving pneumatic transport or moving bed. In this manual the term is restricted to dense-phase or bubbling fluid bed, which is mostly used for combustion of wastewater solids.

One of the most important factors in determining the regime of fluidization is the gas velocity U_0. If the gas velocity is too low, then the bed becomes sluggish, which could lead to severe consequences (such as agglomeration of sand in the vicinity of the fuel guns or even explosion). If the gas velocity is too high, then fresh feeds could bounce prematurely to the freeboard, leading to over-bed burning and lack of combustion in the sand bed. This situation may cause higher fuel consumption and incomplete combustion, leading to higher emissions. For a given bed of solids, the fluidizing velocity of gas U_0 is located between two limits: the minimum fluidization gas velocity, U_{mf}, and the terminal gas velocity, U_t.

2.4.1 *Minimum Fluidization Gas Velocity U_{mf}*

Estimation of the gas velocity U_{mf} at the onset of fluidization is essential because it is the most fundamental design parameter in fluidization. It determines the transition point between the fixed bed and the fluidized bed. Gas velocity U_{mf} can be determined by measuring the bed pressure drop as a function of gas velocity (Figure 5.3). On the pressure drop curve versus gas velocity, U_{mf} is the velocity at which a sudden reduction in pressure drop occurs.

The minimum fluidization gas velocity can also be estimated by the following equation defined by Kunii and Levenspiel (1969) for small particles:

$$U_{mf} = d_p^{\,2} (\rho_s - \rho_g)\, g / 1650 \qquad (5.1)$$

Where

U_{mf} = gas velocity (cm/s),
d_p = particle diameter (cm),
ρ_s = density of solid (g/cm^3),
ρ_g = density of gas (g/cm^3),
g = acceleration of gravity (980 cm/s^2), and
μ = viscosity of gas (g/cm · s).

Per equation 5.1, U_{mf} depends on characteristics of both solid particles and fluidizing gas. It varies greatly with the particle diameter.

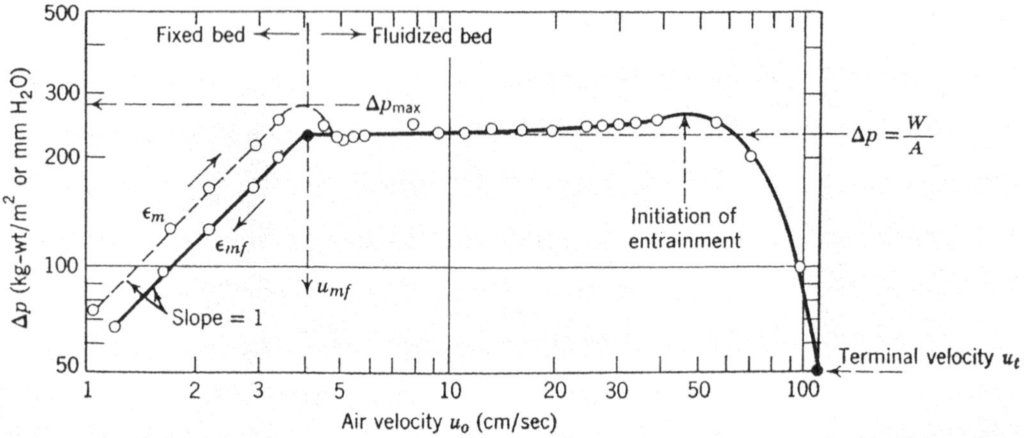

FIGURE 5.3 Pressure drop versus gas velocity for a bed of uniformly sized sand particles (Shirai, 1958).

2.4.2 *Terminal Gas Velocity* U_t

The terminal velocity or free-fall velocity U_t of a particle in a fluid bed of uniform size is the velocity at which entrainment or carry-over of the particle occurs. It is greatly dependent not only on the characteristic of the gas but also on the size, density, and physical shape of the particle (spherical, irregular, rough, or smooth particle surface). It could be determined by measuring the bed pressure drop with gas flow rate (Figure 5.3). It could also be estimated by correlations established from fluid mechanics by Kunii and Levenspiel (1969):

$$U_t = [4gd_p(\rho_s - \rho_g)/3\rho_g C_d]^{1/2} \tag{5.2}$$

where C_d is an experimentally determined drag coefficient.

Depending on the size and density of the particle and also on the characteristic of the fluidizing gas, U_t could range from 10 U_{mf} to 100 U_{mf}. When fluidization is applied to incineration, the bed support material has a wide size distribution. It is composed of fresh sand, eroded sand of smaller size, and fine ash particles. The terminal velocity of the smallest particle in a poly-disperse or mixed-particle assembly limits the operational range of velocity. The terminal velocity of the smallest particle may be just equal to, or even less than, the minimum fluidization velocity of the largest particle. In such a case, the carry-over of fines will take place while the largest particles are fluidized or kept in a fixed bed condition. Particles that fall between the two size extremes will be in a state of fluidization. To avoid carry-over of solids from a bed, the gas velocity for fluidized bed operations should be kept somewhere between U_{mf} and U_t. In calculating U_{mf}, the mean particle diameter for the size distribution actually present in the bed must be used. However, in calculating U_t, the smallest size of sand solid in the bed must be considered.

For WWTP applications, the fluid bed incinerator is typically designed based on a fluidizing gas velocity U_0 equal to 2.5 to 3 U_{mf}.

2.5 Transport Disengaging Height

In a bubbling fluid bed, bubbles once formed in the bed start rising, grow in size, coalesce, reach the bed surface, and finally erupt. When these bubbles erupt at the surface they scatter solids into the region above. Erupting gas bubbles do splash solids far into the freeboard or the region above the surface of the bed. If the gas exit were situated immediately above the top of the bed, then the gas would entrain a considerable amount of solids. With a higher gas exit, the amount of entrainment is smaller, and finally a level is reached above which entrainment becomes approximately constant. The entrainment increases with the gas velocity or when the size of solids decreases. This height of exit above the top of the bed where entrainment becomes approximately constant is known as the transport disengagement height (TDH). Findings from Zenz and Othmer (1960) on TDH at different gas velocity are illustrated in Figure 5.4. The solids used in this research are a combination of sizes between 20 and 150 μm. In this application, depending on the gas velocity, TDH ranges from 0.2 to 0.3 m above the surface of the bed. The TDH increases by approximately 70% for a doubling in gas velocity.

The height of the freeboard of a WWTP fluid bed incinerator is typically 4.6 m (15 ft) minimum (between the fluidized bed surface area and the exhaust gas duct). This height is determined by the freeboard gas residence time of (minimum) 6.5 seconds.

2.6 Freeboard Gas Velocity

Zenz and Othmer (1960) (Figure 5.4) show that for a given solids and vessel, the entrainment is not only sensitive to height of the freeboard, but also to gas spatial velocity. The entrainment increases strongly with gas velocity. It varies as U_f^2 to U_f^4. Freeboard gas velocity U_f is, therefore, an important parameter of design. In WWTP applications, the freeboard gas velocity ranges from 0.64 to 0.76 m/s (2.1 to 2.5 ft/sec), calculated based on combustion gases. The velocity should be small

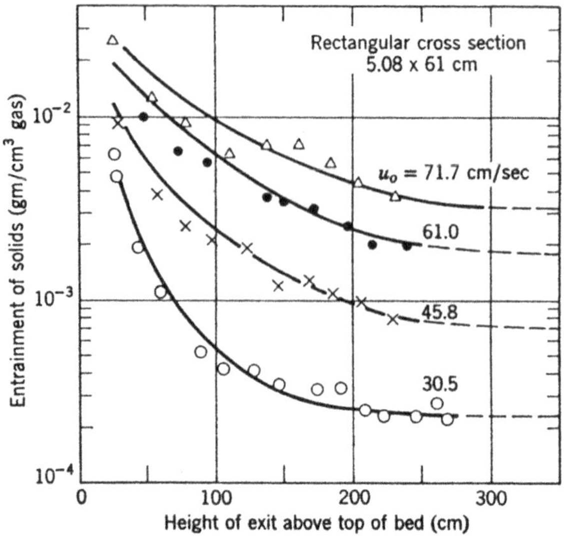

FIGURE 5.4 Entrainment of solids from different heights above the top surface of a dense bubbling bed (as shown in Kunii and Levenspiel, 1969, from Zenz and Othmer, 1960).

enough to minimize sand loss but large enough to allow entrainment of ash. Theoretically, the velocity must be greater than the terminal velocity of the ash, but less than the terminal velocity of the sand.

2.7 Hydrodynamics-Based Groups

In a gas-fluidized bed, the bubbles moving through the dense particulate phase have a strong influence on the quality of fluidization. Quality of fluidization is not the same for all solids. Depending on the mean size of the particle and the difference in density of the gas and the solid, Geldart (1973) has classified powder into four groups and designated them by the letters A, B, C, and D (Figure 5.5). Of these groups, only groups A and B are suitable for fluidization. Of these two groups, group A powders have dense-phase expansion after minimum fluidization but before the commencement of bubbling. Group B powders exhibit bubbling at the minimum fluidization velocity itself. The fluid bed layer in group B expands less than in group A, but exhibits higher mixing. Group C is difficult to fluidize because of its extreme electrostatic effect. Group D can create jetlike upward moving dilute phase, referred to as spoutable. The solid media used in WWTP fluid bed incinerators are of group B.

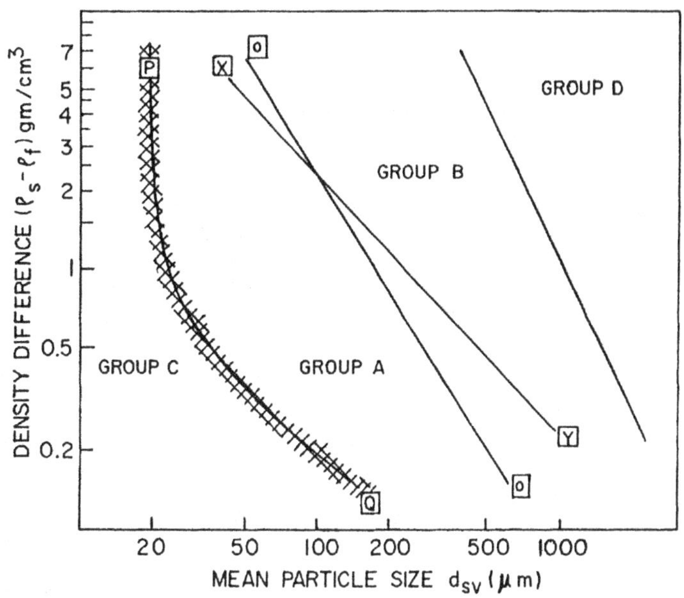

FIGURE 5.5 Geldart classification of powders (Geldart, 1973).

3.0 PRINCIPLES OF COMBUSTION

Combustion can be defined as the instantaneous union of an organic substance with oxygen and is usually referred to as "burning." The reactions always release heat. To achieve complete combustion, five basic requirements must be fulfilled: potential, temperature, residence time, sufficient oxygen, and turbulence.

3.1 Potential

There has to be material available that will burn or be oxidized by oxygen to cause combustion. Combustion will occur if materials are available that have elemental components (such as carbon, hydrogen, and sulfur) at a higher energy state than their products of combustion. Some of the oxidation reactions are shown in Table 5.1.

Gasoline and fuel oil have a large potential for combustion, as do feed streams that are fed to the incinerator. There are factors that will inhibit the potential for combustion of feed streams. These factors must be fully understood so that deviations from design process conditions may be rectified or adjusted before serious operational problems occur. These factors may be generalized into two basic areas.

TABLE 5.1 Oxidation reactions.

Reactions	Heat released, kJ/kg (Btu/lb)
$C + O_2 \rightarrow CO_2$	33 700 (14 500)
$2H_2 + O_2 \rightarrow 2H_2O$	144 000 (62 000)
$S + O_2 \rightarrow SO_2$	10 500 (4 500)
$2CO + O_2 \rightarrow 2CO_2$	10 200 (4 400)
$CH_4 + 2O_2 \rightarrow CO_2 + 2H_2O$	55 600 (23 900)

3.1.1 Heating Value of Combustible Materials

The heating value is the quantity of energy (heat) that can be released per unit mass of solid material being fed to the system and is the prime indicator of combustion potential. The conventional unit of measurement is kilojoules per kilogram (kJ / kg) or British thermal units per pound (Btu / lb). The heating value can be determined by an oxygen bomb calorimeter. For wastewater solids, the heating value is never the same at each test. It varies from plant to plant and from sample to sample and depends on its chemical oxygen demand or on the treatment process used. As an example, primary sludge has higher volatile contents and higher heating values than digested biosolids. Some waste products will digest and evolve organic gases if they are allowed to be stored for an extended period of time, thereby reducing their heating value. Inert solids in process feed streams also will have an effect on their heating value.

Typically, heating values for WWTP solids are in the range of 12 800 to 19 750 kJ / kg (5500 to 8500 Btu / lb) of dry solids. In general, a decrease in heating value and a decrease in total combustible solids content will decrease the capacity of the unit. An increase in heating value and increase in total combustible solids will increase the capacity of the unit.

3.1.2 Water Content of Combustible Materials

The quantity of water in the feed material has a significant effect on the operation of the incineration system. Water has no heating value but requires a large amount of heat to be vaporized and heated to the operating temperature of the incinerator. The heat required to vaporize this water must be supplied by the combustible materials in the feed stream or by auxiliary fuel.

Changes in water content will affect operating temperature and capacity of the incineration system. When feed material is wetter than design condition more fuel will be required to maintain bed temperature and, therefore, decrease feed capacity. To maintain design capacity and optimize operation, the percent total solids should be monitored twice daily.

3.2 Temperature

Combustible materials must be introduced into the fluid bed incinerator when the bed operating temperature is greater than the ignition point of those combustibles. If the bed operating temperature is not high enough to instantaneously ignite combustible material in the bed, then overbed combustion would be excessive and result in incomplete combustion. Municipal WWTP solids can generally be completely combusted with a bed temperature range of 650 to 760°C (1200°F to 1400°F) and a freeboard temperature range of approximately 815°C to 870°C (1500°F to 1600°F).

3.3 Gas Residence Time

Combustible materials must have sufficient time to react. The fluid bed incinerator is designed to allow sufficient time for the feed material and any auxiliary fuels to react with oxygen in the combustion air. The bed section is designed to completely disintegrate the feed and combust some of its volatiles to keep temperatures greater than 650°C (1200°F). The freeboard is designed to completely combust any volatiles that escape from the bed. Typically, the gas residence times are two to three seconds in the bed and six to seven seconds in the freeboard. Although lower freeboard residence times would achieve high combustion efficiencies, sufficient disengagement height is provided in the freeboard to reduce sand carry-over in the exhaust gases. This limitation results in the long freeboard residence times.

3.4 Sufficient Air

Oxygen is supplied to the fluid bed incinerator system in the form of fluidizing and combustion air. This air must be supplied in an amount slightly greater than that theoretically required for complete combustion. The normal indicator for excess air is percent free oxygen released to the atmosphere. Dependent upon the feed material and combustion temperature, the oxygen content in the exhaust gases should be a minimum of 4% by volume on a dry basis, or approximately 2% on a wet basis as measured in the exhaust gas before the scrubber.

For efficient operation, the high range of oxygen must be 5% based on wet gas. Anything greater than 5% will have a quenching or cooling effect on the bed and will reduce overall efficiency.

3.5 Turbulence

Combustible materials and combustion air must be mixed to optimize combustion or to efficiently operate at low excess air. The combustible material must be provided with adequate surface area to contact with oxygen and react. Highly agitated hot sand quickly fragments feed material into small particles, which in turn are quickly heated to volatilization temperature without depressing the bed temperature because of the large heat inventory of the sand bed. With turbulence, feed is better distributed to the bed and every particle of the fluid bed is exposed to the fluidizing combustion air, providing almost infinite extended surface. On the other hand, with lack of turbulence, feed material will be poorly distributed and a larger portion of the volatilized organics will reach the freeboard before being oxidized in the bed. This phenomenon can lead to excessive over-bed burning and subsequent higher emission of hydrocarbon volatiles and other products of incomplete combustion.

If the system is being operated in violation of one or more of the above five main principles of combustion then incomplete or poor combustion will result.

4.0 FLUID BED DESIGN CONSIDERATIONS

To design a fluid bed incinerator that is as efficient as possible with minimum emissions, the system must not operate with scale formation in the exhaust system nor formation of agglomerates (clinkers) in the sand bed at minimum auxiliary fuel consumption. Although scaling and formation of clinkers depends on the characteristics of the feed material, emissions of pollutants and auxiliary fuel consumption are directly linked to the fundamental parameters of design.

4.1 Fundamental Design Parameters

To determine equipment size and characteristics, it is necessary to determine required airflow to the furnace, flue gas quantities, supplementary fuel requirements, and cooling water requirements. First, a mass balance is determined. Then the heat balance can be prepared. Finally, system exit characteristics can be found. Heat and mass balances are discussed elsewhere in this chapter; this section emphasizes the fundamental parameters of design commonly used in the sizing of the incinerator.

TABLE 5.2 Fundamental parameters of design.

Parameter	Values
Bed material size range, μm (U.S. mesh)	1 680–177 (10–80)
U_o gas velocity in bed, m/s (ft/sec)	0.75–1 (2.5–3)
U_f gas velocity in freeboard, m/s (ft/sec)	0.76–0.64 (2.5–2.1)
Excess air, over stoichiometry	40%
Freeboard temperature, °C (°F)	843 (1 550)
Residence time in freeboard, sec	6.5 minimum
Combustion air temperature	Depends on percent total solids

There are several fundamental parameters: excess air, size, bed material, bed gas spatial velocity, freeboard gas spatial velocity, temperature in freeboard, and gas residence time. See Table 5.2.

4.1.1 *Size of Bed Material and Gas Velocities*

In designing the fluid bed system, the selection of bed material is critical. As shown in Figure 5.5, solid particle size has a direct effect on the quality of fluidization. For incineration of WWTP solids, sandlike material with a median size of 550 μm (30 mesh) is typically used. At bed operating conditions, using equation 5.1, the minimum fluidization gas velocity U_{mf} of the 550 μm (30 mesh) particle equals 0.33 m/s (1 ft/sec).

Because optimal bed fluidizing gas velocity is in the range of 2.5 to 3 U_{mf}, U_o is equal to 0.75 to 1 m/s (2.5 to 3 ft/sec) for the selected solid. Gas corrected to bed temperature and pressure should be used in sizing of the bed section.

Because entrainment increases with freeboard gas velocity, U_f is maintained as low as possible. It is typically in the range of 0.76 to 0.64 m/s (2.5 to 2.1 ft/sec).

4.1.2 *Excess Air*

If the ultimate analysis of the feed and feed rates is known, then combustion air can be calculated from the oxidation reactions shown in Table 5.1, with an excess of 40% greater than stoichiometry. Because the combustion of dewatered wastewater solids is a two-phase process (evaporation followed by combustion), almost one-half the volume of gas in the reactor is water vapor. Therefore, designing the system based on an excess air of less than 40% presents the risk of having incomplete combustion. Greater than 40% excess air is not efficient and can generate higher emission of nitrogen oxides (Dangtran and Holst, 2001).

4.1.3 Combustion Temperature and Gas Residence Time

For wastewater solids, a combustion temperature of 843°C (1550°F) is typically used, with a minimum freeboard gas residence time of 6.5 seconds. Although there are no regulatory temperature and residence time requirements for incinerators in the United States, the previous figure is conservative compared with European regulatory requirement of a freeboard temperature of 850°C (1562°F) for a minimum of two seconds (European Commission, 2000).

4.2 Combustion Air Temperature versus Solids Content

One of the most important criteria in designing the incineration system is to minimize fuel consumption. Supplementary fuel consumption is calculated from the heat and mass balances. It depends on two factors: (1) the heat content of the feed material (or specifically its solid content, i.e., percent dry solids); and (2) the heat content of the combustion air (or its temperature).

Solids content of the feed depends on the dewatering equipment and on the quantity of polymer used as a dewatering aid. Heat content of combustion air depends on how intense heat recovery is in the heat exchanger. Typically, a heat exchanger can recover up to 40% of flue gas enthalpy to preheat the combustion air to approximately 675°C (1250°F).

The theoretical curve of supplementary fuel consumption is presented in Figure 5.6. The calculation was based on a combustion gas temperature of 843°C (1550°F) and a throughput capacity of 454 kg (1000 lb) of dry solids per hour. The feed material is typical of that resulting from WWTP with 75% volatile solids and a high heating value of 23 260 kJ/kg (10 000 Btu/lb) of volatile solids. Fuel consumption decreases with either percent dry solids or combustion air temperature increase, or both.

The greater the solids content and the greater the combustion air temperature, the lower the auxiliary fuel requirement. With a wind box temperature of 648°C (1200°F), the feed material is burned autogenously, or thermally self-supporting, at a solid content of 27%.

To avoid emission of nitrogen oxides (NOx emissions increase with dry solids, according to Dangtran and Holtz, 2001) and to lower polymer consumption, the system is typically designed based on autogenous combustion at maximum air temperature and minimum dry solids.

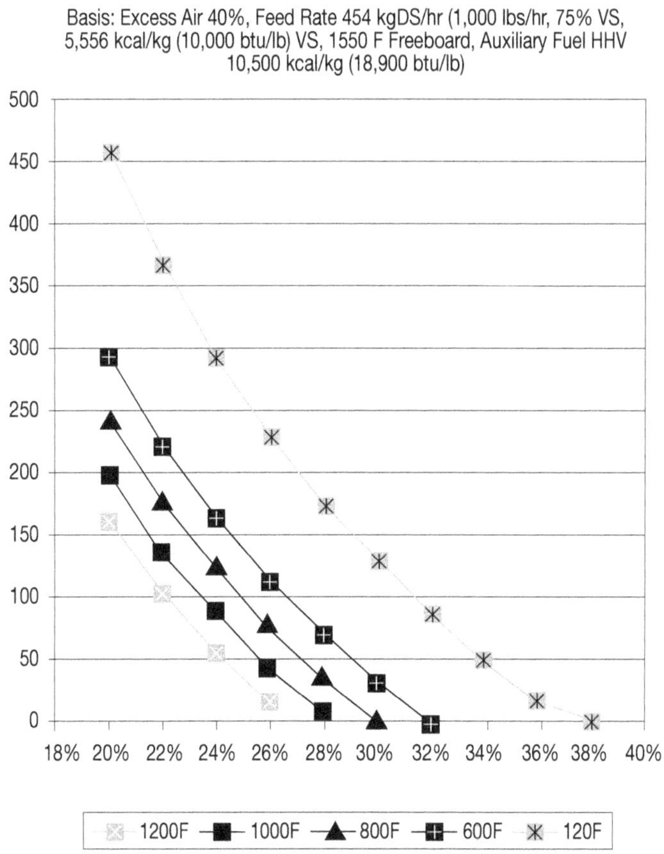

FIGURE 5.6 Auxiliary fuel consumption versus sludge solid content at various wind box preheat air temperatures (kg/h = 2.205 lb/hr; °C = [°F − 32] × 0.555).

4.3 Ash Analysis

In designing a fluid bed incinerator, it is also important to evaluate feed material for its ash chemical composition. Sodium and potassium chlorides have a low melting point. Large quantities of these materials in the feed can lead to glassification of the bed media. The bed media can become sticky, and agglomerates (or clinkers) can form, which results in segregation of bed materials and eventually defluidization of the bed. Furthermore, if the feed contains iron, phosphorus, and chlorides, deposition of iron oxides can occur. Scaling of these materials can obstruct the exhaust gas duct, which can result in excessive backpressure and lead to operating difficulties and shutdown of the plant. These problems can be eliminated by chemical addition (Jeffers et. al., 1999). With new trends in liquid stream biological nutrient removal (BNR), phosphorus concentration in sludge could be higher than before. Caution should be taken when designing sludge incinerators for disposal of waste activated sludge. Increased concentrations of phosphorus in the ash could change the eutectic characteristics and result in lower ash fusion temperatures.

To neutralize sodium and potassium, kaolin clay (a mixture of hydrous aluminum silicates) is used. It is typically available in a very fine powder and is a convenient source of both SiO_2 and Al_2O_3. It will react with sodium and potassium chlorides to form high melting point crystalline sodium and potassium aluminum silicates. These silicates have a melting point of approximately 1100°C (2000°F).

Lime is used to convert iron phosphate to iron oxides in the sand bed at bed temperature. This conversion prevents iron from forming gaseous iron chlorides, which can precipitate and form scales in the freeboard and in the exhaust gas duct.

To calculate the dose of chemical additives, a complete analysis of both soluble and total concentrations of the components in the ash is required. Details of the calculations can be found elsewhere (Jeffers et. al., 1999).

5.0 DESCRIPTION OF MAJOR COMPONENTS

The incineration system typically is composed of three major components: the fluid bed incinerator with subsystems such as for feeding of the wastewater solids and sand and other auxiliary equipment; the heat recovery system; and the air pollution control system. The heat recovery system and the air pollution control system are detailed in Chapters 6 and 7; this chapter is limited to the incineration system and its subsystems.

The fluid bed incineration plant could be located indoors or outdoors. Puerto Nuevo, Puerto Rico is an outdoor plant. It is composed of a hot wind box fluid bed, a heat exchanger to preheat combustion air to approximately 675°C (1250°F), a quench section followed by a cooling tray and multiple Venturi scrubber, a wet electrostatic precipitator, and a stack. An overview of the plant is shown in Figure 5.7.

The storage and feed system remains the same most of the time and is composed of a live-bottom bin and piston pumps for the feed. Incineration, heat recovery, and air pollution control systems, however, could be different from project to project. The fluid bed could be of either the hot or cold wind box types. The heat recovery system

FIGURE 5.7 The Puerto Rico fluid bed incineration plant (6000 lb dry solids per hour) (courtesy of Degremont Technologies – Infilco).

could be composed of heat exchangers for preheating the combustion and for plume suppression or a waste heat boiler for steam generation, or both. The air pollution control system could be either a dry-ash system or a wet-ash system.

A typical process flow diagram of fluid bed incineration system using a wet-ash system with a hot wind box and heat recovery by heat exchangers is shown in Figure 5.8. Solids are dewatered using belt filter presses and pumped to the furnace via piston pumps through two or four feed ports. no. 2 fuel oil or natural gas is used as auxiliary fuel during startup and operation as needed. The freeboard operates at a design temperature of approximately 843°C (1550°F). The reactor offers an expanded freeboard to allow deceleration of the larger particles to minimize sand carry-over and maximize carbon burnout. The bed makeup sand may be fed to the furnace pneumatically during operation, if required. The hot wind box furnace in this example is equipped with a refractory arch that supports the sand bed and evenly distributes the air. To minimize auxiliary fuel use, fluidizing air is preheated to approximately 675°C (1250°F) in an external tube and shell heat exchanger using the exhaust flue gas of the reactor as a heat source. The air pollution control system

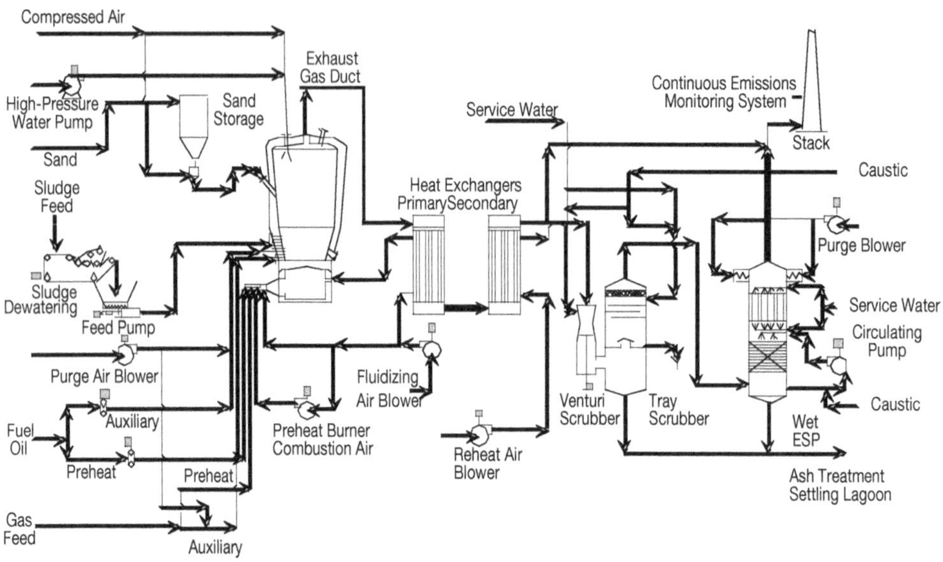

FIGURE 5.8 A typical process flow diagram for a wet-ash system.

includes a Venturi scrubber followed by a tray tower. A wet electrostatic precipitator could be used to eliminate submicrometer particulates. Hot air at 260°C (500°F) is optionally added to the stack gas for plume suppression. Plume suppression air is preheated in a secondary heat exchanger using the exhaust flue gas from the primary heat exchanger. Ash and fine sand particles carried in the flue gas are removed in the high-pressure drop Venturi scrubber where an ash-slurry is produced. The slurry is either pumped or flows by gravity to an outdoor ash settling lagoon system for dewatering. Dry ash at approximately 50% total solids is removed from the drying lagoon approximately once per month, depending on the size of the lagoon.

In the wet-ash system, acid gases such as SO_2 and HCl are removed by water in the Venturi scrubber and cooling tray. These gases are soluble in water, which means that up to 95% of the acids can be removed by effluent plant water alone. With stricter regulation, a solution of caustic can be added to the cooling tray to increase removal further. Mercury and dioxins can be removed from the flue gas in an activated carbon adsorber, which is installed before the stack.

Because of its simplicity and the availability of effluent water and space at most wastewater treatment plants, the wet-ash system is mostly used in North America (more than 90% of existing plants).

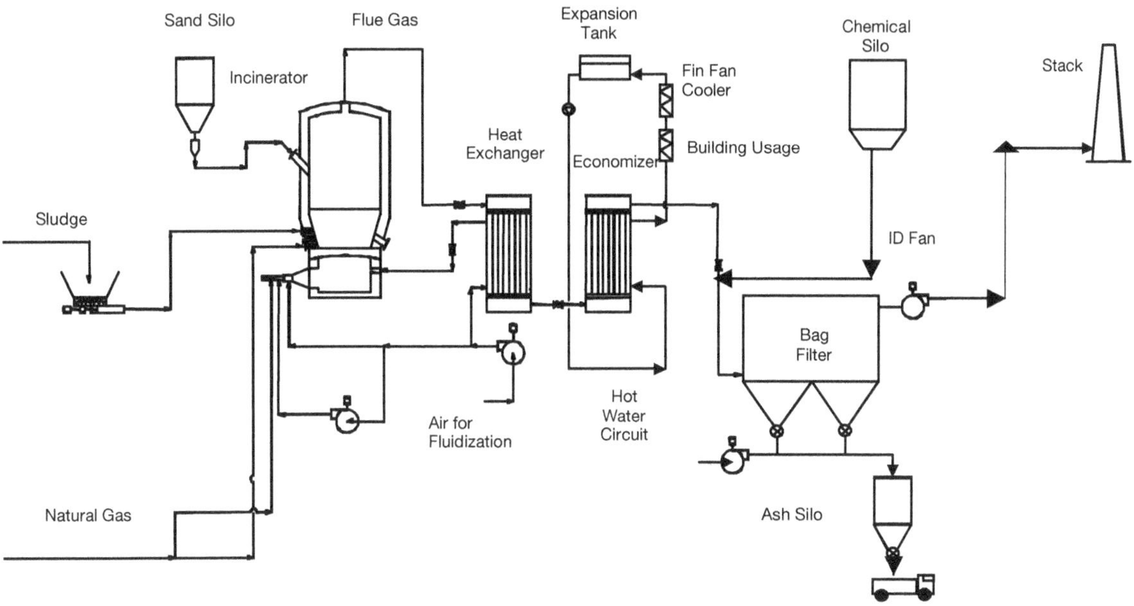

FIGURE 5.9 A typical process flow diagram for a dry-ash system (ID = induced draft).

Other types of heat recovery—such as waste heat boiler and dry type air pollution control systems—could be used. A typical process flow diagram of the dry-ash system is presented in Figure 5.9. In the dry-ash system, the flue gas temperature has to be in the range of 150 to 205°C (300 to 400°F), before its entrance to a bag-filter (or a dry electrostatic precipitator). A waste heat boiler or an economizer can be installed between the fluid bed and the air pollution control system or between the heat exchanger and the air pollution control system to generate steam or hot water. Chemical sorbent can be injected to the duct or in a separate reactor chamber installed between the heat recovery system and the air pollution control system to remove acid gases, mercury, and dioxins.

6.0 FLUID BED FURNACE

The fluid bed furnace can be divided into two types: the hot wind box and the cold wind box (or warm wind box). The hot wind box is typically provided to incinerate wastewater solids that typically have low heat values and that require intensive air preheating to minimize auxiliary fuel consumption. The cold wind box furnace is typically used to burn heat-treated solids, solids that have been dried, primary sludge, scum and grease, or other materials such as wood chips or sawdust that can be burned autogenously without heat recovery or with moderate heat recovery.

6.1 Hot Wind Box Fluid Bed

The hot wind box fluid bed is designed for maximum temperature exposure of approximately 650 to 980°C (1200 to 1800°F) for most applications. It is used when the wind box air temperature is greater than approximately 400°C (750°F). A cross-section of a typical hot wind box fluid bed is shown in Figure 5.10. The unit is a vertical steel shell made of carbon steel. The inside lining is made of refractory and insulating brick. The refractory lining is necessary because of the inside temperature of approximately 980°C (1800°F). The fluid bed is composed of four sections:

- The wind box section—the lower section is the wind box, which is a distribution chamber for fluidizing air and a combustion chamber for the preheat burner. The hot wind box is a refractory-lined plenum in which the hot combustion air is received. The wall of the wind box has openings for fluidizing air supply, a preheat burner, observation port, and instrument ports.
- The bed support and air distributor section—the roof of the wind box separates it from the next compartment. The roof can be constructed of refractory or metal alloys, depending on the service and design temperature requirements. The refractory arch construction is self-supporting because of the special shape of the refractory elements used, and is typically referred to as the dome. It supports the weight of the bed material, when the bed is not fluidized, and is a distributor plate for the fluidizing air. To allow the passage of air without the back drainage of sand, the dome has a number of air nozzles, commonly referred to as *tuyeres*. These are of special shape and material to prevent sand drainage, to provide uniform air distribution, and to withstand operating temperatures. The refractory arch distributor and the refractory-lined wind box are designed for a wind box temperature of approximately 980°C (1800°F). In a hot wind box with refractory arch distributor, the combustion air is typically preheated to approximately 675°C (1250°F).
- The bed section—the section immediately above the distributor that contains the fluidized mass of sand is called the bed section or combustion zone. The air from the distributor causes the bed of sand to fluidize. There is no physical

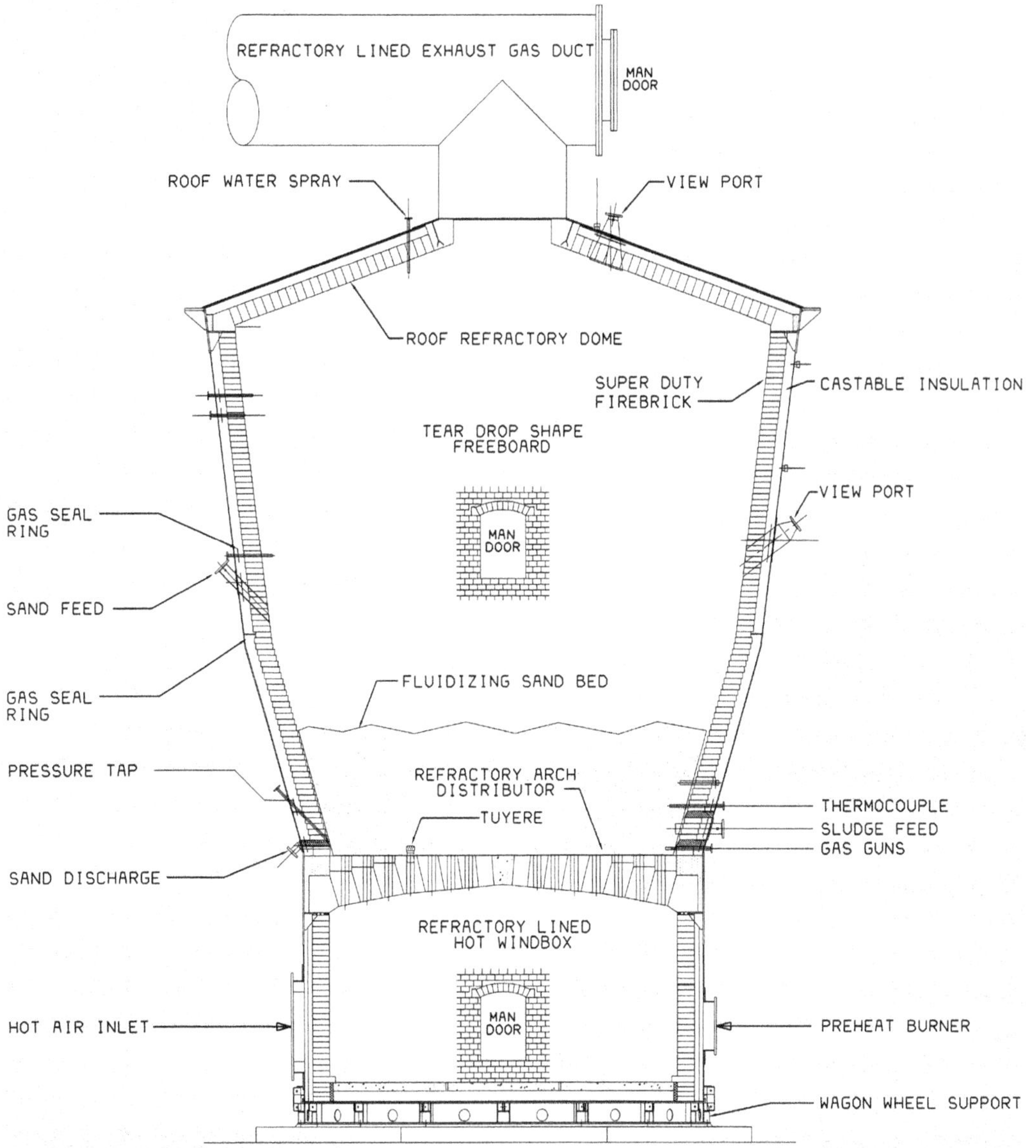

FIGURE 5.10 A typical cross-section of a fluid bed (courtesy of Degremont Technologies – Infilco).

upper boundary. Its height depends on the amount of sand in the bed. For incineration of wastewater solids, an expanded bed of approximately 1.5 m (5 ft) is typically used. The sidewalls slope outwardly from bottom to top of the bed to ensure water vapor expansion and to keep the gas velocity within acceptable limits. The walls are also equipped with nozzles and ports for sludge and auxiliary fuel injection and for various instruments.

- The freeboard section—the space above the bed is called the freeboard or disengagement zone. It acts as a retention chamber for the combustion gases to ensure a sufficient residence time at the temperature required and as a separation chamber for the bed material particles to disengage from the gas. To ensure complete combustion of any escaping volatile hydrocarbons (from the bed), the freeboard must be sized to provide approximately 6.5 seconds of gas residence-time minimum. It is typically 4.6 m (15 ft) high. The shape of the freeboard could be a cylindrical straight shape or a conical teardrop shape. The cylindrical straight shape is generally designed based on a gas velocity of 0.76 m/s (2.5 ft/sec). In the conical teardrop shape, the freeboard is expanded laterally along its height to maximize residence time and to reduce the gas velocity further. The gas velocity at the top of the teardrop shaped freeboard is 0.64 m/s (2.1 ft/sec). The exhaust gas duct is installed in the center of the roof dome to minimize gas bypassing and dead zone and to maximize residence time in the freeboard. The gradient of decreasing gas velocity in the freeboard and a lower gas velocity at the exhaust minimizes sand losses.

6.2 Cold or Warm Wind Box Fluid Bed

A cold wind box (or warm wind box) incinerator is used for feed material that can be incinerated without heat recovery (or with moderate heat recovery). Wind box air temperature is typically limited to less than 400°C (750°F). As the hot wind box, the cold (or warm) is also composed of four sections, and the designs in general are quite similar, except as follows:

- The wind box is not refractory-lined and the bed support and air distributor can be a metal alloy plate. To sustain the high temperature of the bed, the top of the plate is usually refractory lined.
- The preheat burner is installed in the freeboard, angled downward to heat the top of the fluidized sand bed during startup.

7.0 FLUID BED INCINERATION SUBSYSTEMS

As shown in the process flow diagrams in Figure 5.8 and Figure 5.9, the incineration system can be divided into subsystems, which are described in the next several sections.

7.1 Air System

Air supplied to the incineration system could be divided in three categories: fluidizing air, purge air, and atomizing air.

7.1.1 Fluidizing Air

Combustion air (same as fluidizing air) is typically supplied by a multistage centrifugal blower. The system shown in Figure 5.8 operates under positive pressure and is referred to as a push-type system. Complete gas tightness of the system is essential because the pressure in the system is greater than atmospheric and the temperature of the air and gases is high. This type of system is common in North America, where heat recovery is aimed at autogenous operation and plume suppression and wet-ash types of air pollution are typical. In contrast, the system shown in Figure 5.9 operates at both positive and negative pressure, with the zero point typically at the furnace outlet. This type of system is referred to as a push-pull type system. In this case, an additional induced draft fan is required to ensure that negative pressure is maintained in the exhaust gas system. Push-pull systems are typically used where waste-heat boilers and dry-ash type systems—which are more difficult to make gastight—help avoid any risk of leakage of hot gases and dry ash.

In a hot or warm wind box incinerator, the air is preheated in a shell and tube heat exchanger, which uses the reactor offgases as the hot medium. Heat recovery by combustion air preheating reduces supplemental fuel and increases capacity of the plant. In a cold wind box, combustion air from the centrifugal air blower is supplied directly to the metal wind box.

The fluidizing air blower is designed with several stages to provide the required outlet pressure. The main control of the incineration process capacity is regulated by the quantity of fluidizing air entering the wind box. The airflow is controlled by a damper on the blower inlet side, and measured by a flow meter.

7.1.2 Purge Air

Purge air is used in all ports to the incinerator and also to all expansion joints and pressure taps in the ductwork to keep them cool and free of sand and ash. Purge air can be either high pressure or low pressure. For example, high-pressure air is supplied from a compressed system for the roof spray nozzles and pressure ports to keep them cool and free of deposits. Purge air from the fluidizing air blower can be used in annular sleeves of all ports, including site ports, bed oil guns, sand inlet/outlet, sludge inlet, and roof sprays.

7.1.3 *Atomizing Air*

Atomizing air at approximately 48 to 55 kPa (7 to 8 psig) is required when fuel oil is selected as auxiliary fuel. Atomizing air can be provided by an injection-purge air blower.

7.2 Feed System

Solids from the wastewater process are dewatered first. Dewatering can be achieved in a plate filter press, a belt filter press, or a centrifuge decanter. A belt press typically can dewater undigested sludge to approximately 25 to 30% dry solids. High-performance centrifuges typically can dewater to approximately 30 to 35% dry solids. Polymer is used in both the belt press and centrifuge decanter as a dewatering aid.

Continuous, even transport and distribution of the dewatered cake to the furnace is an important adjunct to stable and economical operation. Past practice was to use screw extrusion feeders for dry cake and progressing cavity pumps for wet cake. Today, hydraulic piston pumps are mostly used to convey the cake from dewatering equipment to the furnace. The piston pump is preferable to others because of its flexibility and insensitivity to feed quality. Recent developments have been made in the design of progressing cavity pumps for pumping cake.

Two types of feed operations can be found in the literature: overbed feeding and in-bed feeding. Overbed feeding consists of dropping the feed cake either by gravity or by air spraying on to the bed from the freeboard sidewall or from the roof of the furnace. In-bed feeding consists of conveying the feed cake at high pressure directly to the bed of sand at a height of approximately 1.2 m (4 ft) under the bed surface area.

Overbed feeding is simple but is prone to bypassing of uncombusted feed cake particles into the exhaust. Overbed feeding is more commonly used in other applications such as fluid bed boilers burning coal or other solid waste-fuel, in which a cyclone is installed at the exhaust of the furnace to return the unburned carbon back to the bed.

In-bed feeding is mostly used for incineration of wastewater solids because the combustion process is slower and in two stages (evaporation and combustion). The feeding location, at 1.2 m (4 ft) under the bed surface or 30 cm (1 ft) above the distributor, ensures that maximum possible retention time of the feed cake particles is obtained before they reach the bed surface where carry-over may occur. It is important that the feed cake release its maximum energy to the sand bed to counteract the quenching effect of water evaporation.

Depending on the diameter of the furnace, there are usually two or four feed injection points to ensure that the cake is evenly distributed throughout the bed.

7.3 Sand System

Sand-like material is used as bed media. The furnace is typically filled with sand to a static height of 0.9 m (3 ft). When the fluid bed is in operation, the bed material will expand, because of fluidizing air, to a height of about 1.5 m (5 ft). With time, the sand gets abraded, and makeup is required. Makeup sand can be pneumatically fed into the furnace during normal operation. The feed system is generally of the dense-phase type pneumatic conveying.

Bed removal systems have been provided with some existing systems to cool the sand. It's critically important to remember that removing sand from a hot bed is unsafe. As a safety precaution, sand should be removed from the bed when the furnace has been cooled and the sand bed is at about 38°C (100°F).

Hydrodynamics of the fluidized layer depends on size and density of the media because the furnace is sized based on gas flow rate. The media should have a bulk density of approximately 1600 kg/m^3 (100 lb/cu ft), with typical particle size analysis as shown in Table 5.3.

The sand must be angular, dry, and free of sodium and potassium. It must not grind into fines at an operating temperature of 870°C (1600°F) or fuse at 980°C (1800°F).

Two types of fine bed media can be used: silica or olivine sands. Although silica sand is lower in unit cost, abrasion of the silica is higher than with olivine sand.

Table 5.3 Typical sand particle size distribution analysis.

Particle size μm (U.S. mesh)	Distribution, %
2 380–841 (8–20)	0–20
841–500 (20–30)	10–30
500–350 (30–40	20–25
350–295 (40–50)	20–25
295–210 (50–70)	0–5

Abrasion results in sand particle size being reduced and in the sand exiting the furnace with the flue gases, commonly referred to as elutriation. This sand must be replaced. Depending on the quality of the feed, the selection of media needs to be carefully planned. Olivine sand is not desirable in two particular cases. First, if the feed contains grit, then the bed material can build up and removal of excess bed material will be required, resulting in unnecessary shutdowns. Second, if the feed contains high alkali metals, then accumulation of these elements on the olivine bed can lead to low melting eutectics (Jeffers et al., 1999).

7.4 Auxiliary Fuel System

Auxiliary fuel is used at startup in a preheat burner or during normal operation by direct fuel injection into the fluidized sand bed via fuel guns. A wide variety of fuels including coal, saw dust, and digester gas can be used as supplementary fuel to the fluid bed incinerator. The only stipulation is that the fuel must be suitable for feeding in a reliable and controllable manner. Among these fuels, natural gas and no. 2 fuel oil are most common.

7.4.1 Preheat Burner

When starting up from cold temperatures, it is necessary that a proper burner system preheats the fluid bed to the temperature at which it is possible to inject fuel to the bed itself. It is important that proper ignition takes place. In a hot wind box, the preheating is done by a standard industrial oil (or gas) burner, which is placed in the sidewall of the wind box. The fluidizing air from the heat exchanger (pre-heater) is heated by mixing with the hot gases from the burner; the resulting hot air then fluidizes the sand bed. Heat from the air is then intimately transferred to the fluidizing sand. In a cold or warm wind box where the preheat burner is installed in the freeboard, the heat transfer from the gas to the bed is less efficient. Therefore, fuel consumption is greater and length of time to startup is higher in a cold wind box than in a hot wind box.

The air supply to the preheat burner is taken from the outlet side of the fluidizing air blower. The air is then pressurized further by the combustion air blower to ensure a burner supply pressure that is approximately 14 kPa (2 psig) greater than the pressure inside the wind box.

7.4.2 Bed Fuel Injection

Once the fluidized bed is sufficiently hot to support instantaneous ignition of the fuel, the preheat burner can be shut down; preheating is continued by injection of auxiliary fuel directly to the fluidized bed. This mode of operation is much simpler than with the preheat burner and is typically preferred to such an extent that a bed once heated will be kept hot for an extended period by reheating. Another reason for keeping the bed hot at all times is the extension of the refractory life.

Oil injection takes place through oil guns located at the periphery of the bed at approximately 0.3 m (1 ft) above the distributor. To prevent fouling of the injector guns, the oil is mixed with purging air at the supply end of the gun and blown into the bed in the form of a coarse mist. Natural gas injection takes place through gas guns located at the distributor level. To homogenously distribute the gas throughout the sand bed, gas guns are delivered in different lengths.

It is important to maintain airflow on the oil or gas guns whenever they are inserted to the bed. Therefore, the air supply line is equipped with a flow indicator.

7.5 Water System

To protect the heat exchanger from excessive temperatures in a hot or warm wind box, the fluidized bed is equipped with water quench spray nozzles. The nozzles are installed through the roof of the furnace above the freeboard. The nozzles operate at a high gauge pressure (approximately 2100 kPa or 300 psig) to create a fine water mist that will evaporate and quench the exhaust gas as fast as possible. The evaporation takes place close to the freeboard exhaust gas duct, limiting the cooling effect of the flue gas leaving the incinerator. The nozzles are used in sequence, depending on the temperature of the gas at the heat exchanger inlet.

The spray nozzles have small orifices and swirl grooves. To maintain clear internal passages and provide some cooling for the nozzle when it is not spraying, it is purged with air from the compressed instrument air system at a gauge pressure of approximately 410 kPa (60 psig). Both air and water supply have check valves installed to prevent backflow of one medium into the pipes of the other.

The water supply includes the water pump, pressure regulator, filter, and relief valve.

7.6 Ductwork and Expansion Joints

All hot gas and air ducts are furnished with internal refractory lining and internal thermal insulation and expansion joints. The refractory lining must resist abrasion because the hot gas contains abrasive ash and elutriated sand. Internal thermal insulation is selected and sized for a typical steel temperature of 93°C (200°F).

Expansion joints are constructed of high-temperature metal. The expansion joints are refractory lined and of metallic bellows construction to allow for duct and equipment movement caused by temperature fluctuations.

Purge air is provided to the bellows to prevent sand and fly ash from passing through and restricting movement of the joint.

7.7 Process Control System

The control system is designed with alarms and interlocks to ensure safe operation. In general, interlocking is based on a few fundamental philosophies. None of the combustion operations can be started until the various safety checks are cleared. Safety checks include airflow rates per design conditions, water flow rates to the Venturi scrubber, and so forth. Typically, the control system consists of programmable logic controllers (PLC) and programmable controllers with a screen control monitor as interface. All process information recorded by the instrument and control equipment is displayed on the operator's graphics computer screen for plant monitoring.

For the safety of the operation, the plant is fitted with temperature elements called thermocouples. The thermocouples are used to give control signals to the various combustion control loops associated with the incinerator operation and to determine the bed temperature span, which is an indication of fluidization quality. The incinerator is also fitted with pressure taps. The differential pressures in the bed indicate bed height and are also used to monitor the quality of fluidization. A wide span of bed pressure differential generally indicates a well-fluidized bed. Thermocouples and pressure taps are also used in the heat recovery and air pollution control systems. Water flow and airflow are measured by a mass flow meter.

The fluidized bed exhaust is supplied with an oxygen sampling and monitoring system to assist the operating personnel in monitoring combustion and function as a source of interlocks and alarms.

The operation of the fluid bed and its performance depend on the feed rates of the three major flows to the incinerator—air, cake, and supplementary fuel. Airflow can be set constant, although the composition of the dewatered cake, especially the solids content and, therefore, the cake feed rate, varies with time. The variation of cake feed rate is the primary reason for the need to observe and control the process continuously. The control is simple and is based on only two parameters of control: temperature and excess air (or oxygen). These parameters are continuously monitored during the process.

7.7.1 Bed Temperature

Thermocouples placed in the sidewall of the furnace bed measure the bed temperature. The thermocouples are averaged to get the bed temperature for control. The operating range for bed temperature is typically from approximately 675 to 790°C (1250 to 1450°F) during stable operation. The temperature of the fluidized sand bed must be high enough to ensure that ignition of the cake and of the fuel takes place. A sufficient part of the cake combustion must take place in the bed itself so that the heat liberated is available for evaporation of water and subsequently to reduce supplementary fuel.

The bed temperature must also be kept sufficiently high to ensure that the freeboard temperature is greater than the required minimum of 843°C (1550°F) to ensure complete combustion of hydrocarbons. Alternatively, the bed temperature should not be kept higher than necessary to minimize fuel consumption and to respect the upper limit to the heat exchanger gas inlet temperature of 870°C (1600°F).

Interlocks are installed to stop cake feed if the bed temperature drops below the ignition temperature of sludge and fuel.

7.7.2 Oxygen

An analyzer installed in the hot gas duct, which connects the heat recovery system and the air pollution control system, measures oxygen. It is recorded in real time and displayed in the operator console.

To ensure complete combustion, it is necessary to have excess oxygen available. The design excess air of the fluidized bed incinerator is 40%, which corresponds to approximately 3.5 to 4.0% oxygen (by volume on a wet basis) in the furnace offgas.

To ensure good fluidization, fluidizing airflow rate must be maintained constant at design airflow rate. Excess air is, therefore, a function of the total amount of combustible material (dewatered cake and fuel) being fed to the combustor.

If the oxygen content is high, the combustion air is not being efficiently utilized for combustion, which can mean that the capacity of the unit is less than optimal.

If oxygen concentration in the reactor offgas falls to less than 2% (by volume wet basis) for two minutes, alarm and interlock will be activated and interrupt cake feeding.

8.0 ADVANTAGES OF THE TECHNOLOGY

As outlined above, fluid bed incineration is the most efficient thermal technique for wastewater solids disposal. The technique offers several advantages over other techniques, including

- Flexibility, suitable to intermittent operation. The inventory of hot bed solids acts as a thermal reservoir, thereby causing only a small change in temperature upon shutdown. This situation permits quick startup following daily or weekend shutdown.

- Flexibility, suitable to feed variability. Within a defined size range, the fluid bed can easily handle feeds with varying chemical properties, moisture, and volatile contents, particularly during short periods of time. This characteristic is primarily a result of the large quantity of heat stored in the sand bed and to the fact that the feed represents approximately only 1% of total bed material.
- Ease of control and automation. The rapid mixing of solids caused by the turbulence in a fluid bed provides uniformity of bed temperature. Temperature control is reduced to basically one point of control, which is the average of all measurements.
- Lower auxiliary fuel usage. Much less fuel is required, mainly because of relatively low excess air requirements (only 40% excess over stoichiometry compared with typically 100% excess over stoichiometry in other thermal technologies) and possibility of heat recovery from flue gas and preheating of combustion air to approximately 650°C (1200°F).
- Reduced maintenance costd. There is no moving part exposed to the combustion section. Absence of thermal shock from the thermal reservoir of the bed solids, results in slow temperature changes and leads to longer refractory life and lower maintenance cost.
- High efficiency of combustion. High turbulences of the sand bed increase the contact surface of cake particles and oxygen, increasing combustion efficiency. Combustion is basically instantaneous, with low emissions of CO and THC.
- Low NOx emissions, primarily because of low excess air and good distribution of temperature throughout the bed.

9.0 MULTIPLE-HEARTH FURNACE

The MHF was developed in 1888 for roasting ores that contained metal sulfides to isolate metals from further refining. Nichols Engineering—a New Jersey firm that designed and built Hereshoff-type metallurgical furnaces in the early part of this century—realized that the extended holding time and positive material flow the furnace offered was well suited to the needs of countercurrent drying and burning wastewater filter cake. This product of dewatering must be dried to approximately 50 to 55% moisture before it can sustain its own combustion; but it is sticky and rapidly fouls indirect heating surfaces. It was found that the plowing action, called rabbling, that occurred in the MHF resulted in uniform drying. The firebrick walls and hearths were amply resistant to combustion temperatures (WEF, 1992).

Nichols began to help municipal consulting engineers use this design in the early 1930s. Twenty-three units were installed in 12 cities before World War II in the Northeast or Midwest. Feed solids handled included raw primary, digested primary, and digested activated. Ferric and hydrated lime were the primary chemicals used in solids conditioning for dewatering. In the late 1940s and 1950s, many other cities adopted this approach, resulting in the installation of 44 more units by 1960.

Since then, several factors have affected the use of MHFs. In late 1973, the first oil embargo by the Organization of Petroleum Exporting Countries caused many furnaces to shut down because of the cost and scarcity of oil and gas. Improvements in dewatering technologies and development of polymers as dewatering aids resulted in dryer feed solids and reduced fuel costs. Of the approximately 400 MHF units installed in municipal wastewater plants, 250 to 260 remain in operation. The last new unit was constructed in 1993. Since then, a number of units have been upgraded to meet the Title 40 Part 503 of the Code of Federal Regulations (40 CFR Part 503) or have been replaced with fluid bed incinerators.

New MHFs are unlikely to be constructed at wastewater plants. However, the many operating MHFs have several years of significant useful life remaining. To maximize the existing capital investment, many MHFs will require upgrades to achieve their maximum useful life and to satisfy regulatory requirements. This section describes MHF combustion technology and potential upgrades that are available to achieve maximum life expectancy of existing facilities while meeting regulatory requirements.

9.1 Process Design Considerations

The MHF is designed for continuous operation. Startup fuel requirements and the extended time needed to bring the hearths and internal equipment to the correct temperature from a completely cold condition typically preclude intermittent operations. The MHF is a vertical, cylindrical, refractory-lined steel shell containing a series of horizontal refractory hearths, one above the other (Figure 5.11). A central shaft, hollow to allow the passage of cooling air through it, runs the height of the furnace and rotates within the furnace at roughly one rotation per minute carrying the rabble arms above each hearth with it. There are two or four rabble arms per hearth. Each arm contains rabble teeth or plows that rake the dewatered cake spirally across the

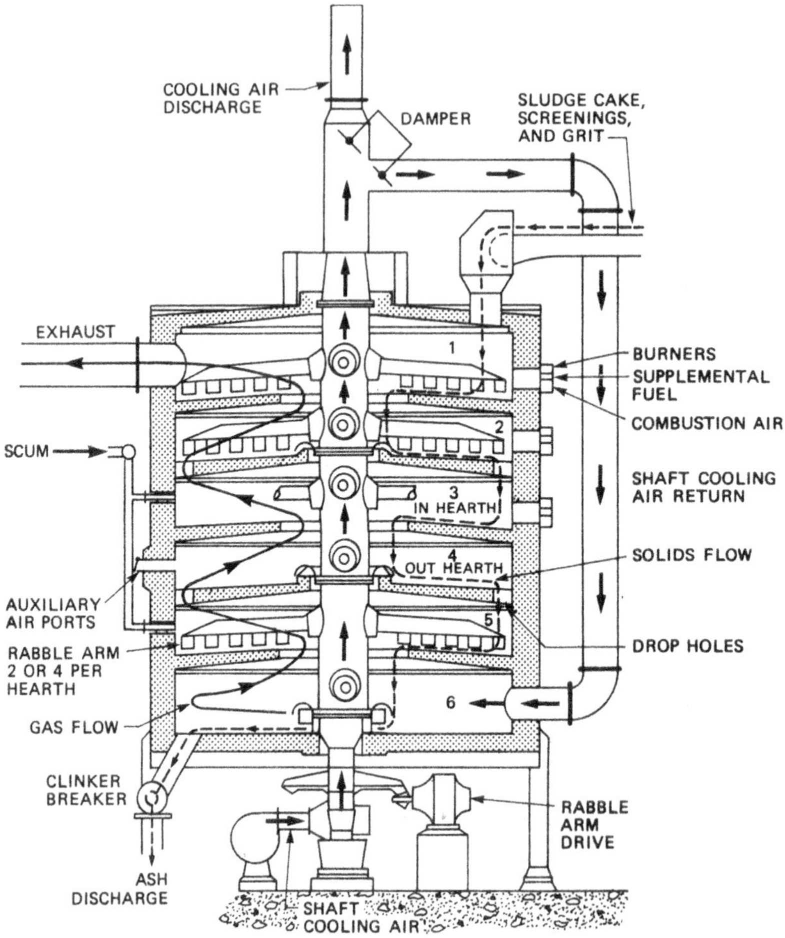

FIGURE 5.11 A typical cross-section of an MHF (WEF, 1992).

hearth as the arm rotates above each hearth. The cake typically is fed to the feed hearth, and it is raked either toward the center or toward the periphery, where the cake drops to the hearth below. When the cake is raked toward the center, the hearth is referred to as an *in-hearth*. When the cake is raked outward to holes in the periphery, the hearth is referred to as an *out-hearth*. The bottom hearth is always an out-hearth to facilitate removal of bottom ash from the periphery. The alternating drop hole locations on each hearth and the countercurrent flow of rising exhaust gases and descending cake provide contact between the hot combustion gases and the cake feed solids to ensure complete combustion.

The MHF can be divided into five zones during incineration (Figure 5.12). The first zone is the afterburning zone. Afterburning is typically required to meet hydrocarbon emission requirements of the Part 503 regulations. The afterburning zone can be located on the upper hearths or can be external to the incinerator. The second zone is the drying zone which consists of the upper hearths (below the afterburner, if it is internal), where most of the water is evaporated. The third zone, generally consisting of the central hearths, is the combustion zone, where temperatures reach approximately 760°C to 930°C (1400 to 1700°F). The fourth zone is the fixed carbon burning zone which oxidizes carbon to carbon dioxide. The fifth zone is the cooling zone in which ash is cooled by the incoming combustion air. The sequence of these zones is always the same, but the number of hearths in each zone is dependent on the quality of the feed, design of the furnace, and operational conditions.

When the heating value of the feed cake is insufficient to sustain autogenous combustion, the additional heat required is supplied by adding supplemental fuel to

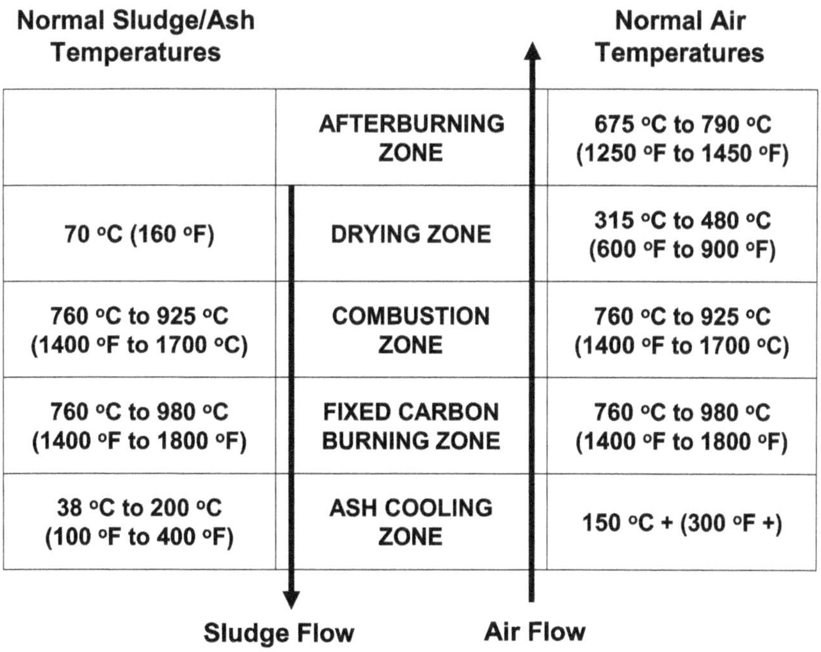

FIGURE 5.12 Multiple-hearth furnace process zones (adapted from U.S. EPA, 1979).

burners located at various points in the furnace. Burners may operate either continuously or intermittently and on all or selected hearths.

A measure of the quantity of water evaporated from the cake during burning is the drop in temperature of the hot gases as they pass between the combustion zone and the gas outlet. In an MHF, gas temperatures in the combustion zone may exceed 925°C (1700°F). These gases sweep over the cold, wet cake fed into the drying zone, giving up considerable portions of their heat in evaporating water. While the temperature of the solids is only marginally increased in the drying zone, the gas temperature is drastically reduced, typically to approximately 315 to 480°C (600 to 900°F). Exhaust gas temperatures should be maintained at less than 480°C (900°C) by controlling airflow to prevent distillation of odorous greases and tars from the drying solids. However, afterburning is still likely to be required to meet hydrocarbon limits (U.S. EPA, 1979).

Excess air of 100 to 125% must be provided to ensure adequate cake burnout. Some 10 to 20% of the ash is airborne, and gas cleaning equipment must be provided for its capture. Occasional odor problems may require installation of afterburning equipment.

Chapter 4 discusses combustion theory and practice for the MHF; Chapter 10 provides information on operating a MHF.

9.2 Description of Major Components

The MHF has been supplied in diameters ranging from approximately 2 m (6 ft 6 in) o.d. to 7.85 m (25 ft 9 in) o.d., with anywhere from 6 to 12 hearths. Some furnaces have been supplied with purpose-built internal afterburning hearths, typically the top hearth, which has a larger volume than typical hearths. Other MHFs have been modified to provide internal afterburning by removing hearths to create a single system with larger volume or by dedicating more than one hearth.

Multiple-hearth incinerator systems consist of the MHF and the air pollution control system. Figure 5.13 illustrates a typical multiple-hearth incinerator system process flowsheet. The system consists of an MHF with an internal afterburner, a waste heat boiler, a Venturi scrubber, and a wet electrostatic precipitator (ESP). Combustion air is supplied to the burners and the furnace by a combustion air burner blower. Shaft cooling air is supplied to the central shaft by a shaft cooling air blower. The furnace draft and pressure drop through the exhaust gas components is provided by an induced draft fan located upstream of the stack. An emergency bypass stack (not shown) is provided for emergency relief.

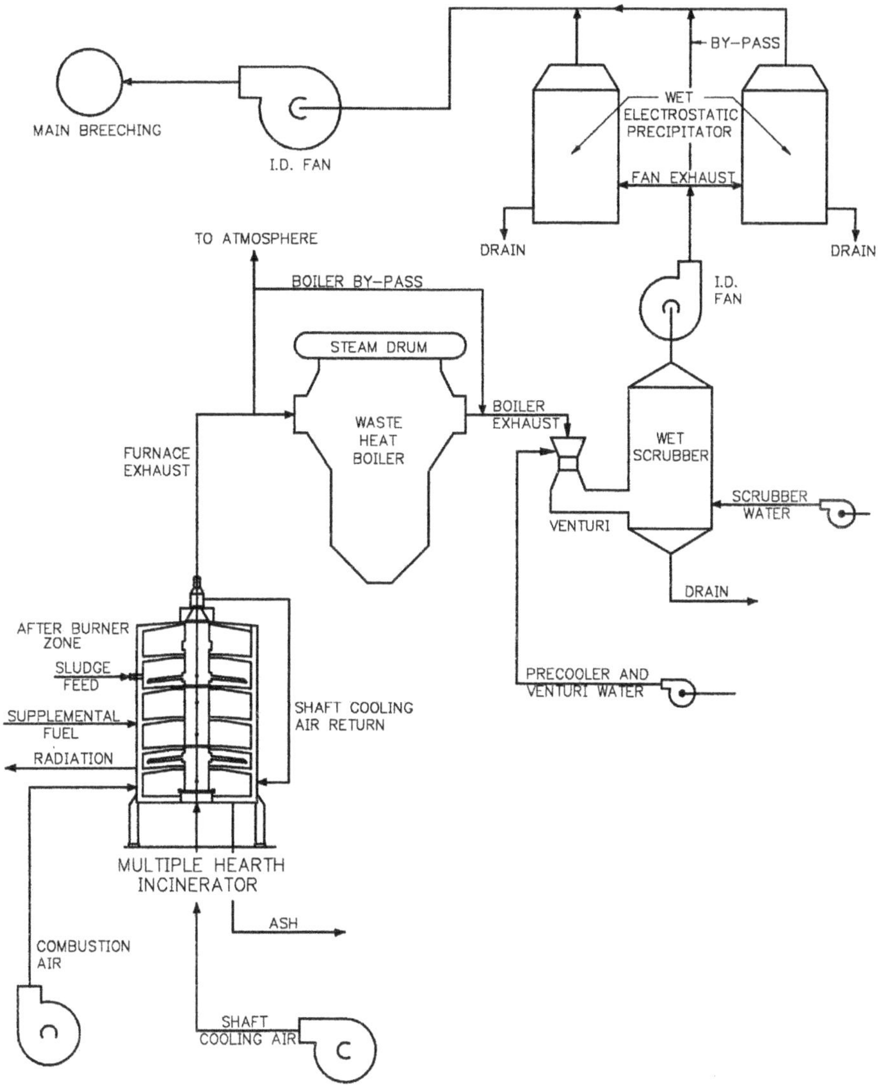

FIGURE 5.13 A typical multiple-hearth incinerator process flowsheet (I.D. = induced draft).

In the system shown, the waste heat boiler and the wet ESP are optional depending on the owner's desire for heat recovery and regulatory requirements. Also, the afterburner could be located where the waste heat boiler is shown. Heat recovery and air pollution control are discussed in detail in Chapters 6 and 7.

9.3 Composition and Construction

The shell of the MHF is constructed of welded or bolted steel plate, although older furnaces have been provided with rivets for joining the steel. The shells are provided with reinforcing bands at each hearth to support the lateral forces from the refractory hearths. Openings are provided for several components: access doors; burners; feed ports; exhaust gas duct; combustion air ports; observation ports; and ports (lance tubes) for cleaning clinkers and instrumentation. Openings are reinforced as necessary. The exhaust gas outlet is located on the wall of the top hearth and is a rectangular opening with a flanged stub that is refractory lined. The floor is constructed of steel plate supported on a steel frame with structural steel columns to provide headroom between the furnace floor and the building floor for the central shaft drive and to support bearing and the shaft cooling air housing. The furnace floor and frame support the entire weight of the furnace. The floor includes an opening for the central shaft and sand seal and openings for ash discharge. The roof is constructed of steel plate, reinforced to support the top bearing and sand seal for the central shaft. An opening is provided for the central shaft, feed ports, and the emergency exhaust.

The shell, floor, and roof are lined internally with refractory. The wall refractory typically consists of approximately 230 mm (9 in) thick high-heat-duty refractory brick, backed by approximately 155 mm (4.5 in) thick insulating brick or block. The shell may be coated with a corrosion-resistant lining. The floor and roof are typically lined with castable refractory.

9.3.1 *Hearths*

Each hearth is constructed of high-heat-duty refractory firebrick arranged in successive rings from the outside to the inside. The outer ring is constructed of a special firebrick, referred to as a skewback, having an angled internal face for supporting the hearth. The brick is approximately 300 mm (12 in) square and much heavier than a single, standard firebrick. This brick also supports the wall refractory between each hearth. Each ring is constructed with regular and special shapes with angled faces to form a tight circular structure. Successive inner rings have slightly different angles to match the planned diameter of the ring. The hearth starts out at the wall approximately 230 mm (9 in) thick.

Out-hearths are provided with holes that allow cake solids that are rabbled from the inside of the hearth to drop to the hearth below and gases to pass from the hearth below. These holes are constructed with special refractory shapes that form the opening. The inner ring is constructed with a lip to prevent cake solids from getting between the hearth and the central shaft and to keep air from short-circuiting between hearths. The central shaft has a hearth cap that fits over the lip to complete the seal.

In-hearths are constructed with an annular space between the hearth and the central shaft. This configuration allows cake solids that are rabbled inwards to discharge to the hearth below and for the gases from the hearth below to enter the in-hearth.

9.3.2 Central Shaft and Rabble Arms

The central shaft, also known as the center shaft, supports the rabble arms at each hearth. It rotates to provide mixing and movement of cake solids across the hearth and to provide a duct for air that cools the shaft and rabble arms. The central shaft is driven by a motor and gear reducer through a bevel gear mounted on the bottom of the shaft and a pinion gear mounted on the driven end of the gear reducer. The variable-speed drive allows central shaft speed to vary from 0.3 to 3 rpm. Several types of variable drives are available, including electric drives with mechanical variable sheaves, direct current electric motors, electric motors with variable-frequency drives, and hydraulic drives.

The central shaft is constructed of a cast alloy metal (heat-resisting cast iron), suitable for the temperature and environmental conditions within the furnace. It is constructed in multiple flanged sections joined end-to-end with two annular rings. The inner ring provides shaft cooling air; the outer provides heated (shaft return) air. Each central shaft section contains four socket holes for mounting rabble arms. The rabble arms are mounted in pairs and pinned to the shaft section. The holes are plugged when there are less than four arms in a hearth. The lower part of the shaft is provided with air inlet openings for the cooling air. A fixed air housing constructed of steel with seals to prevent air leakage between the fixed housing and rotating shaft is mounted over the openings (Figure 5.14). The shaft is supported by a lower thrust bearing which is mounted on the floor. The top of the shaft is supported by the top bearing; a fixed steel housing with seals is provided over the heated air outlet.

The rabble arms are constructed of a cast alloy metal (chrome-nickel cast stainless steel) suitable for the temperature and environmental conditions within the furnace. Each rabble arm is hollow; an air tube allows shaft cooling air to enter the inner

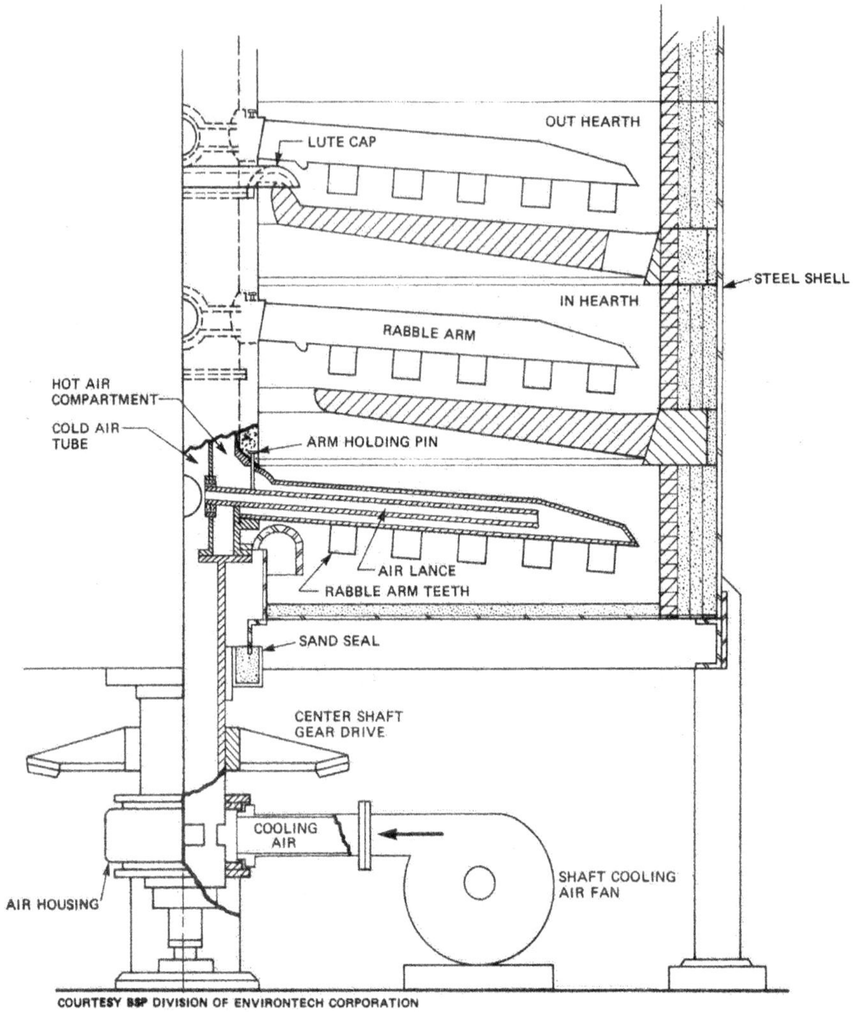

FIGURE 5.14 Shaft cooling air arrangement in an MHF (U.S. EPA, 1979).

end of the arm and discharge into the cavity at the outer end. The air is heated while the arm cools and discharges to the outer annulus in the shaft. Shaft cooling air enters at the bottom, proceeds through the rabble arms, and discharges at the top.

Each rabble arm contains a slot in the bottom for inserting the rabble teeth, which are constructed of similar material to the rabble arm (Figure 5.15). Spacers are inserted between the teeth and the teeth are prevented from sliding out by a

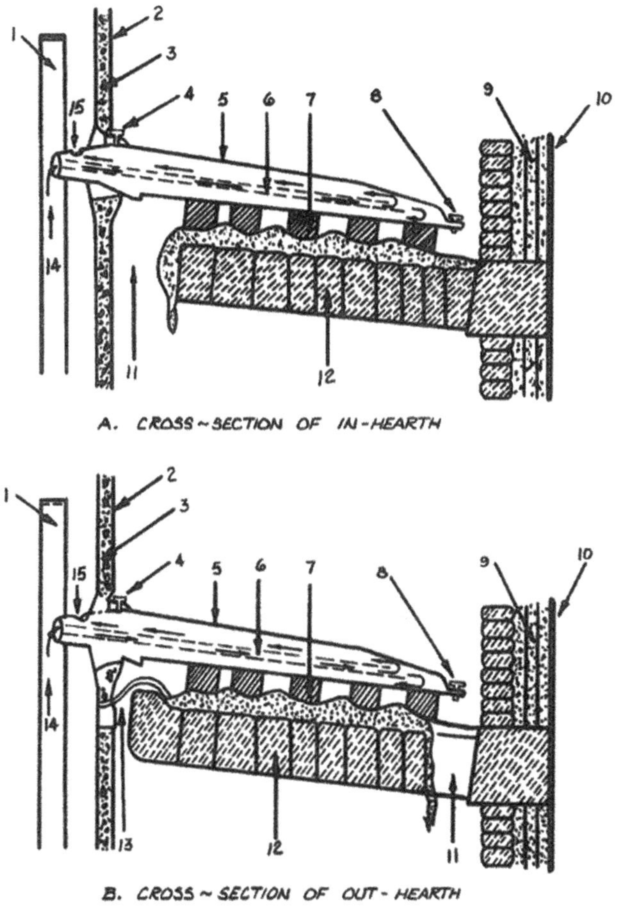

FIGURE 5.15 Typical rabble arm arrangement: 1 = shaft cold air tube; 2 = shaft castable insulation; 3 = rotating shaft; 4 = arm holding pin; 5 = rabble arm; 6 = inner tube; 7 = rabble tooth; 8 = tooth-holding pin; 9 = mineral wool insulation; 10 = steel furnace shell; 11 = drop hole; 12 = hearth; 13 = lute cap; 14 = cold air; and 15 = heated air.

tooth-holding pin inserted to the end of the arm. Rabble arms are arranged at the same angle as the hearth to provide a uniform space between the teeth and the hearth. Rabble teeth are angled to move the cake solids in the desired direction. Teeth can also be installed to provide movement in the opposite direction to increase mixing and residence time on the hearth. This arrangement is referred to as back rabbling.

The central shaft is lined with castable refractory to protect it from high temperatures. The castable refractory is anchored to the metal surfaces by V-shaped stainless steel anchors welded to the metal surface. Shaft refractory is typically 140 mm (5.5 in) thick and rabble arm refractory is typically 50 mm (2 in) thick.

9.3.3 Burner Systems

Burners are mounted on steel structures called burner boxes. They may be solidly bolted in place or have a swing-out feature that allows for rapid cleaning. Burner boxes are refractory lined and contain a monolithic burner tile composed of either ceramic or castable refractory. The burner is arranged with an angled hole, typically convex to allow the burner flame to propagate into the hearth and in the desired firing axis. Good burner design provides for the burner flame to propagate into the hearth without impinging on the rabble arms or the sludge bed. The location and quantity of burners per hearth varies.

Depending on burner type, an MHF can use a variety of fuels including natural gas, digester gas, fuel oil, and bunker oil. Each burner has connections for main and pilot fuel(s), combustion air, control piping, flame rod or UV sensor, spark plug, and sightglass. A description of the auxiliary components associated with burners—including fuel trains, safety controls, combustion air blowers, and fuel pumps—are described in Appendix B.

9.3.4 Central Shaft Return Air

Ducts with motorized dampers in the central shaft allow heated air to be either returned to the furnace as auxiliary heat or to be vented. Control systems allow adjustment of the amount of air returned to the furnace and vented.

9.3.5 Auxiliary Combustion Air Ports

Earlier MHFs were equipped with auxiliary combustion air ports located on the walls of lower hearths. These combustion air ports were equipped with manual or motorized dampers. The desired damper opening was set by the operator, either locally or remotely, and auxiliary combustion air was drawn into the furnace by the draft created by the induced draft fan.

Later versions of MHF were equipped with auxiliary combustion air ports on multiple hearths to allow air to be injected to hotter hearths to control temperatures. These ports were equipped with motorized dampers that were controlled automatically.

In many cases, the auxiliary combustion air was supplied by an air system that included fans or blowers. This type of system has been retrofitted on some older MHFs.

9.3.6 *Access Doors*

Hinged access doors are constructed of ductile or cast iron. The door is lined with castable refractory. Each door has a sliding or hinged plate that acts as an observation port. The hinged door is mounted on a cast iron frame that is bolted to the incinerator shell. The door is kept closed by bolted clamps. There are typically two doors per hearth located 180° apart. The door sizes vary but are large enough to permit entry of workers to a cold furnace.

9.3.7 *Emergency Bypass*

An emergency bypass damper is provided to vent furnace gases in the event of a power failure or loss of induced draft fan. The purpose is to prevent unburnt hot gases from being blown into the building through furnace doors because of pressure buildup that would occur when the induced draft fan stopped. It also prevents the internal furnace from overheating because of cake in the furnace continuing to burn.

The emergency bypass system includes an emergency bypass damper and stack. The emergency bypass damper is connected to the stack by a duct that is either connected to the top of the furnace or to the exhaust gas duct. The damper is arranged to fail open, which means that upon loss of power or signal, the damper opens. The damper is typically of the butterfly type and is refractory lined. Guillotine-type, refractory-lined dampers (single and double) also have been used. Ducting between the furnace and the damper is typically refractory lined. The stack is typically constructed of stainless steel and does not have refractory lining but may be insulated externally for personnel protection.

9.3.8 *Instrumentation*

The MHF is equipped with pressure, temperature, and oxygen monitoring instrumentation to allow the operator to supervise operations. Instrumentation also provides process measurements that are incorporated into automatic control systems, such as hearth temperature control through burners and auxiliary combustion air dampers and furnace draft control through induced draft fan damper modulation or speed adjustment.

Temperature is typically measured by Type K thermocouples mounted in thermowells in the furnace wall. Thermocouples typically extend up to 300 mm (1 ft) into the furnace. The signal from the thermocouple is connected to a remote transmitter or a PLC terminal unit by instrumentation wiring. At least one thermocouple is provided on each hearth, with multiple thermocouples placed on some hearths, depending on the furnace control system. A thermocouple is also provided at the outlet of the incinerator to measure exhaust gas temperature.

Pressure is measured by pressure transmitters through a pressure tap. The transmitter transmits the signal to the control system through instrumentation wiring. Pressure taps may be short pipes or tubes connected to the furnace wall and the transmitter, or may be provided with purge air to prevent the pressure taps from being plugged. Pressure measurements are typically made on some of the lower hearths and the top hearth or the exhaust air duct.

Oxygen is measured either remotely or by an in-situ oxygen analyzer. Where a remote oxygen analyzer is provided, a sample of exhaust is withdrawn through a stainless steel tube and conveyed to the oxygen analyzer system. Oxygen is typically measured in the exhaust gas duct.

9.4 Furnace Subsystems

The MHF includes several subsystems:

- Dewatered cake conveyance and feed system—this includes belt conveyors, screw conveyors, dewatered cake pumps, and piping systems.
- Burner fuel system—this includes the burner control system, the fuel valve and safety trains, pumps, and piping.
- Burner combustion air system—this includes combustion air blowers and ducting systems.
- Auxiliary combustion air system—this includes fans/blowers and ducting systems.
- Shaft cooling air system—this includes shaft cooling air fans and ducting.

Subsystems are described in Appendix B.

9.5 Multiple-Hearth Furnace Combustion Enhancements

Several combustion enhancements have been developed for MHFs to assist in meeting requirements of the 40 CFR Part 503 and other regulations. These include a patented afterburning system called the RHOX process, flue gas recirculation, and oxygen injection.

9.5.1 RHOX Process

The RHOX process (Figure 5.16) provides a practical and economical means of producing a high-quality exhaust gas while maintaining conditions for fuel combustion of wet solids (RHOX, 1989). In the RHOX process, a regenerative thermal oxidizer (RTO) afterburner is installed downstream of the scrubber to reduce THC, carbon monoxide, dioxin, and furan emissions. The RTO is equipped with a low NOx burner. The RTO uses the high-heat-transfer efficiency of the regenerative thermal oxidizer. Therefore, the RTO has to increase exhaust gas temperature from the scrubber exit by approximately 38°C (100°F) to approximately 110 to 140°C (230 to 280°F). The RTO provides greater fuel efficiency than a conventional external or internal afterburner. In a conventional afterburner, the mass load includes the water evaporated in the MHF. Thus, if a furnace has a gas exit temperature of approximately 480°C (900°F) and an afterburner temperature of 675°C (1250°F), the temperature of the entire mass of exhaust gas has to be increased approximately 195°C (350°F).

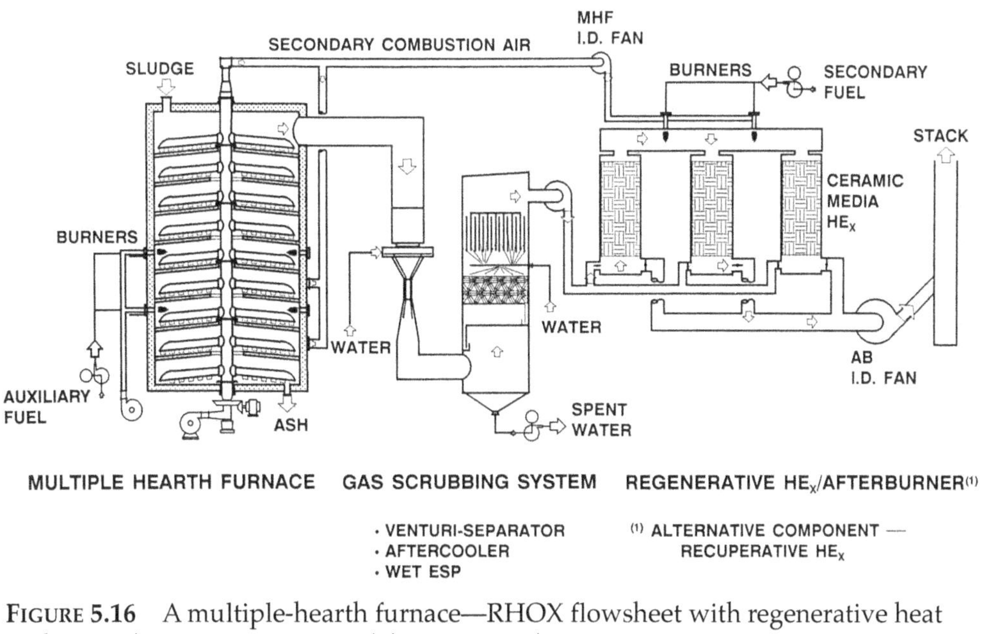

FIGURE 5.16 A multiple-hearth furnace—RHOX flowsheet with regenerative heat exchanger (I.D. = induced draft) (RHOX, 1989).

Saturated gas, which contains a relatively small amount of water, has to be raised from a temperature of 60 to 140°C (140 to 280°F) to approximately 80°C (140°F). The mass loading of the exhaust gas after the scrubber is approximately 70% of the mass loading out of the furnace.

The RTO in the RHOX process requires gas with low particulate concentrations to prevent fouling of the ceramic surfaces of the heat exchangers. Therefore, existing scrubbers may have to be augmented by installation of a wet ESP or replacement with a scrubber.

9.5.2 Flue Gas Recirculation

Flue gas recirculation (FGR) in MHFs (Figure 5.17) recirculates flue gas from the hearth onto which the dewatered cake is fed to a hearth below the burning hearth. Typically, the dewatered cake is fed below the afterburning hearth(s) (referred to as AB in Figure 5.17). The gas is recirculated through ducting and fans. Typically, two sets of ducts and fans are provided. The flow rate and temperature of recirculated gas is measured by a flow meter in the FGR duct and is controlled either through dampers or by varying fan speed. Cooling air is provided to the FGR to control temperature. The FGR process provides many benefits, including

- Increased stability of operation.
- Reduction in NOx emissions (elimination of yellow plume).
- Reduced slag formation in the MHF.
- Increased furnace throughput capacity (because of lower downtime from slag removal).
- Reduced THC emissions (because of more stable furnace operation).
- Complete ash burnout (because of higher temperatures in the lower hearths because of the recirculated gases).

9.5.3 Oxygen Injection

An oxygen-enriched MHF demonstration was carried out at the Frank E. Van Lare Sewage Treatment Plant by Praxair, Inc., under a New York State Energy Research and Development Authority (NYSERDA) grant (NYSERDA, 1998).

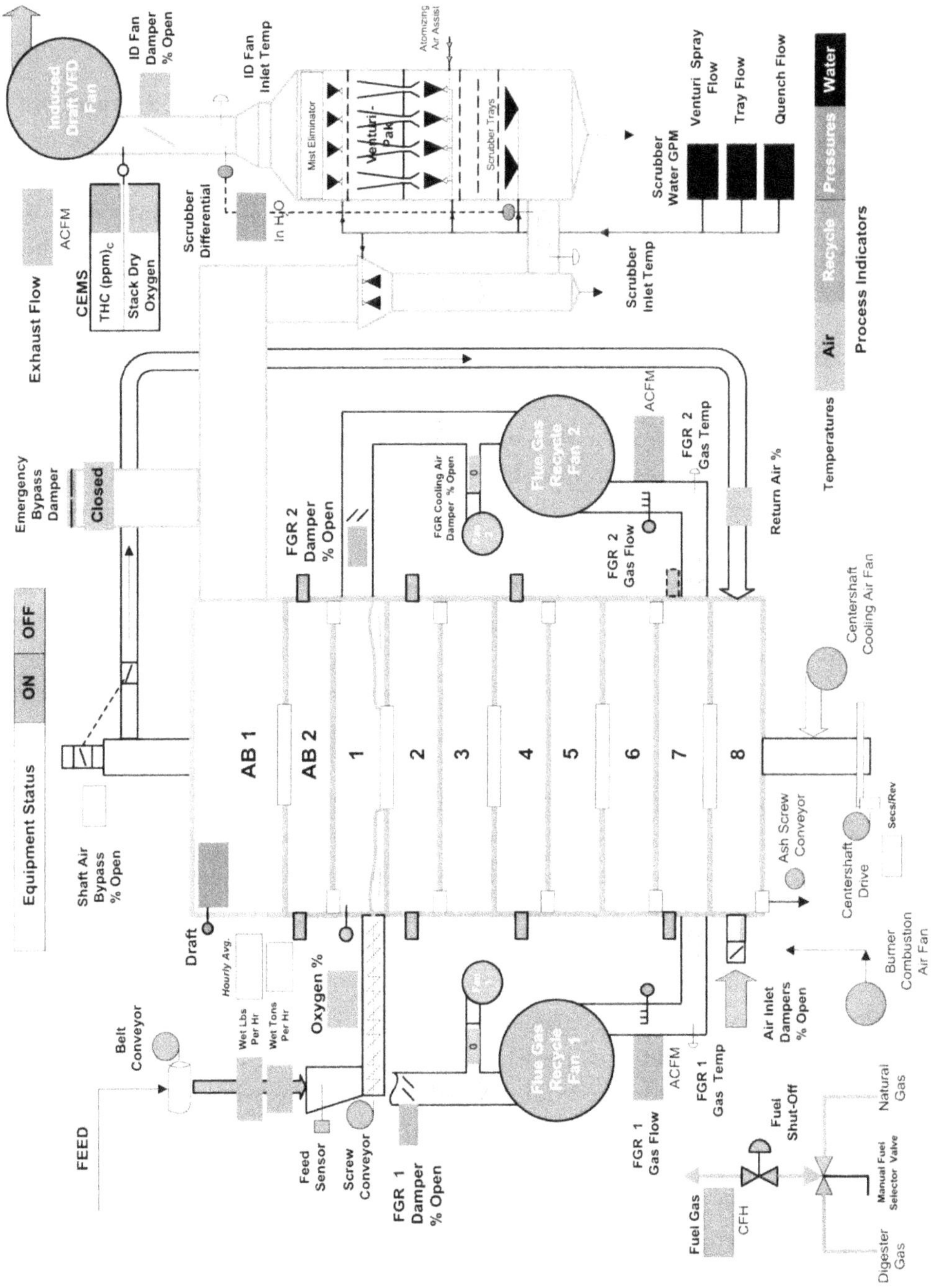

FIGURE 5.17 A multiple-hearth furnace flue gas recirculation system (CEMS = continuous emission monitoring system; ID = induced draft) (Porter et al., 2002).

10.0 OTHER THERMAL PROCESSING TECHNOLOGIES

Several emerging thermal processing technologies can be used for solids treatment. Some of these technologies, such as vitrification, are commercially available for processing WWTP solids. Some technologies, such as gasification and plasma arc, have been used commercially for other feedstocks but not WWTP solids; others, such as SlurryCarb™, have been developed but not applied commercially. The following briefly describes these technologies.

10.1 Vitrification

Vitrification, defined as a thermal process for converting minerals into glass, is an emerging technology in the area of treating WWTP solids. Vitrification has a well-established track record in other industrial processes, especially in furnaces used in the glass manufacturing industry and slagging furnaces used in coal-fired power generation.

Wastewater treatment plant residuals possess characteristics common to both glass manufacturing and power generation and, therefore, play two important roles in the vitrification process. First, the organic fraction provides the thermal energy required to complete vitrification. Second, the mineral fraction (ash, clays, and mineral fillers) melts into a glass aggregate product with multiple beneficial construction and industrial applications.

Japan has practiced vitrification of WWTP solids for a couple of decades. Until recently, however, commercial vitrification of waste materials in the United States has been limited to hazardous waste applications. Commercial-scale vitrification of high-volume industrial WWTP residuals emerged in 1998 with Minergy's Fox Valley Glass Aggregate Plant (FVGAP) located in Neenah, Wisconsin. The FVGAP vitrifies approximately 270 000 metric tonne/a (350 000 ton/yr) of wastewater solids from several local paper mills into approximately 45 000 metric tonne/a (50 000 ton/yr) of glass aggregate that is sold and used locally. Minergy subsequently developed a second generation vitrification technology, GLASSPACK®, applicable for individual on-site use.

The GLASSPACK® system is a patented, modular, closed-loop oxygen-enhanced combustion process. It uses enriched oxygen to achieve temperatures that promote vitrification, provide complete destruction of organic compounds, and reduce emissions. The high process temperatures completely melt the inorganic fraction into an inert, beneficially reusable glass aggregate product. The process train comprises the melter, a heat exchanger, the air cleanup system, and an exhaust recycle fan (Figure 5.18).

The closed-loop oxygen enhanced process uses pure oxygen (greater than 90%) injection as the source of oxidizer, eliminating use of atmospheric air and, therefore, the ballast and diluent nitrogen. The melter, composed of separate but interconnected refractory lined chambers surrounded by a water jacket to dissipate heat lost, provides a three-zone operation. Feedstock that has been predried to at least 85% solids or greater is injected along with synthetic air to the zone 1 chamber. In this zone, the organic component of the feedstock is completely combusted, liberating a significant amount of heat energy and resulting in temperatures between 1315 and 1482°C (2400 and 2700°F). At these high temperatures, the mineral (ash) component of the feedstock melts to form a pool of molten glass at the bottom of the zone 1 chamber.

Phase separation of the molten glass and exhaust gas occurs by gravity draining the molten glass from zone 1 through a drain port on the bottom of the zone 2 chamber. The molten material drops into a water quench tank and is cooled into the glass aggregate product. Hot combustion gases are directed out of zone 2 through a refractory-lined duct to zone 3. Hot exhaust gas is cooled through dilution mixing with lower temperature gases obtained by the recirculation of cooled exhaust gas.

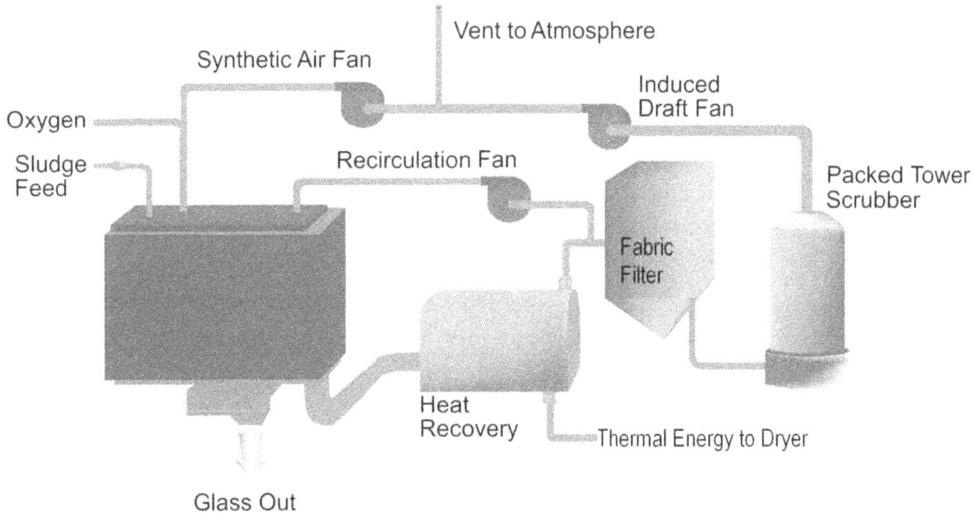

FIGURE 5.18 The GLASSPACK® closed-loop oxygen enhanced vitrification process (courtesy of Minergy Vitrification LLC).

The temperature of the exit gas varies depending on the temperature and quantity of the dilution gas, but is typically in the range of 370 to 870°C (700 to 1600°F).

Exhaust gases are ducted into a heat exchanger to recover thermal energy and generate low temperature dilution gas. Several options are available for recovering the heat, including steam generation, thermal oil, and hot gas heat exchangers. After the exhaust gas exits the heat recovery unit, it is split into two directions. Approximately 70% of the exhaust gas flow enters the exhaust gas recirculation (EGR) fan and is injected back into zone 3. The remaining 30% is directed to a fabric filter to capture and remove particulate from the exhaust stream.

After exiting the fabric filter, the exhaust is further cooled and water vapor, produced during combustion, is condensed in a packed tower condenser. The exhaust gas is cooled to 32 to 49°C (90 to 120°F) and directed to the gas recycle header where a portion of the exhaust gas—approximately 10%—is vented out of the process to advanced air pollution control equipment. The remaining 20% of recycled gas is boosted in pressure through a recycle fan and enriched with oxygen. The end result is synthetic air which is injected back to zone 1.

Heat energy recovered from the system can be transferred directly to a feed drying circuit. In most cases the dry feed cake provides enough thermal energy to eliminate the need for an additional energy source for drying. Figure 5.19 illustrates the thermal energy balance for the process installed at the North Shore Sanitary District's Sludge Recycling Facility, Zion, Illinois.

Oxygen can be provided to the process from either on-site liquid storage or generation with either adsorption or cryogenic air separation technologies. Liquid oxygen is vaporized and warmed to ambient conditions to meet process demands.

Markets for the glass aggregate product are large and diverse. More than 2.25 mil. metric tonnes (2.5 mil. tons) per year of the material is currently produced in a similar industry known as slag marketing. The process of water quenching the molten glass as it exits the system results in the formation of an environmentally inert aggregate. During quenching, heavy metals that may be present are physically sequestered in the glass matrix resulting in low leaching.

10.2 Miscellaneous Technologies

10.2.1 Gasification

Gasification is a process that uses heat, pressure, and steam to convert materials to a gas composed primarily of carbon monoxide and hydrogen (California Integrated

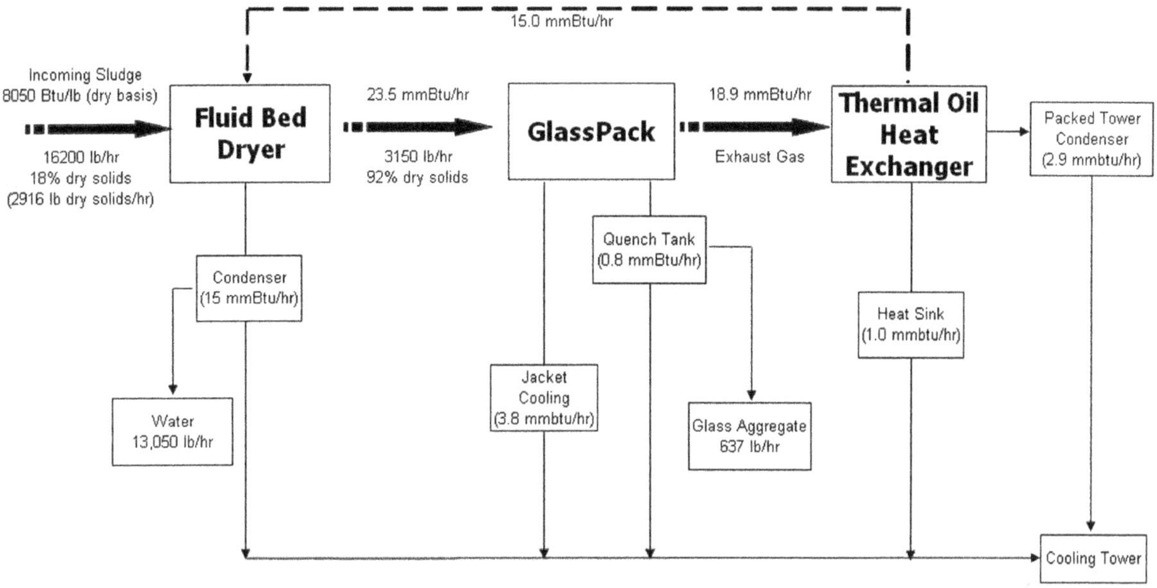

FIGURE 5.19 Thermal energy balance for GLASSPACK® application at North Shore Sanitary District, Zion, Illinois (Btu/hr × 0.2931 = W; Btu/lb × 2.326 = kJ/kg; lb/hr × 0.4536 = kg/h) (courtesy of Minergy Vitrification LLC).

Waste Management Board, 2001). There are many variations in process operating temperatures and pressures that will affect the byproducts, which may be in the form of a syngas, char or slag, oils, and reaction water (Figure 5.20). Operating temperatures may be in the range of 815 to 1815°C (1500 to 3300°F) and pressures may be 2800 kPa (400 psi). Process dynamics and products vary considerably depending on the type of feed. Pilot testing is typically required to determine yields of the offgases and residues. With WWTP solids, the process has proven to be expensive. Typically, the economics of energy recovery are not positive because of low calorific value and high moisture content of the feed cake, which must be heat dried before gasification. The syngas produced generally has a relatively low heating value of approximately 4000 to 8000 kJ/m^3 (105 to 210 Btu/cu ft) and needs to be combined with other higher quality fuels, such as natural gas, before it can be used. The char and oils produced will have less heating value than those produced in a pyrolysis system because of partial combustion of the organics in the gasification process.

Gasification systems have been used more widely in Europe and Asia, both of which use high calorific value feed stocks such as wood wastes. Advantages over

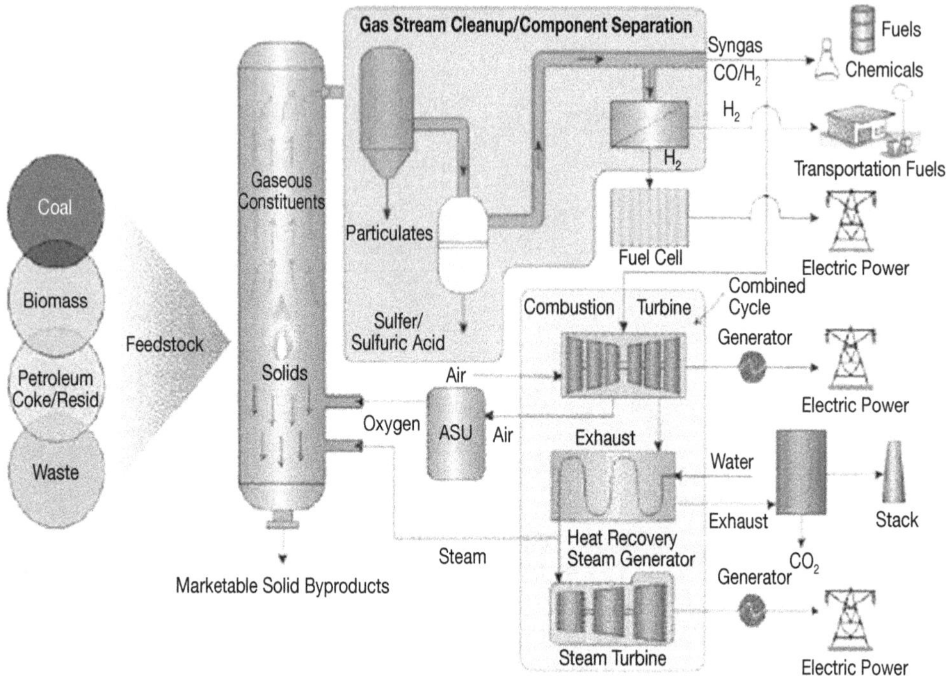

FIGURE 5.20 A typical process schematic for gasification.

incineration include the ability to control air emissions and, because of the production of products with energy value, the process is seen as an energy-recovery technology. Incineration often is considered a disposal or destructive technology. In addition, gasification has not received the negative public perception that incineration has.

10.2.2 Plasma Arc

Plasma arc technology is a non-incineration thermal process that uses extremely high temperatures in an oxygen-starved environment to completely decompose waste into simple molecules (CIWMB, 2001). The extreme heat and lack of oxygen results in pyrolysis of the feed cake. Plasma arc technology has been used for many years for metals processing. The heat source is a plasma arc torch, a device that produces a very high-temperature plasma gas. A plasma gas is the hottest sustainable heat source available, with temperatures ranging from approximately 1480 to 6650°C (2700 to 12 000°F). A plasma arc system is designed specifically for the type, size, and quantity of feed solids to be processed. The high temperature profile of the plasma gas provides an optimal processing zone in the reactor vessel through which all input material is forced to pass. The reactor vessel operates at atmospheric pressure. The process requires a minimum feed calorific value of 20 000 kJ/dry tonne to be autothermal at such low organic concentration. Typical digested biosolids have a calorific value of 13 700 kJ/dry tonne and would, therefore, need to be dewatered or dried to a higher solids concentration for the process to be autothermal. The inert material remains as an ash that needs to be disposed; condensate from the vapor stream will need to be returned to the sewer or treated. The process has been pilot tested on various waste streams, including paper and pulp solids and manure. There are no full-scale operating facilities treating WWTP solids.

10.2.3 SlurryCarb™ Process

The SlurryCarb™ Process is a thermal treatment process to improve dewaterability of biosolids. EnerTech developed the process, which has been tested with feed solids concentrations up to 20% (Orange County Sanitation District, 2003). The feed cake is diluted if the solids concentration is greater than 20% and then macerated to ensure particle size is smaller than 12 mm (0.5 in). The resulting feed slurry is then pumped up to the required pressure setpoint of approximately 7 to 10 kPa (1000 to 1500 psi) and passed through heat exchangers to raise the temperature to approximately 200 to 230°C (400 to 450°F), as shown in Figure 5.21. During the thermal decomposition reactions, organics in the slurry are broken down and carbon dioxide gas is separated from the solids. Simultaneously, any chlorine in the slurry is converted to hydrochloric acid, which is neutralized by the inherent buffering strength of the slurry before heat treatment. Chlorine is a precursor to dioxins and furans, and this process enables the chlorine to be washed out of the treated solids in the form of aqueous salts. The treated solids or char is passed through recovery heat exchangers used to heat the feed slurry and then washed and dewatered in a centrifuge to a solids concentration between 50 and 55%. Depending on the reuse options for the final product, this dewatered cake is either used directly or dried. As a result of the chemical changes that occur during thermal treatment, the viscosity of the resulting slurry is greatly reduced and the dewaterability is increased. The product can be further dried to 95% solids concentration if required by the reuse options. The product has a heating value of approximately 15 100 kJ/kg (6500 Btu/lb) at a solids concentration of 95% if undigested sludge is used.

The process does produce an effluent waste stream that is high in ammonia and organics, which can be controlled by conventional treatment techniques such as

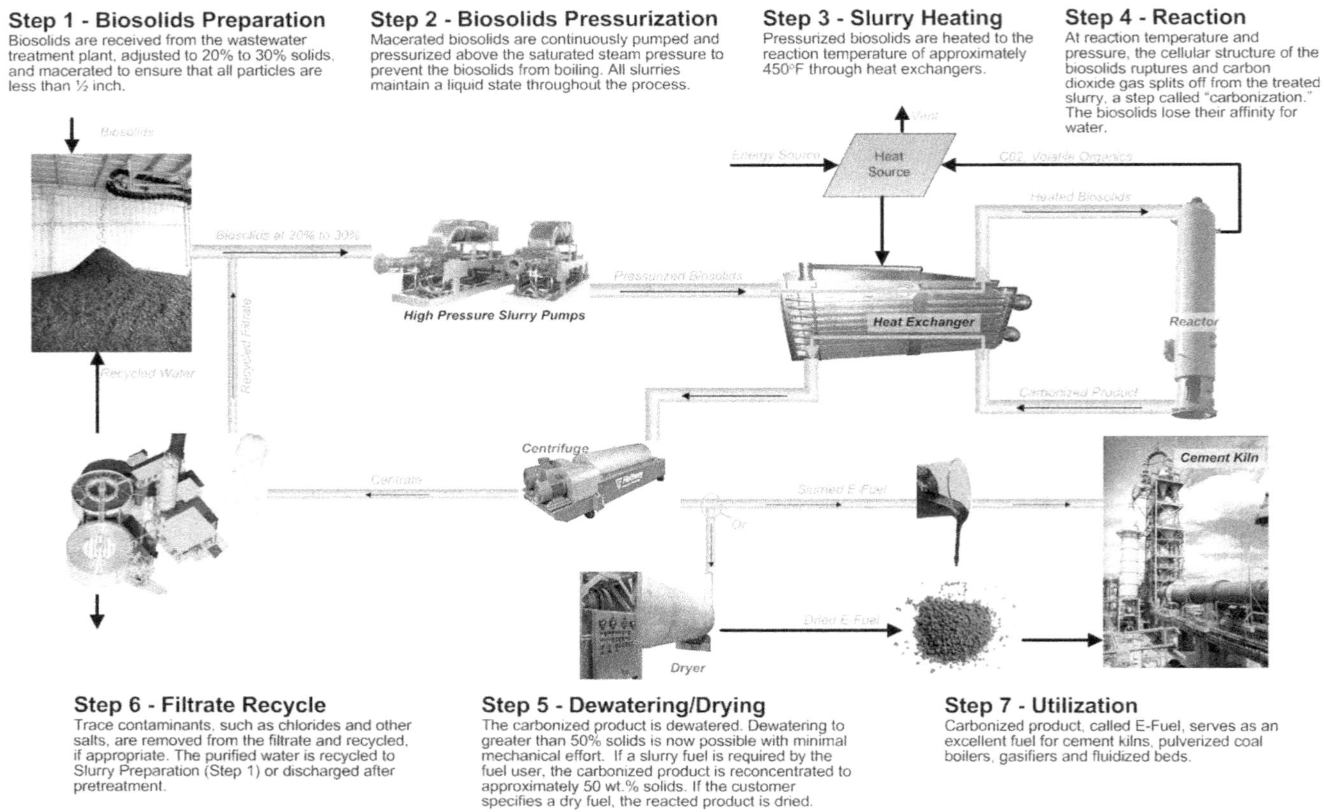

FIGURE 5.21 Flow scheme of the SlurryCarb™ process (BOD = biochemical oxygen demand; COD = chemical oxygen demand) (courtesy of EnerTech Environmental, Inc.).

membrane filtration, anaerobic digestion, and air floatation, to meet required effluent specifications. The CO_2 byproduct stream and dryer exhaust streams are directed to a recuperative thermal oxidizer to oxidize any volatile organic compounds in the gases.

EnerTech has operated one plant in Japan, which only operated on slurried organic solid waste and had a capacity of 20 metric tonne/d (22 ton/d) of dry solids. A demonstration-scale facility has been operated in Atlanta, Georgia using residuals from several WWTPs.

In southern California, EnerTech is developing a regional facility at a site adjacent to the Rialto wastewater treatment plant. The site is in a heavily industrialized area. The facility will process approximately 178 metric tonne/d (196 dry ton/d) from five municipalities and produce approximately 150 metric tonne/d (167 ton/d) of product.

EnerTech states that the process is a net producer of power, generating an excess of 1.25 to 1.5 times the energy requirement, although the understanding is that this is based on the use of undigested sludge. Feed solids concentration is limited by the ability to pump the feed cake before heat treatment. Several cement kilns within an 80-km (50-mile) radius of the proposed site in Rialto have expressed interest in the fuel product. Tests conducted by General Electric show that NOx emissions from the char is similar to coal and that SOx and other emissions are lower, providing a cleaner burning fuel overall. The ash from burning the fuel is used in the cement process and, therefore, does not require disposal.

11.0 REFERENCES

California Integrated Waste Management Board (2001) *Conversion Technologies for Municipal Residuals.* Background Primer for the *Conversion Technologies for Municipal Residuals Forum*; Sacramento, California, May 3–4; CIWMB: Sacramento, California.

Dangtran, K.; Holst, T. (2001) Minimization of Major Air Pollutants from Sewage Sludge Fluid Bed Incinerators. *Proceedings of the 74th Annual Water Environment Federation Technical Exposition and Conference* [CD-ROM]; Atlanta, Georgia, Oct 13–17; Water Environment Federation: Alexandria, Virginia.

European Commission (2000) *Directive 2000/76/EC of the European Parliament and of the Council of 4 December, 2000 on the Incineration of Waste.* Brussels, Belgium.

Geldart, D. (1973) Types of Gas Fluidization, *Powder Technol.*, **7**, 285–292.

Jeffers, S.; Mullen, J. F.; Cohen, A. J.; Dangtran, K. (1999) Control Problem Waste Feeds in Fluid Beds, *Chem. Eng. Prog.*, **May,** 59–63

Kunii, D.; Levenspiel, O. (1969) *Fluidization Engineering.* Wiley & Sons: New York.

New York State Energy Research and Development Authority (1998) *Oxygen-Enriched Multiple-Hearth Sewage Sludge Incineration Demonstration.* Final Report 98-9; New York State Energy Research and Development Authority: Albany, New York.

Orange County Sanitation District (2003) *Long-Term Biosolids Master Plan.* Job no. J-40-7; Orange County Sanitation District: Huntington Beach, California.

RHOX International, Inc. (1989) *RHOX Process.* Technical Bulletin MHF-1; RHOX International: Salt Lake City, Utah.

Shirai, T. (1958) *Fluidised Beds,* Kagaku-gijutsu-sha, Kanazawa.

U.S. Environmental Protection Agency (1979) *Process Design Manual for Sludge Treatment and* U.S. Environmental Protection Agency: Washington, D.C.

Water Environment Federation (1992) *Sludge Incineration: Thermal Destruction of Residues,* Manual of Practice no. FD-19; Water Environment Federation: Alexandria, Virginia.

Water Pollution Control Federation (1988) *Incineration;* Manual of Practice no. OM-11; Water Environment Federation: Alexandria, Virginia.

Zenz, F. A.; Othmer, D. F. (1960) *Fluidization and Fluid-Particle Systems.* Reinhold Publishing Corp.: New York.

12.0 SUGGESTED READING

Porter, J.; Lill, W.; Mansfield, W. (2002) Reviewing Multiple Hearth Furnaces: The Atlanta Experience. *Proceedings of the 16th Annual Water Environment Federation Biosolids Conference* [CD-ROM]; Austin, Texas, Feb 27-29; Water Environment Federation: Alexandria, Virginia.

Chapter 6

Heat Recovery and Reuse

1.0 INTRODUCTION

Heat recovery and reuse is an integral part of a modern incineration system. In most cases, feed cake to the incinerator is not sufficiently dry to be fully autogenous or thermally self-supporting during combustion. Therefore, it is necessary to provide additional energy to the process, either using auxiliary fuel (natural gas, fuel oil, etc.), thermal energy recovered downstream of the incinerator, or both, to properly sustain the combustion process.

Most high-temperature thermal processes, such as incineration, offer a variety of opportunities for energy recovery. Every stack that exhausts hot flue gases to the atmosphere represents irretrievable thermal energy. Similarly, quenching of hot flue gases in the wet scrubber represents energy "down the drain," not to mention the demands this places on plant water systems. In a typical incinerator operating at 760 to 870°C (1400 to 1600°F), heat loss to the stack or scrubber drain can be significant. The incinerator exhaust gases represent a valuable energy resource that, when effectively recovered, can have a positive effect on operating costs. If this energy is recovered as useful heat, then it translates to energy cost savings and possible reduction in capital costs. Heat recovery from hot flue gases will reduce spray-cooling requirements and reduce volume flow through downstream gas cleaning systems. However, a careful technical and economic evaluation must be conducted before embarking on expensive and complex schemes. Although inclusion of heat recovery systems does require some capital investment and routine operation and maintenance effort and cost, this approach can often offset capital and operation and maintenance cost savings of downstream equipment.

Recovered energy may or may not suffice to completely eliminate the need to purchase auxiliary fuel for incineration; however, it can greatly reduce the fuel requirements and is typically cost-effective. Moreover, recovered energy can be used to serve other portions of the plant through the generation of steam, hot water or heated thermal fluids or, taken a step further, for the generation of electrical power. Whenever there is a market for energy recovered in the wastewater treatment plant (WWTP), there is an opportunity for savings. Taking advantage of the energy properties of feed solids is what makes incineration with energy recovery a true form of beneficial use.

Waste heat recovery can take many different forms. In general terms, it can be classified as *primary recovery* and *secondary recovery,* based on ultimate the use of the recovered energy. Primary recovery refers to energy recovery that is used specifically to improve performance of the combustion process by reducing or eliminating the need for auxiliary fuel. This process can include use of recovered energy to preheat combustion air to the incinerator or use of recovered energy to thermally dewater or dry incoming feed cake, or both. Thermal dewatering or drying systems may use energy in the flue gases either by direct contact or indirect, where steam or thermal fluids heated by energy in the flue gases is then used to thermally treat incoming feed cake prior to incineration. Either of these approaches can be used to reduce the need for auxiliary fuel.

Secondary recovery uses heat in the flue gases for beneficial purposes outside of the combustion envelope. Heat recovery could be in the form of preheating an external medium or generation of power. For seasonal space heating, an economizer may be added as a source of hot water for space heating or in-plant process purposes. Steam generated in a boiler can be used throughout the WWTP, sold to neighboring users, used for production of electricity in a cogeneration system, or used directly in steam turbine drives. The exhaust flue gases can also be used, directly or indirectly, to preheat flue gases downstream of the scrubbing system before they are exhausted to the atmosphere to provide plume suppression. When recovered energy is beneficially used, energy is used that would otherwise be wasted to displace an equivalent (or greater) amount of energy that would have been required from other sources.

From a practical perspective, heat transfer technology can be classified as either direct or indirect. Direct heat transfer refers to a process in which the heat source comes in direct contact with the material or flowing stream being heated. An example would be direct injection of a stream of hot air to the exhaust stack to reheat flue gases before discharge for plume suppression. Another example would be a process by which feed cake is preheated or dried by direct contact with flue gases. This approach has gained limited acceptance as a means of preheating or thermally dewatering the feed cake. Most municipal WWTP incineration systems use indirect heat transfer.

Indirect heat transfer refers to a process in which a physical barrier separates the heat source from the material or flowing stream being heated. For example, indirect heat transfer occurs when incinerator flue gases are used to preheat combustion air in a heat exchanger. This example is referred to as recuperation and is the most common and economical approach. In a fluid bed incinerator, the fluidizing air is typically preheated to temperatures from 200 to 650°C (400 to 1200°F). In a multiple-hearth furnace, the utility of combustion air preheat is practically limited to a maximum of approximately 200°C (400°F), as the nature of the process and equipment precludes operation with high combustion air temperatures. In some variations of the multiple-hearth incineration process, wet scrubber exhaust is reheated to 760 to 870°C (1400 to 1600°F) in a secondary combustion unit. Here the energy in the furnace exhaust could be used to preheat the scrubber exhaust flue gas, thus reducing auxiliary fuel requirements for secondary combustion. In this case, such recovery allows high-temperature treatment with lower energy costs.

No matter what form heat recovery takes, the goal should be economical plant operation, both in fuel efficiency and operating and maintenance costs. Energy recovery should first address the primary objective: efficient and effective combustion of waste residue. Remaining energy can then be used for secondary purposes.

This chapter covers the key concepts and considerations in heat recovery and reuse applications. Finally, design and operational details of major types of equipment used for heat recovery are described.

2.0 CONSIDERATIONS IN HEAT RECOVERY AND REUSE APPLICATIONS

The source of recoverable energy is the hot furnace exhaust flue gas. As a resource, these flue gases contain most of the heat energy input to the system, including the heating value of the feed cake and auxiliary fuel and the sensible heat of the combustion air. A small portion of this energy is lost to the surrounding environment through radiation and convection, and some is lost because of the unburned combustibles. These losses, however, are relatively minor when compared with the total heat energy input to the furnace.

The inherent nature of fluid bed systems provides several primary heat recovery opportunities to reduce auxiliary fuel requirements. Fluid bed systems are able to accommodate relatively high levels of combustion air preheat, typically as high as 650°C (1200°F) and sometimes greater. Furnace exhaust temperatures may be as high as 870°C (1600°F) under normal operating conditions. There is always sufficient energy available in the exhaust gases to preheat the typical amount of combustion air (stoichiometric amount plus 40% excess air) to as much as 650°C. In fact, there is typically a meaningful surplus of recoverable energy in the exhaust flue gases that can be applied to other secondary uses.

One unique attribute of exhaust flue gas is its relatively high moisture content. Essentially, all moisture in the furnace feed cake is contained in the exhaust gases as water vapor. In most cases, water vapor accounts for 25 to 45% of the furnace exhaust

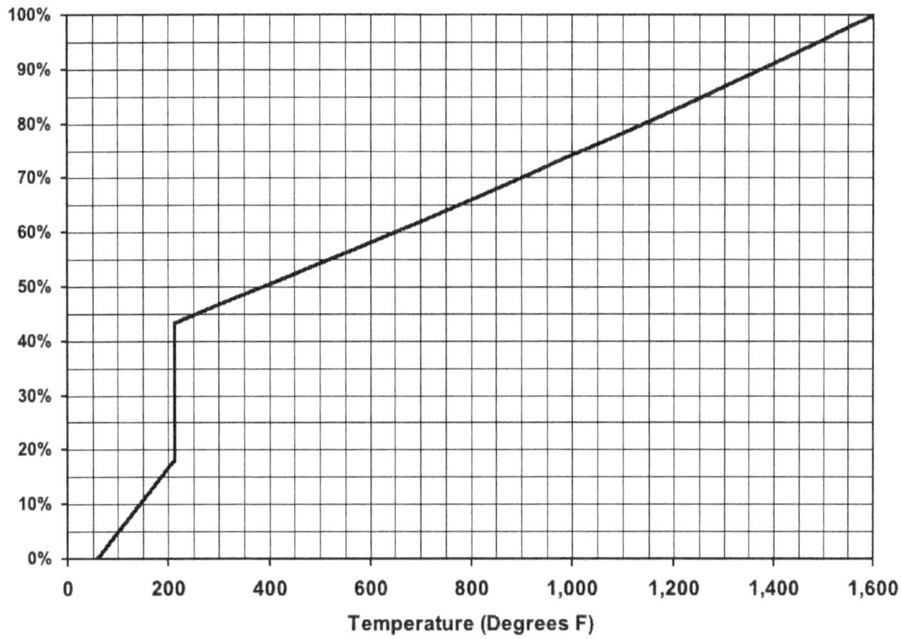

FIGURE 6.1 Flue gas energy versus temperature (0.5556 [°F − 32] = °C).

gas on a molar or volume basis. Because of high moisture content, a large fraction of the furnace exhaust energy is the latent heat of vaporization of this water, which is generally not recoverable in this type of application. Figure 6.1 shows a typical graph of the fraction of total flue gas energy versus temperature, with 100% occurring at 870°C (1600°F) and 0% occurring at 16°C (60°F).

The steep segment of the line occurs at 100°C (212°F) and represents the latent heat of vaporization of moisture in the flue gas. In a real-world application, condensation would occur over a range of different temperatures, corresponding to changing flue gas saturation properties. This simplified example, however, is for illustration purposes only. In a typical system, the latent heat of water vaporization in the flue gas would equate to approximately 25% of total flue gas energy.

In WWTPs, few, if any, heat recovery systems are designed to recover energy from flue gas less than approximately 175°C (350°F). High acid gas content of typical flue gases makes acid dewpoint corrosion a significant concern at lower temperatures and makes low-temperature heat recovery problematic. Allowing for a reasonable safety factor greater than the acid dewpoint, the practical lower limit for heat recovery systems is approximately 175 to 200°C (350 to 400°F). As illustrated in Figure 6.1, typical recoverable energy above this limiting temperature range is approximately 50% of the total energy in the flue gas at a starting temperature of 870°C. This proportion may change slightly depending on site-specific factors such as furnace exhaust temperature, actual flue gas composition, and moisture content, among others.

2.1 Potential Uses for Primary Energy Recovery

The most important use for recovered energy is primary recovery to reduce or eliminate auxiliary fuel requirements for combustion. The most common form of primary recovery is for preheating of the combustion air to the system. As noted earlier, this is more appropriate for fluid bed incinerators because these systems can take advantage of preheat temperatures as high as 650°C (1200°F) or more. Another form of primary recovery is using excess energy to thermally dewater incinerator feed cake. The feed cake is dried to the point that little or no auxiliary fuel is required, which typically requires two steps. First, the system either generates steam or heats a thermal fluid. Second, an indirect dryer uses that steam or fluid in the drying process.

Nearly all fluid bed systems have some form of heat recovery for combustion-air preheat purposes. To illustrate the value of preheating combustion air, Figure 6.2 shows a graph of auxiliary fuel requirements as a function of preheat temperature for a 25% solids feed cake to a fluid bed operating at 40% excess air and an exhaust temperature of 870°C (1600°F). Volatile solids levels of 50, 60, and 70% are shown for illustration purposes.

Figure 6.2 shows the wide range of potential auxiliary fuel requirements and the dramatic reduction that can be effected by preheating combustion air to the system. In this example, a feed cake having 70% volatile solids is nearly autogenous (requires zero fuel) at an exhaust temperature of 870°C and a combustion air preheat level of 650°C (1200°F). As a practical matter, most fluid bed systems operate at a slightly lower exhaust temperature. So this air preheat temperature would typically reflect autogenous combustion in the range of 840 to 870°C (1550 to 1600°F), which would be acceptable for design purposes. In this example, preheating combustion air to 650°C reduces auxiliary fuel requirements to levels from 5 to 35% of those without preheat.

To put this into perspective, it is useful to consider the practicality of preheating combustion air to 650°C using the available energy in the furnace exhaust flue gases. The fraction of total flue gas energy required to achieve various combustion air preheat levels is presented graphically in Figure 6.3, using the same example as in Figure 6.2.

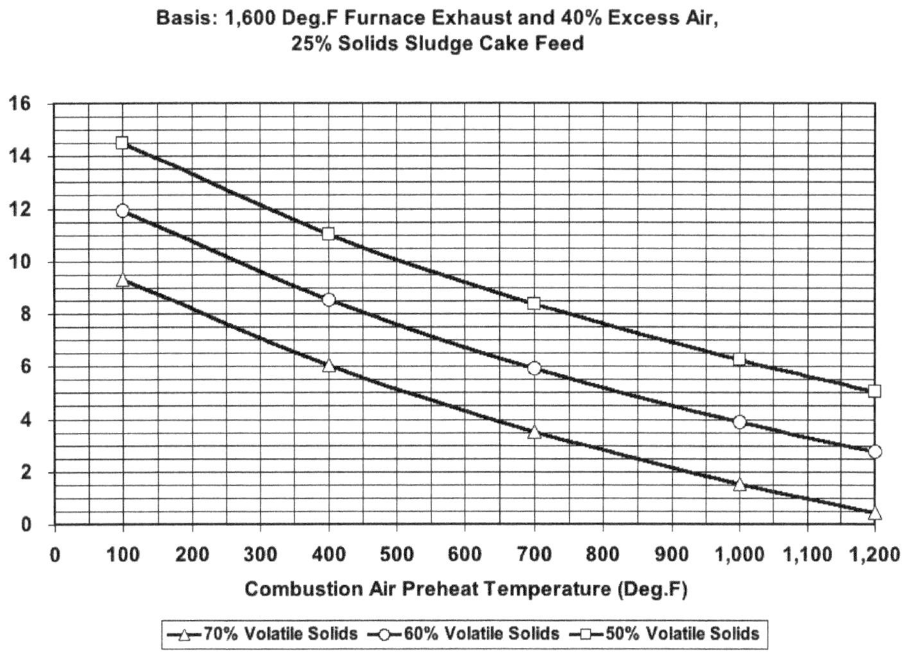

FIGURE 6.2 Auxiliary fuel requirements for a typical fluid bed incinerator (Btu/ton × 1.163 = J/kg; 0.5556 [°F − 32] = °C).

As indicated in Figure 6.3, it takes approximately 23.5% of the energy in the furnace exhaust flue gases at 870°C (1600°F) to preheat combustion air to 650°C (1200°F). Figure 6.1 shows that this can be achieved by cooling exhaust flue gases to approximately 540 to 600°C (1000 to 1100°F). This range is well within the capability of the equipment and current heat exchanger designs.

2.2 Potential Uses for Secondary Energy Recovery

It was noted earlier that the amount of heat energy that can be practically recovered from exhaust flue gases equates to approximately 50% of the total energy starting at a temperature of 870°C (1600°F). This point is set by limiting the flue gas temperature at the outlet of the last heat recovery unit to a minimum of about 180 to 200°C (350 to 400°F). As illustrated earlier, combustion air preheating typically will require from 15 to 25% of the total flue gas energy, depending on the design level of preheat. Therefore, approximately 25 to 35% of the total flue gas energy remains available for fur-

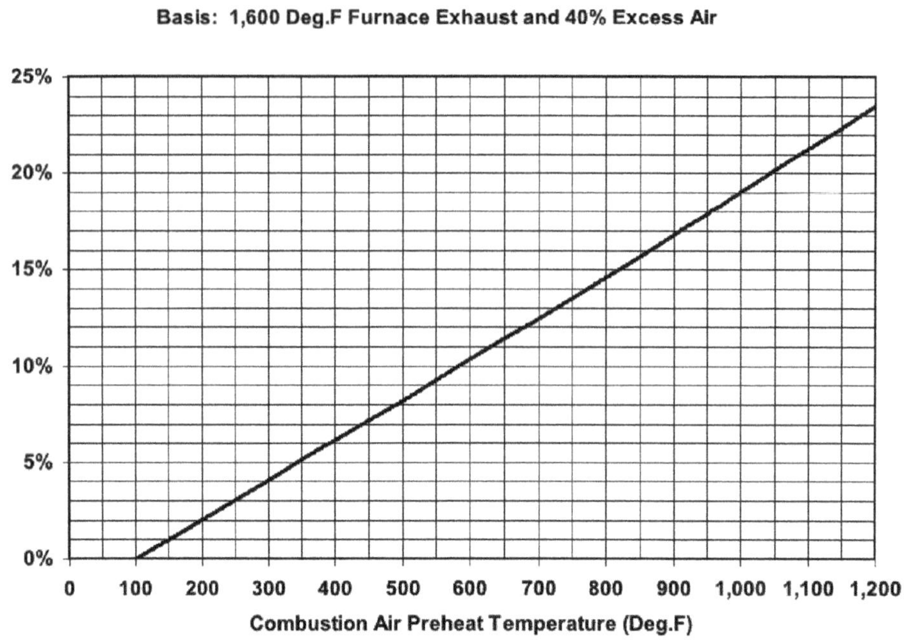

FIGURE 6.3 A fraction of flue gas energy required for combustion air preheat (0.5556 [°F − 32] = °C).

ther heat recovery and reuse downstream of the primary air preheater. Accordingly, once the desired combustion air preheat is achieved, other uses for energy recovered from the exhaust flue gas should be considered. These options fall into the general category of secondary energy recovery.

A variety of heat recovery systems exist that can take advantage of energy in exhaust flue gases from fluidized bed systems downstream of the primary air preheater:

- Several types of waste heat recovery boilers that can produce steam for process uses or for power generation.
- Heat exchangers that can produce hot thermal fluids for process use.
- Heat exchangers to produce hot water for process or building heating use, or both.
- Heat exchangers to produce hot air or flue gas for exhaust gas reheating.

These and other applications will be discussed in"Heat Reuse Applications" section of this chapter. As fluidized bed systems become more integrated with other solids processing systems—such as anaerobic digestion, dewatering, or thermal drying systems—more direct markets for this energy will develop.

The focus up to this point has been on fluid bed systems because they offer the greatest potential for primary heat recovery. Multiple-hearth systems, on the other hand, cannot typically make effective use of combustion air preheat temperatures higher than approximately 200°C (400°F). Moreover, furnace exhaust temperatures from multiple-hearth systems are typically no higher than 760°C (1400°F) when high-temperature afterburning is provided and are more often in the range of 480 to 650°C (900 to 1200°F). Further, multiple-hearth systems typically operate at 75 to 150% excess air, so there is a higher mass flow of flue gases per unit of feed cake compared with fluid bed systems. In general, exhaust gases from a multiple-hearth system are nearly equivalent in resource value to exhaust gases from a high-temperature air preheater on a fluid bed system. If afterburning is provided on the multiple-hearth system, energy recovery potential is even higher. Therefore, regardless of the type of furnace used, the potential for secondary heat recovery and reuse is approximately the same when exhaust flue gases are cooled to 180 to 200°C (350 to 400°F).

2.3 Application Considerations

Several issues must be considered when designing or operating primary and secondary energy recovery equipment and systems. These issues include addressing the unique needs and composition of flue gases from the particular type of incinerator (fluid bed or multiple hearth) and accommodating the myriad process and feed variations. Many systems have been designed to operate well for a continuous feed at a maximum feed rate and may be severely challenged under conditions such as turndown or variations in the thermodynamic properties of the feed cake.

2.3.1 Gas Composition

Exhaust flue gases have many undesirable traits including high moisture content and fairly high acid gas content upstream of the wet scrubber. In fluid bed systems, essentially all of the ash in the feed cake exits the furnace with the flue gas. In multiple-hearth systems, this discharge is usually lower. Flue gases, however, still carry 5 to 15% of the ash from the feed. The potential fouling characteristics and particulate loading should be addressed during the system design phase. Table 6.1 lists typical composition values for exhaust flue gases from fluid bed and multiple-hearth incinerators.

TABLE 6.1 Typical composition of wet flue gas (volume basis).

Parameter	Fluid bed systems	Multiple-hearth systems
Oxygen (O_2)	3.5 to 6%	6 to 10%
Nitrogen content (N_2)	45 to 55%	50 to 65%
Carbon dioxide (CO_2)	7 to 10%	5 to 8%
Water vapor (H_2O)	35 to 45%	25 to 35%
Sulfur dioxide (SO_2)	0 to 1500 ppm	0 to 1000 ppm
Hydrogen chloride (HCl)	0 to 1000 ppm	0 to 700 ppm
Acid dewpoint*	50 to 90°C (125 to 200°F)	50 to 90°C (125 to 200°F)

*Add 56°C (100°F) design safety factor to account for potential presence of SO_3.

Values are typical ranges of parameters of interest from a heat transfer and corrosion perspective only and are for illustration purposes; no individual incinerator is "typical." Each application must be evaluated using the specific thermodynamic characteristics of the feed cake and the design and operating parameters of the case being analyzed.

2.3.2 Process Considerations

One of the key process considerations when designing and operating a heat recovery and reuse system is the real-world variability in operating conditions that the system must be able to accommodate. Incineration systems often must operate with a wide range of feed cake characteristics. Sometimes key characteristics, such as cake solids and volatile solids content, are subject to sudden fluctuations on a daily or hourly basis. Further, most WWTPs experience seasonal variations in the amount and quality of the feed cake. Some systems are operated on campaigns of hours, days, or weeks and are then either placed in a "hot" standby condition or are temporarily shutdown. Other systems operate more or less continuously; however, the feed rate to the system reflects the actual solids load at the plant, which may be significantly less than the design point.

Most incineration systems are designed to be able to operate over some range of turndown and to accommodate variations in characteristics of the feed cake. It is important that equal attention be paid to designing the heat recovery and reuse systems with appropriate features and providing ways to handle these variations, which can sometimes be a challenging endeavor.

Intermittent operation can be of greater importance to secondary heat recovery and reuse than it is to primary heat recovery. The simple reason for this situation is that if the incineration system (the market for reuse of the primary energy recovered) is out of service, then the need for the recovered energy is nonexistent. On the other hand, if the secondary energy recovered is used, for example, to heat processes or buildings or to generate power, then demand will continue independent of the incinerator operation. Hence, an alternate means must be provided to ensure that energy demands are met when the incineration system is not in service or when operating at partial load. In the latter case, the recovered energy may not fully satisfy the load it serves when the incineration system is operating at 50 to 60% of the design feed rate. Finding a creative way to address the issues of turndown and cyclic operation in design and operation of heat recovery and reuse systems can be the key factor in success.

Operation of a high-temperature air preheater at turndown is a significant concern, particularly when design temperatures approach design limits of the materials of construction or of the equipment manufacturer's warranty. If a preheater is designed to produce 650°C (1200°F) combustion air temperature at the 100% design point, then it will tend to produce higher temperature air when the system is turned down to 60 to 70% of the design point. The typical recuperator has the same amount of heat transfer area, regardless of operating throughput. Further, fluid bed systems may experience prolonged temperature excursions for a variety of reasons, thereby changing the flue gas inlet temperature to the air preheater to an "off-design" condition. At the extreme, metal temperature limits can be approached, leading to potential equipment damage; at the least, equipment life may be reduced.

A second consideration with respect to combustion air preheat relates to changing thermodynamic characteristics of the feed cake to the incinerator. For example, the incineration system may have been designed for a 25% solids feed cake, with 70% volatile solids. During the design phase, a combustion air preheat level of 650°C (1200°F) may have been selected for this application to yield autogenous operation at slightly less than 870°C (1600°F) furnace exhaust. For one reason or another, this system may be called on to process 27% solids feed cake, with 75% volatile solids, which may only require the combustion air to be preheated to 480°C (900°F). If flexibility was not built into the system, then furnace exhaust and preheated air temperatures would rise above design levels, which are unacceptable conditions. In the extreme, the operator may have to turn down the feed cake rate to the system and operate at an artificially inflated excess air level to limit furnace exhaust temperatures to within satisfactory limits and to protect the air preheater equipment. De-rating of the system capacity is rarely an acceptable approach to address these "off-design" conditions. Fortunately, several creative approaches have been developed to address such conditions without sacrificing system capacity under abnormal feed cake characteristics or system operating efficiency when conditions are normal. One approach is to partially bypass a controlled amount of cold air from the inlet to the outlet of the air preheater when less air preheat is required. Another approach is to control the bleed of hot air from the discharge of the air preheater to the atmosphere (or stack).

Several incineration systems have waste heat recovery boilers or economizers to capture available energy in flue gases for use elsewhere in the plant, or both. Rarely is there a perfect match between available energy and independent energy demand; therefore, these systems are often equipped with bypass ducts around the heat recovery equipment. The bypasses can be throttled to match energy demands of the plant or to isolate the heat recovery unit from the flue gas stream when energy demand is not present or the energy recovery equipment is out of service. Despite good intentions, these bypass ducts do not typically provide a complete barrier and seal against flowing hot flue gases; maintenance of the heat recovery equipment typically requires a complete system shutdown. Bypasses can be used to isolate the equipment from the flue gases to allow the incinerator to operate until shutdown is convenient for heat recovery system maintenance and repair. Care should be taken in design of the system to protect the off-line heat recovery equipment from damage from the ever-present leakage around the bypass dampers. Protection may require some form of supplemental pressurized air purge for the isolated equipment or evaporative cooling sprays in the ductwork upstream of the unit to limit flue gas temperatures to acceptable levels to protect off-line equipment.

2.3.3 *Safety Considerations*

Safety is a significant concern in the operation of complex, high-temperature equipment such as those found in incinerators and energy recovery systems. Expansion joints are required at key points in the system because of thermal expansion and contraction of equipment. Several good expansion joint designs exist, such as joints with multiple barriers and pressurized air purge features. In the long run, though, potential for rupture and leakage through these expansion joints is a significant concern.

Many earlier simple fluid bed systems were equipped only with a combustion air preheater and not with secondary energy recovery devices. Flue gases passed directly from the outlet of the air preheater through appropriate ductwork to the inlet of the wet scrubbing system. In many cases, the air preheater was "hard-connected" to the furnace and the only expansion joint was at the inlet to the wet scrubber. Many of these systems operated under pressure developed by the fluidizing air blower all the way to the stack discharge. These configurations are referred to as "push" systems. The disadvantage of this system is that if a leakage develops then hot flue gases may be released into the operations building, resulting in a potentially severe safety hazard.

Systems with more extensive heat recovery equipment, particularly those having waste heat boilers or economizers, often are designed to operate under slight negative pressure downstream of the furnace. In this design, an induced draft fan downstream of the wet scrubber pulls flue gases from the furnace through the system and discharges them to the stack. The fluidizing air blower is only designed to develop the pressure necessary to deliver the combustion (fluidizing) air into the furnace and through the bed zone. These designs are referred to as "push-pull" systems. The disadvantage of this approach is that two fans must be operated and maintained. Air from any leakage that develops, however, will be drawn into the hot flue gases from the operations building area, which is safer.

3.0 HEAT REUSE APPLICATIONS

Although a variety of equipment is available to recover energy from furnace exhaust flue gases, extensive recovery has rarely been practiced. The only exception is that most fluid bed systems use combustion air preheaters for primary energy recovery. Multiple-hearth systems do not have a meaningful demand for primary energy recovery. In the past, the potential benefits of secondary energy recovery were largely ignored in favor of a simpler system with no linkages to other plant processes and demands.

Newer incineration systems generally are based on fluid bed technology. Many of these systems contain more extensive primary and secondary heat recovery and reuse systems. Although many older incinerators use multiple-hearth technology, these too provide opportunities for retrofitting with heat recovery systems, most often involving secondary recovery only.

3.1 Primary Energy Recovery Systems

Primary recovery systems are almost exclusively related to preheating of combustion air, using flue gas to air heat exchangers, before its introduction to the furnace. Although combustion air preheating is rarely used with multiple-hearth systems, there are some installations that are designed for a modest preheat level of 200°C

(400°F), which can be achieved using a high-pressure (3110 kPa [450 psig]) steam coil. These applications typically have a high-pressure waste heat boiler to recover energy from hot flue gases in the form of steam. Part of the steam generated is used to preheat combustion air; the balance is available for other purposes.

A second form of primary recovery involves the use of energy from furnace exhaust flue gases—either as steam or a heated thermal fluid—to indirectly heat the thermal dewatering system. This type of system can be modulated to remove additional moisture from mechanically dewatered feed cake, up to the point where the cake is autogenous (or nearly so) in the combustion system. Typically, this process involves thermally dewatering the cake to 27 to 33% solids, depending on the particular characteristics of the cake being processed. Integration of a thermal dewatering system to the incineration process provides enhanced flexibility to accommodate changes in feed cake characteristics and turndown operations in an efficient manner.

3.2 Secondary Energy Recovery Systems

Most secondary recovery systems are an add-on to the primary recovery systems. Occasionally, the primary and secondary recovery systems are integrated into a single unit that can serve both primary and secondary energy demands. In general, energy in flue gases can be converted to heated air or flue gas using gas-to-gas heat exchangers, to high- or low-pressure steam using firetube or watertube boilers, or to heated water or thermal fluids using gas-to-liquid heat exchangers or economizers. The choice of what equipment to use is based on a review of available energy and the nature and magnitude of energy demands to be served.

There are several uses for common secondary energy recovery and reuse applications:

- Reheating of exhaust flue gases downstream of the scrubber before discharge for plume suppression.
- Generation of high-pressure steam to operate turbine-driven equipment or electrical power generators.
- Generation of steam or hot water for building or structure heating, potable water heating, or hot flushing operations.
- Heating water or a thermal fluid to meet process demands in the plant, such as heating of digesters or preheating the feed to centrifuge dewatering equipment.

Steam turbine equipment includes backpressure and condensing turbine designs. With either style of turbine, it is advisable to provide some amount of steam superheat upstream of the turbine to reduce or eliminate moisture formation at the low-pressure end of the turbine. When power generation is used, the maximum amount of power that can be generated through secondary energy recovery is much greater than the total power demands for the incineration system itself. Thus it is possible to support part of the electrical load of the balance of the WWTP in this manner. It is unlikely that the amount of power generated will exceed the balance of plant demands, so the potential for sale of power to the utility is minimal.

In general, most applications to date use exhaust gas reheat for aesthetic reasons, such as for plume suppression. Although it does accomplish this goal, it also provides enhanced dispersion of stack exhaust gases and can often help to meet emerging strict regulations on air toxics found in many jurisdictions. It is expected that more incineration systems will incorporate this feature in the future.

There are several ways to reheat exhaust gases. One method involves providing a second flue gas-to-flue gas (or flue gas-to-air) heat exchanger downstream of the main combustion air preheater on fluid bed systems or at a convenient point downstream of the furnace in multiple-hearth systems. If an induced draft fan is available, then exhaust can simply be routed through the heat exchanger to pick up heat before discharge through the stack. For purely "push" systems, with no induced draft fan, a smaller fan may be provided that discharges fresh air through the heat exchanger and blends it with the flue gas upstream of or within the stack itself.

Another technique for reheating exhaust gas in fluid bed systems involves making the fluidizing air blower and the combustion air preheater slightly larger than required for the combustion air alone. A relatively small side stream of hot (540°C [1000°F] or greater) pressurized air is then blended with exhaust flue gas upstream of or within the stack itself. With this method, the operator has the advantage of greater control and can suspend operations in an emergency to provide additional combustion air. In addition, extra air passing through the combustion air preheater can reduce tube metal temperatures, providing some additional protection for the preheater. At turndown, if combustion air temperatures rise too far, some excess air can be bled to the stack to serve as "extra" reheat air while providing added air preheater protection.

3.3 Typical Energy Recovery Flow Sheets

Figure 6.4 shows a generic flow sheet illustrating the typical arrangement of energy recovery equipment in an integrated fluid bed system. Both primary and secondary

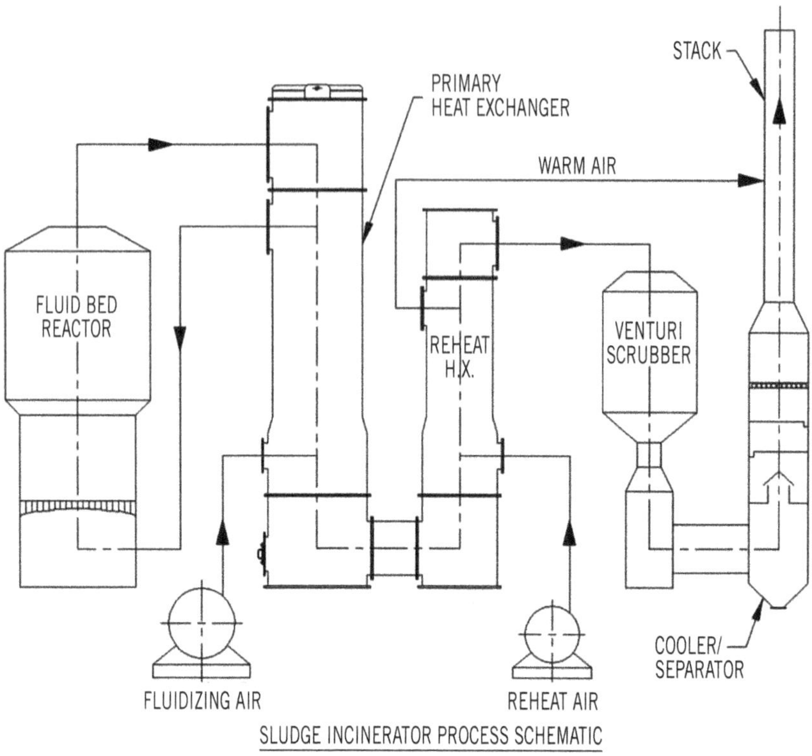

FIGURE 6.4 Primary and secondary heat exchangers in a fluid bed system.

heat exchangers are shown. Additional heat recovery, in the form of a waste heat boiler, could be added to this basic system either upstream or downstream of the secondary heat exchanger, depending on the needs of the application.

4.0 HEAT RECOVERY TECHNOLOGIES

This section addresses the most common types of equipment and systems used for energy recovery and reuse. Included are discussions of recuperative air preheaters, plume suppression heat exchangers, economizers, thermal fluid heaters, and waste heat recovery boilers.

4.1 Recuperative Air Preheaters

Fluidized bed combustion was commercially produced starting in 1963. Early units required co-firing of fuel (oil or natural gas) to sustain combustion. To reduce the cost of auxiliary fuel, heat recovery through recuperation was attempted. Fuel savings were 35 to 75% at 540°C (1000°F) air preheat and combustion could be self-sustaining at 650°C (1200°F) and a well-dewatered feed cake (25 to 27% solids). Recuperation is the most basic and cost-effective form of heat recovery.

In most recuperators, a hot, dirty flue gas stream flows through tubes, while combustion air passes over in multiple, cross-counterflow passes. The axial (straight) flow of dirty gas through the tubes solves several problems. Because particulate matter in gas is carried parallel to the tube wall, abrasive impingement and erosion is minimized. Further, vertical tubes do not offer areas on which ash can collect and they minimize damage from thermal expansion. The so-called flue-gas-through-tube (FGTT) design is illustrated in Figure 6.5.

4.1.1 Air Preheater Operation

Figure 6.4 provides a schematic of a modern fluid bed system, showing the process location of primary and secondary heat exchangers. Typical operating temperatures

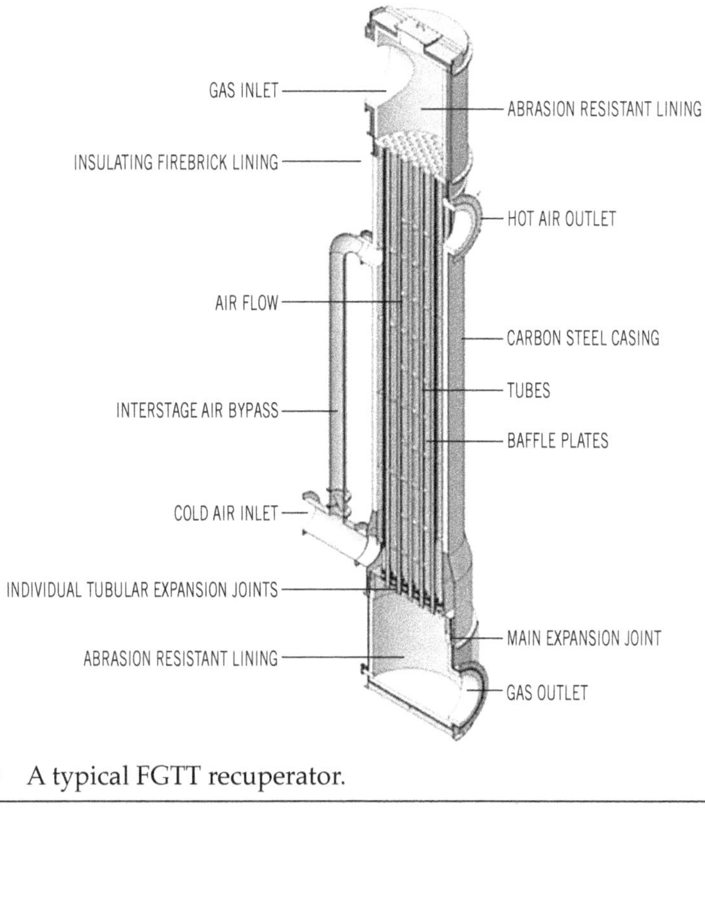

FIGURE 6.5 A typical FGTT recuperator.

of the fluid bed are 820 to 870°C (1500 to 1600°F), with excursions to 930°C (1700°F) or higher. A typical temperature range of preheated air is 540 to 650°C (1000 to 1200°F); however, this can rise even higher during furnace temperature excursions. Although the feed cake is dewatered before incineration, fluid bed furnace exhaust flue gases still contain approximately 35 to 45% water vapor by volume.

Fluid bed incinerators must, by function, operate as an oxidizing environment with excess air rates of 40% or greater; however, occasional upsets may create reducing environments. This situation not only aggravates corrosion but also introduces the risk of after-combustion in the heat exchanger and connecting ductwork. Chlorides in the feed cake are another common source of corrosion attack. The typical composition of flue gases was discussed in section 2.3 of this chapter.

Many incinerators are run on a demand basis rather than continuously. A common operating cycle follows the normal working week: 8 hours per day; 5 days per week; or, in some cases, 24 hours per day, 5 days per week. Such cyclic operation can be damaging, both from mechanical stress and materials wear. Repeated transitions through the dewpoint range are particularly harmful because this can cause corrosive condensation, deposits, and potential plugging problems. If lime is introduced to the feed cake for sulfur control, then plugging can be further aggravated, which can lead to corrosion, erosion, and mechanical problems.

4.1.2 Air Preheater Design

With an FGTT design, each tube is fitted with an expansion joint to allow differential thermal movement between individual tubes. As shown in Figure 6.5, large main expansion joints compensate for differential movement between the tube bundle and the cool, refractory-lined shell. Most modern expansion joint bellows are made of Alloy 625.

A typical FGTT recuperator may have 91 tubes of type 304 stainless steel or alloy 20, arranged on a triangular pitch. Tube (pipe) size may be 9 m long × 89 mm diameter × 3 mm wall thickness (30 ft × 3.5 in × 0.120 in). Units have been built with as few as 19 and as many as 162 tubes. Pipe sizes range from 50 to 200 mm (2 to 8 in) in diameter, illustrating the range of the preheater supply and capacity.

Tubes are welded directly to the tube sheet. Alloy or stainless steel is typically specified for the upper tube sheet, which is exposed to the full temperature of the hot air and supports the entire tube bundle. The shell, flue gas plenums, and lower tube sheet are carbon steel. Insulating refractory linings are typically required for the entire casing and hot face on the tube sheets. Dense, abrasion-resistant refractories help avoid erosion by the particulates in the flue gas. Vapor barriers or coatings on interior carbon steel surfaces minimize acid attack on flue gas plenums.

Although problems and failures have been greatly minimized because of proper design, heat exchangers still experience several material and mechanical failures. Some of these are related to upset conditions, such as overheating incidents, which can cause rapid oxidation, distortion, and cracking of the hottest parts, such as tubes, tube sheets, and expansion joints.

4.1.3 Tubes

Austenitic Cr-Ni stainless steel such as type 304 typically possesses adequate resistance to corrosion by hot flue gases, even those with high sulfur content. Alternating oxidizing and reducing conditions may greatly accelerate corrosion by destroying the protective oxide films that rapidly form on stainless steel in excess-air environments. Reducing conditions also create the risk of afterburning of incompletely combusted substances inside the bundle, which can lead to tube failure. This problem is best addressed by better controls and tighter operating practices. Alloy 625 and alloy 20 provide added protection against chlorides and are often specified. Although more costly, the added life is considered worthwhile and should be evaluated and used if no standby capacity is available.

A common mode of tube failure is stress corrosion cracking (SCC) caused by chlorides in the flue gas. Many stainless steels are susceptible to SCC; maximum susceptibility occurs when nickel content is approximately 8%. Using an alloy with no nickel or one with a high nickel content (greater than 30%) results in significant improvement to resistance.

Stress corrosion cracking occurs in austenitic stainless steels only when tensile stress, chlorides in solution, and elevated temperature are present. During cool down (shutdown), water condensation occurs and promotes chloride concentration. During startup, moisture from the flue gas will condense on cooler surfaces. As the equipment heats up, the liquid boils away and concentrates dissolved salts until the liquid is saturated. Because it is impractical to prevent condensation formation, especially in cyclic operations, the solution typically lies in upgrading materials.

4.1.4 Tubesheets

The main concern with tubesheets is cracking of the highly stressed upper tubesheet, particularly in the area of the tube attachment weld. Cracking may result from thermal fatigue caused by repeated cycling with high thermal stresses or from high temperature creep, or both. As with tubes, upgrading materials to alloy 625 or alloy 20 is justified in many instances.

4.1.5 *Expansion Joints*

The most common location of heat exchanger failure in earlier incinerator service was at expansion joints. The bellows in such joints must not only face repeated cycling loads, but are thin to permit flexing and are highly stressed. The most prevalent failure mode has been stress corrosion cracking from chlorides. As with tubes, it is virtually impossible to prevent the formation of corrosive condensate throughout the entire operating cycle. Consequently, the only practical solution with chloride-rich feeds is to use construction materials of higher quality. Because bellows materials have been upgraded, failures are now uncommon. Design improvements minimize leakage in the event of bellows cracking so that the effect on operation is minimized.

4.1.6 *Materials Recommendations*

As discussed in the preceding section, upgrading materials from conventional stainless steels of the type 300 series may be necessary in special situations, most often to provide improved resistance to corrosion cracking from chloride stress. Unfortunately, it is difficult to formulate strict guidelines. In general, experience indicates that the frequency of problems with stainless steels increases considerably when chloride levels in the flue gas reach approximately 100 ppm and becomes progressively worse with increasing amounts of chloride.

For chloride levels greater than 100 ppm, intermediate alloys such as alloy 20 800H, and type 825 have been used with success. For higher chloride levels, greater than 1000 ppm, alloy 625 is required. Hot tube sheets follow the same pattern and should be compatible with welding the tubes to the sheets. The lower tube sheet is typically composed of carbon steel. It is sufficiently cool and protected from flue gas exposure by refractory and insulation so that failures are rare.

Expansion joints are now almost exclusively furnished with alloy 625 bellows, which effectively resist corrosion cracking from chloride stress because of their high nickel content. Although this alloy is costly, the overall increase is limited by the small amount used in the thin bellows. This extra cost is easily justified when compared with the high total cost of a typical heat exchanger and the additional life that can be obtained by using this alloy.

4.1.7 *Long-Term Operation*

Unfortunately, nothing lasts forever. The typical operating environment for an FGTT air preheater is severe: extreme high temperatures, a corrosive and erosive environment, and cyclic operation. For a typical air preheater designed for a 650°C (1200°F) air outlet temperature, the normal operating temperature of the tubes and associated parts at the hot end is approximately 760°C (1400°F). This situation means the tubes and tubesheet will be glowing red hot in normal operation. At this temperature, the short-term strength of type 304 stainless steel is only approximately one-fourth of that at room temperature. But at 760°C, short-term strength cannot be used for design purposes because of creep.

Creep is the effect of slow extension of a metal under a tensile load at high temperatures. Crack-like voids are created in the microstructure as grain boundaries slowly move, and eventually the material breaks or ruptures at loads much less than the tensile strength or yield strength. Creep strength is reported as a stress at a given temperature to produce rupture in a certain period of time.

In general, once the temperature is greater than approximately 540°C (1000°F) and a tensile load is present, the metal has a finite life and will eventually fail. The creep life of the material is used up with time at high temperature. Similarly, the fatigue life of a material is expressed as the number of cycles to failure at a given level of alternating stress. Cracks grow in the metal with each cycle until the part eventually fails. Unfortunately both creep and fatigue create crack-like defects in the microstructure, and the combined effect is greater than either one alone. Every hour at high temperature and every cycle consumes some of the lifetime of the equipment. Corrosion and erosion compound the problem.

Excursions or operation at higher temperatures have an exponential effect on both creep rupture life and fatigue life. For an air preheater designed for 650°C (1200°F) preheat, if it is operated at 700°C (1300°F), then the creep rupture life of the material will be typically consumed at a rate more than ten times faster. This means the creep lifetime is reduced to only 10% of the creep lifetime at 650°C preheat. Figure 6.6 illustrates this reduction in life versus temperature.

Failure eventually will occur by cracking. Cracking has been observed in tubes near the tubesheet weld, expansion joints, and, more recently, in the support. A specific pattern of failure by type or area does not exist.

Timely inspection and repair of cracks by welding can extend the useful life of the equipment, except for expansion joints. A typical unit can last approximately 10 years, but some units have lasted much longer and others have lasted a shorter time. Units that operate with higher preheat temperatures than the design preheat temperature usually have shorter lifetimes.

Newer units can be provided with an integral air-side bypass to control the air preheat temperature. Cold air is bypassed around a portion of the tube bundle but

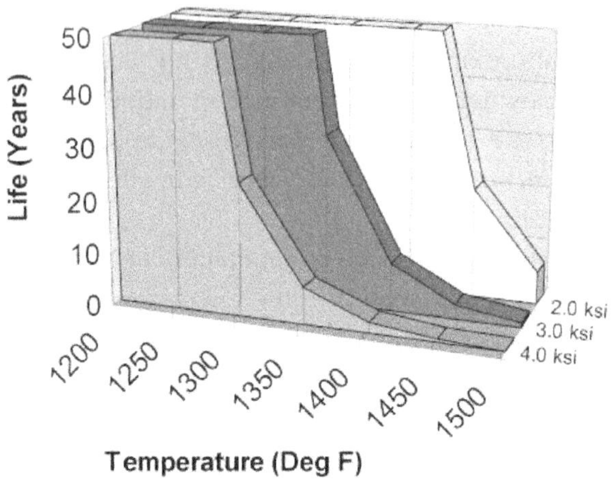

FIGURE 6.6 Creep rupture life versus temperature at several stress levels (0.5556 [°F − 32] = °C).

not the entire tube bundle. If airflow were bypassed around the entire tube bundle, then the air-preheat and metal temperatures inside the bundle would be much higher and failure would result. Without cooling airflow over the tubes, the metal temperatures would rise to the flue gas temperature. By maintaining 100% of the airflow in the hot end of the tube bundle, the hottest parts are effectively cooled by airflow. The partial bypass reduces the air temperature at the hot end, resulting in longer equipment life.

4.2 Plume Suppression Heat Exchangers

Since the early 1990s, plume suppression heat exchangers have been a common accessory to the heat recovery system. Their function is to preheat scrubber exhaust or to clean air for blending into the exhaust, eliminating a visible plume at the stack.

The construction is similar to that of the primary heat exchanger. Warm exhaust gas flows through the tubes (FGTT), and the preheated medium (air or scrubber exhaust) flows over the tubes in multiple cross-counter-flow passes. Tube materials range from nickel alloys to stainless steels, depending on operating parameters.

Because operating temperatures are much lower than in the primary heat exchanger, individual expansion joints are not required. Inlet and outlet plenums provide transitions for the flue gas flow and access for inspection and service. The plenums are refractory lined prior to shipment. The entire assembly (plenums and heat exchanger) can be shipped as a single piece, simplifying field erection.

4.3 Economizers

Flue gas exiting the primary heat exchanger (recuperator) still has considerable sensible heat. At 480 to 590°C (900 to 1100°F), the opportunity exists for further heat recovery. One option is to install an economizer to generate hot water. Existing installations have used both watertube and firetube designs.

The watertube construction is the more conventional approach for economizer design. Flue gas flows over the tubes, presenting the potential for erosion of the tube walls. The tubes must not be placed directly in the path of the gases leaving the primary heat exchanger tubes or severe erosion/corrosion occurs. Auxiliary cleaning in the form of soot blowers may also be necessary to maintain the heat transfer performance.

Most recent installations have used the firetube design (FGTT). Flue gas flowing within the tubes can be cooled to 180 to 230°C (350 to 450°F). Hot water is produced at 90 to 150°C (200 to 300°F) for use in seasonal space heating and boiler feedwater.

The FGTT design, which is used for the primary air preheater, eliminates many maintenance concerns. Despite the high speed at which the flue gas travels through these units, there is no erosive impingement because flow is parallel to rather than across the tube walls. Additionally, this design eliminates the need for auxiliary cleaning (soot blowing) because of its self-scrubbing tendency. Tubes are vertical to avoid surfaces where ash can accumulate.

Replaceable ferrules of stainless steel at the tube inlets further inhibit erosion and help prevent the possibility of unwanted steam forming in the hot end because of high local heat-transfer rates.

The primary heat-transfer surface consists of a matrix of carbon steel tubes, which are rolled, seal welded, and lightly rerolled on each end into 13 mm (0.5 in) thick tube sheets. The gas inlet tubesheet is insulated with an abrasion-resistant castable on its hot face. The carbon steel casing is self-supporting, with external blanket insulation and jacketing. The casing and tubes are operated at temperatures sufficiently low that expansion joints are not required.

4.4 Thermal Fluid Heaters

A thermal fluid heater warms thermal fluid (synthetic or natural heat transfer oils) to use in an indirect dryer or other heat transfer device. It typically follows the primary air preheater, operating within a temperature range in which overheating and breakdown (cracking) of the heat transfer oil can be avoided. For common furnace sizes, the heater can generate as much as 1.47 to 5.9 MW (5 to 20 mil. Btu/hr) of 206°C (500°F) oil, while cooling the flue gas to approximately 200 to 260°C (400 to 500°F).

The thermal fluid heater unit is generally an all-convective tubular type, but either watertube (as with economizers) or firetube types can be used. With the watertube approach, the design is similar in concept to a serpentine tube economizer. The hot exhaust gas exiting the primary air preheater flows vertically across the horizontal tubes contained within the casing. Hot oil is circulated within multiple passes in the tubes. Because of the potential for fouling, it is important that the tubes are arranged on a rectangular pitch to accommodate the use of online cleaning equipment. For the same reason, bare tubes (no finning) are used. To avoid erosive attack, 15 m/s (50 ft/sec) is considered a conservative inlet velocity for the flue gas. Internal insulation of the casing is provided. Tube sheets at both ends support the tubes.

With the firetube approach, the design is similar in concept to the air preheater. The hot exhaust gases exiting the primary air preheater flow vertically up and through the vertical tubes. Hot oil is circulated within the shell, over the carbon steel tubes. As with the air preheater, flue gas velocity within the tubes is high enough that the unit is self-cleaning. An auxiliary sootblower cleaning system is not required. The tube bundle is contained within a carbon steel shell. External insulation of the shell is required. Top and bottom tube sheets at both ends support the tubes. Tubes are connected by rolling, seal welding, and lightly rerolling into the tube sheets.

With any design operating at pressures greater than 103 kPa (15 psig), the bundle will be designed, fabricated, tested, and stamped in accordance with the latest version of the ASME Code, Section VIII, Division 1.

4.5 Waste Heat Recovery Boilers

A waste heat recovery boiler for WWTP incineration service falls into a narrow, specialized category in the overall family of boilers, which includes small boilers for residential heating and power boilers for large utilities.

The usual inlet temperature of waste gases is between 540 and 980°C F (1000 and 1800°F), depending on the type of incinerator, the presence of an afterburner, and whether a combustion air preheater is incorporated to the system. The expected dust loading for a fluid bed system can be as great as 45 to 70 g/m^3 (20 to 30 gr/dry sq ft). The boiler exit temperature must be maintained at greater than the acid dewpoint of the flue gas, which can be as high as 120 to 180°C (250 to 350°F). The actual acid dewpoint depends on how much water vapor and acid gases, such as SO_2, SO_3, and HCl, are in the flue gas. To prevent corrosion, the metal surface of the casing and tube walls must be above the dewpoint. This is accomplished by keeping the fluid temperature above the dewpoint. The range of boiler sizes in this service ranges from approximately 2300 to 23000 kg/h (5000 to 50000 lb/hr) of steam, with pressures from 410 to 4100 kPa (60 to 600 psig).

4.5.1 Firetube Boilers

In a firetube boiler, hot waste gases pass through straight tubes surrounded by a cylindrical shell holding water that is heated to the boiling point. The construction is simple, and the design offers several advantages. Because the shell is inherently tight, the unit is well suited for gases under pressure. No internal refractory is required for insulation because the shell is completely filled with water. Units are supplied as packages that require minimal space.

The water circulation is self-induced. The less-dense mixture of steam bubbles in water rises up through "risers" to the steam drum where it is separated from the water. The separated water is denser than the steam-water mixture and sinks back down through "downcomers" to the heated steam generating surfaces. If a boiler design cannot generate enough natural circulation, then a pump may be used to increase and control the flow to protect the heating surfaces. In some horizontal, packaged firetube boilers, the shell acts as the steam drum, and risers and downcomers are just flow paths inside the shell. Feedwater is added to the water in the steam drum to maintain a normal water level in the drum.

Firetube boilers can be designed for steam pressures up to approximately 1720 kPa (250 psig), but cannot provide superheated steam or steam above the saturation temperature. Consequently, their applications are limited to small capacities. Horizontal tubes also will allow some particulate to settle out in the tubes and reduce performance if velocities are too low. Use of FGTT designs with vertical tubes avoids this problem.

4.5.2 Watertube Boilers

The design of the watertube boiler is the reverse of the firetube boiler. Hot gases are in contact with the outside surface of the tubes and boiler water and steam are in contact with the inside surface of the tubes. Figure 6.7 is a schematic diagram of a typical watertube boiler.

Because the outside surfaces of the tubes are accessible to soot blowers for cleaning, this type of boiler is better suited for flue gases with high solids loading.

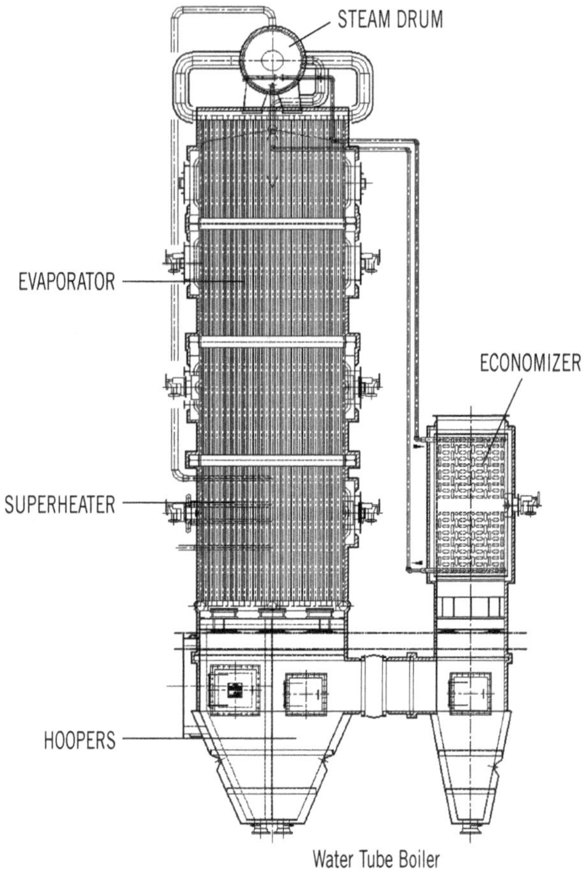

FIGURE 6.7 A typical watertube waste heat recovery boiler.

Boiler tubes are arranged horizontally or vertically, with vertical or horizontal gas flow arranged to allow solids to drop out of the gas stream. At the top, a steam drum separates the steam from the rising water. The steam drum is fitted with feedwater inlet and steam outlet connections and other trim. Some boiler designs have one or more bottom "water drums," sometimes referred to as "mud drums," which can be blown down periodically for cleaning settled solids out of the steam system.

Drums and tubes are supported directly by the casing. Expansion and contraction have no effect on the insulation of the casing. The casing is externally insulated and built to withstand the normal operating pressures of the system. It is desirable to have the boiler at less than slightly negative pressure to reduce the nuisance of dust leaking from the casing flanges and seams. If positive pressure is expected, casings must be welded gastight.

Ash is collected in hoppers as flue gases make low-velocity turns within the boiler. The sides of the hoppers must be steep to prevent ash buildup.

4.5.3 Watertube Boiler Design

Waste heat boilers are typically designed for natural circulation, although some forced-circulation designs have been used. The fluid in the downcomers is mostly water at or slightly below saturation temperature and is of greater density than the water and steam mixture in the risers. Heat transfer depends on tube spacing and diameter and the total heating surface. Tubes for waste heat boilers typically have approximately a 50 mm (2 in) diameter, spaced approximately 100 mm (4 in) apart.

Some boilers also include special banks of tubes not used directly to produce steam. A superheater is a section of tubes that carries only steam so that the heat transferred raises the temperature of the steam to greater than saturation (superheat). Superheated steam is used for steam turbines to prevent excessive amounts of moisture (water droplets) from forming in the turbine. This increases the amount of useful energy that can be taken from steam by the turbine before water creates operating problems. An economizer is also used to increase the temperature of the feedwater and is not designed to produce steam. Because the feedwater is at lower temperatures than the boiling water (saturation temperature), the flue gas can be cooled to lower temperatures for maximum energy recovery from the flue gas.

The usual arrangement for waste heat boilers is to include a superheater ahead of the boiler and an economizer after the boiler. In some designs, the superheater section is located downstream of the boiler inlet, nested in the middle of the boiler evaporator tubes.

4.5.4 Feedwater

Raw water cannot be used in a boiler without chemical treatment to remove scale-forming materials, dissolved oxygen, and acids. Dissolved oxygen will attack boiler steel at saturation temperature. Treatment requirements are sometimes dictated by the ultimate use of the steam, such as when steam turbines are used at the facility.

Not all dissolved minerals in the water are removed by most water treatment systems. A portion of the drum water is removed to reduce concentrated impurities that tend to settle in the mud drum. Low point blowdown connections are provided for periodic removal of settled solids. Dissolved solids and solids suspended in the circulating water are removed continuously from the steam drum. Continuous blow down often represents 3 to 5% of the total feedwater to the boiler.

4.5.5 Soot Blowers

Boiler efficiency is affected by the buildup of ash or soot on boiler tubes. Tubes can be arranged to make cleaning easier. A high-pressure spray of air or steam can be used to periodically blast each tube area that might be expected to show ash buildup. For waste heat boilers, steam is the usual cleaning medium because it is available from the boiler. A combination of inspection doors, automatic soot blowers, and hand lance ports for manual cleaning are recommended. The frequency of soot blowing operations must be determined for each installation and be timed to preserve heat transfer capability and performance of the boiler. This is determined on a trial-and-error basis. Soot blowers are intended to be used to keep tubes clean rather than to remove baked on ash deposits after they have accumulated excessively and hardened on tube surfaces.

5.0 SUMMARY

Although there are many options available for heat recovery, the most effective method is recuperation and air preheating. Further, recovered energy as steam or heated thermal fluids to thermally dewater feed cake to the incinerator is another form of primary recovery and reuse that can dramatically reduce auxiliary fuel requirements. After air preheating or thermal dewatering, options for secondary heat recovery include heat exchangers for plume suppression, economizers, waste heat boilers, or thermal oil heaters. Optimization of heat recovery is essential to efficient energy operation of a plant; it is also what makes incineration with energy recovery a bona fide form of beneficial use.

6.0 SUGGESTED READINGS

American Iron and Steel Institute (1979) *High-Temperature Characteristics of Stainless Steel;* American Iron and Steel Institute: Washington, D.C.

American Petroleum Institute (2003) *Calculation of Heater Tube Thickness in Petroleum Refineries,* 5th ed.; API Standard 530; American Petroleum Institute: Washington, D.C.

Homoki, G. G.; Angel, J. H.; Fedorka, W. R. (2003) Bioenergy Options: Heat Recovery for Thermal Processing. *Proceedings of the Water Environment Federation/American Water Works Association Joint Residuals and Biosolids Management Conference and Exhibition 2003* [CD-ROM]; Baltimore, Maryland, Feb 19–22; Water Environment Federation: Alexandria, Virginia.

Lundberg, L. A. (2004) The Future of Fluidized Bed Incineration. *Proceedings of the Water Environment Federation Residuals and Biosolids Management Conference and Exhibition 2004* [CD-ROM]; Salt Lake City, Utah, Feb 22–25; Water Environment Federation: Alexandria, Virginia.

Lundberg, L. A.; Lewis, F. M. (1993) Integration of Thermal Dewatering and Sludge Incineration Systems—A Marriage of Energy Recovery and Thermodynamic Efficiency. *Proceedings of the American Water Works Association/Water Environment Federation Joint Residuals Conference;* Phoenix, Arizona, Dec 5–8; Water Environment Federation: Alexandria, Virginia.

Lundberg, L. A.; Marchese, N. J. (1991) Integration of Sludge Incineration and Energy Recovery Systems. *Proceedings of the American Water Works Association/Water Pollution Control Federation Joint Residuals Management Conference: Residual Management After 1991;* Research Triangle Park, North Carolina, Aug 11–14; Water Pollution Control Federation: Alexandria, Virginia.

McIntyre, D., Ed. (1997) *Forms of Corrosion: Recognition and Prevention;* National Association of Corrosion Engineers International: Houston, Texas.

Water Environment Federation (1992) *Sludge Incineration: Thermal Destruction of Residues;* Manual of Practice FD-19; Water Environment Federation: Alexandria, Virginia.

Chapter 7

Emissions Control and Monitoring

(continued)

Control of air emissions has become an increasingly important and crucial part of design and operation of incineration facilities. Federal, state, and regional air pollution control boards have imposed increasingly more stringent regulations during the last several decades as environmental awareness has grown. More effective and sophisticated air pollution control (APC) systems have been developed to comply with these regulations.

This chapter covers several topics related to APC systems:

- Formation of pollutant emissions.
- Factors affecting their generation and release.
- Technologies and operating modes to control emissions.
- Advantages and disadvantages of various systems.
- Devices used to monitor emissions.

There are three main types of pollutants: solid, liquid, and gaseous.

1.0 SOLID AND LIQUID POLLUTANTS

1.1 Particulate Matter

According to the definition in the U.S. Environmental Protection Agency (U.S. EPA) New Source Performance Standards (NSPS) contained in 40 CFR, Part 60, Subpart A, particulate matter is "any finely divided solid or liquid material, other than uncombined water, as measured by the reference methods specified" (1991a).

Thus, particulate matter can be any solid or liquid material—excluding water, which is not chemically bonded—that is measured by the method 5 particulate test method (40 CFR, Part 60, Appendix A, Method 5) (1991b). Although small amounts of liquids may be present, particulate matter, or total suspended particulate (TSP) mostly consists of incinerator ash. Total suspended particulate is primarily a mixture of minerals and oxides of several elements: silicon, aluminum, calcium, iron, magnesium, and phosphorus. Although most particulate matter can cause respiratory harm, the much smaller quantity of metals causes the greatest toxic and health effects.

Particulate matter is commonly classified by size and by filterability. For instance, PM_{10} has a mean diameter equal to or less than 10 μm. U.S. EPA Test Method 201 or 201A determines this classification. The text consists of extracting a gas sample at a constant flow rate through an in-stack sizing device such as a cyclone or cascade impactor. Because particulate matter larger than 10 μm is relatively easy to capture, it is generally assumed that all controlled emissions from an incinerator are PM_{10}. A newly determined size distinction is $PM_{2.5}$, which refers to the small fraction of particulate matter that has a mean diameter of 2.5 μm or smaller. The designation of $PM_{2.5}$ is significant because smaller particulates are respired more deeply into the lungs and can have significant health effects. U.S. EPA is evaluating the health effects of $PM_{2.5}$ and may impose emission criteria for large combustion sources, including incinerators.

Particulate matter is also classified as having a filterable portion and a nonfilterable (or condensable) portion. Although these terms are often used in a general sense, they can take on specific meaning when applied to a particular test procedure. For example, U.S. EPA Test Method 5 (40 CFR, 60, Appendix A, Method 5) for the Determination of Particulate Emissions from Stationary Sources is typically used to determine particulate matter from incinerators (U.S. EPA, 1991b). The test involves withdrawing a hot, wet sample of flue gas through a glass fiber, 0.3-μm filter maintained at 121°C (250°F). The mass of particulate captured on the filter is the filterable particulate matter. Following the heated filter, the sample is drawn through a series of four impingers (condensers) in an ice bath that are used to condense out the water vapor in the sample so that the moisture content of the flue gas can be determined. The impingers, however, also will capture any condensable matter in the flue gas. U.S. EPA and most states regulate only the filterable fraction of the particulate catch (front half). Some states, however, require sampling of both the filterable and condensable portions and regulate the total combined catch (front and back halves). Including the condensable fraction can more than double the total amount of particulate matter captured, so it is important that the state testing and reporting requirements are fully understood during the permitting process.

Particulate matter is the portion of incinerator ash that is carried out in the flue gas. The uncontrolled particulate emission rate can readily be calculated using the percent inert fraction of the feed solids on a dry basis and the type of incinerator. In a multiple-hearth furnace (MHF), approximately 85% of the ash exits the incinerator as bottom ash; the remaining 15% is carried out as particulate matter in the flue gas. Thus, a 45-metric tonne/d (50-dry ton/day) MHF with a feed cake that is 75% volatile and 25% inert on a dry basis would have an uncontrolled particulate matter emission rate of 71 kg/h (156 lb/hr, i.e., 50 dry ton/d × 2000 lb/ton × day/24 hr × 0.25 lb inert/dry lb × 0.15). In a fluid bed incinerator, all of the incinerator ash is blown out of the incinerator (with some of the fluid bed sand); hence, the uncontrolled particulate matter emission rate in the above example would be 473 kg/h (1042 lb/hr, i.e., 50 × 2000 × 1/24 × 0.25).

The primary emission criteria for particulate matter are the new source performance standards (NSPS) for municipal incinerators (40 CFR, Part 60, Subpart O), which require that particulate emissions be controlled to not more than 0.65 g/kg (1.3 lb/dry ton) of solids incinerated. This standard sets minimums that all incinerators have to meet. The NSPS also delineates another particulate emission rate, namely 0.37 g/kg (0.75 lb/dry ton) of solids. If an incinerator can demonstrate particulate matter emissions equal to or less than this limit, the incinerator is exempt from certain monitoring and reporting requirements. Hence, the lower particulate matter limit is a typical design objective for new incinerators.

Control methods for particulate matter include cyclones, Venturi scrubbers, impingement tray scrubbers, wet and dry electrostatic precipitators (ESPs), and fabric filters (or baghouses). On existing installations, Venturi scrubbers, tray scrubbers, and wet ESPs are most commonly used. On newer installations, however, dry ESPs and fabric filters have been used.

1.1.1 Opacity

The NSPS for incinerators also limits visible emissions to 20% opacity. According to the NSPS, opacity is "the degree to which emissions reduce the transmission of light and obscure the view of an object in the background."

Hence the greater a plume's opacity, the more it will obstruct light and the more difficult it will be to see through. It should be noted that white plumes as well as black plumes can exhibit high opacity. Opacity measurements can be made by a certified opacity observer or electronically by an opacity monitor. Condensed water vapor (i.e., steam plume) will have high opacity but because condensed water vapor is not a pollutant, it is excluded from opacity measurements. Most incinerators use wet scrubbers, which means that stack gases are wet and contain condensed water vapor. An opacity monitor cannot be used on these sources. If the wet flue gas is reheated, use of an opacity monitor is possible.

Particulate matter from inorganic ash as well as smoke particles from incomplete combustion can contribute to opacity. It is important to note that particulate matter and opacity are not synonymous. A plume with a low, acceptable level of fine particulate can exhibit high light-scattering properties, hence have high opacity.

1.1.2 Metals

High levels of metals in liquid and solids of a wastewater treatment plant (WWTP) are attributable to a number of factors, namely

- Industrial wastewater discharges (particularly from metal finishing and electroplating industries).
- Infiltration of ground water with high metals concentrations into sewer systems.
- Leaching of metals from old piping systems
- Chemicals added at the WWTP.
- Urban runoff entering the sewer system.

Concentration of metals in solids can vary widely (as great as 100- to 1000-fold) from plant to plant. As a result of this extreme variability, there are no typical metal emission rates for incinerators. Even at a given plant, daily metals concentrations can vary by two- to threefold. Primary metals of interest because of their detrimental health effects are: arsenic, beryllium, cadmium, chromium, copper, lead, mercury, nickel, selenium, and zinc.

The vapor pressure of a metal or the compound in which it is contained will determine the fate of a metal during the incineration process. In simple terms, the vapor pressure of a metal is a measure of how easily it can be volatilized into a gaseous state. Several metals and compounds containing these metals (particularly metallic chlorides) are considered volatile, including arsenic, cadmium, mercury, selenium, and zinc. Lead is primarily a nonvolatile metal but it can be converted to a volatile compound, namely lead chloride. Arsenic, cadmium, selenium, zinc, and, to some degree, lead will volatilize at incineration temperatures. These metals will condense onto the fine particulate matter during wet scrubbing of the flue gas and will be collected with the particulate matter in the wet scrubber. Because of its extremely high vapor pressure, mercury is typically assumed to be 100% vaporized during the incineration process and 100% emitted in the flue gas from a conventional incinerator with only a wet scrubbing system. Nonvolatile metals (beryllium, chromium, copper, and nickel) are mostly emitted from the incinerator as fly ash and, as a result, effectively are controlled with a medium- to high-pressure drop wet scrubbing system (Gerstle and Albrinck, 1982).

In general, with the exception of mercury, the same control devices used to control particulate matter are used to control metal emissions. Methods to control mercury are discussed later in this chapter. Typical metal control efficiencies are presented in Table 7.1. Note that the metal control efficiency relates the amount of metal in the feed cake to the incinerator to the amount of metal emitted from the stack. Thus, if the metal feed rate to an incinerator is 100 mg/h and a metal emission rate from the incinerator stack is 1 mg/h, the metal control efficiency is 99%. U.S. EPA metal control efficiencies were compiled from a broad database of both MHF and fluid bed sludge incinerators with different types of wet scrubbing systems, including Venturi scrubber and impingement tray scrubber, Venturi and packed

TABLE 7.1 Metal control efficiencies.

Metal	Classification	U.S. EPA metal control efficiencies (%)* Range	Average	Typical metal control guarantee (%)
Arsenic	Volatile	93.90–100	98.62	98
Beryllium	Nonvolatile	99.9–100	99.99	98.5
Cadmium	Volatile	40.25–99.98	88.54	97.5
Chromium	Nonvolatile	88.92–100	99.16	99
Copper	Nonvolatile	92.28–100	99.39	99
Lead	Intermediate volatility	34.22–99.97	92.24	99
Mercury	Highly Volatile	Negative control	0	0
Nickel	Nonvolatile	89.15–100	98.68	99
Selenium	Volatile	99.4–100	99.81	98
Zinc	Volatile	87.58–100	98.45	97.5

*U.S. EPA (1989).

tower, impingement tray scrubber alone, and Venturi alone. Despite the divergence in type of incinerator and APC equipment, relatively high metal control efficiencies were recorded for all metals except mercury. U.S. EPA metal data are representative of many of the older, existing incinerators equipped with only wet scrubber systems. Recent APC systems which include mercury control are presented at the end of this section under "Current APC Systems."

The last column in Table 7.1 lists typical metal control efficiencies that an incinerator manufacturer would provide as a performance guarantee for a new fluid bed incinerator equipped with just a Venturi and impingement tray scrubber. The guaranteed metal control efficiencies are conservative and reflect the inherent risks involved with providing emissions guarantees for a new facility.

For new facilities, state regulatory agencies often will require low metal emissions as a result of air dispersion modeling and health risk assessments. Ensuring low metal emissions is a two-step procedure for the incinerator design engineer. First, representative maximum metals concentrations in the incinerator feed cake must be determined for each metal. Second, guaranteed metal control efficiencies must be selected for each metal. To account for the high variability of the feed cake metal concentration, it is recommended that the last two to three years of metals data be assessed and the maximum concentrations for each metal used as the design basis. Metals data should be readily available from the WWTP, which are required by Part 503 regulations (U.S. EPA, 1993), to obtain monthly composite solids samples and have them analyzed for metals. The typical metal control guarantees can be used as a starting point for determining control efficiencies (see Table 7.1). If lower emissions are required by the regulatory authority, slightly greater control efficiencies than those shown in Table 7.1 are possible for some metals. However, it is essential that agreement from the incinerator supplier be obtained before going to higher control efficiencies.

If a higher metal control efficiency is required than can be obtained with a conventional Venturi and impingement scrubber, additional particulate control would have to be added to the system. The additional control might be obtained with a higher-pressure drop Venturi scrubber or wet ESP. A dry collector such as a fabric filter or dry ESP also could be considered. In this case, however, the flue gas handling system would have to be significantly reconfigured so that the incinerator exhaust is first cooled in a heat recovery boiler before the dry collector. A heat recovery boiler adds significantly more cost and complexity to the incineration system, but this option is becoming more justifiable and attractive as energy costs keep increasing.

1.2 Gaseous Pollutants

Gaseous pollutants result from the oxidation and volatilization of sulfur, nitrogen, and chlorine in the feed cake and from incomplete combustion of hydrocarbons and other organic compounds. The gaseous pollutants of primary concern are the acid gases, sulfur dioxide (SO_2), and hydrogen chloride or hydrochloric acid (HCl); carbon monoxide (CO); volatile organic compounds (VOCs); polycyclic organic matter (POM); nitrogen oxides (NOx); and greenhouse gases (GHGs).

1.2.1 Acid Gases

Incineration of the feed cake results in two acid gases (SO_2 and HCl). Sulfur can exist in three forms: sulfate sulfur (SO_4), pyritic sulfur (S^-), and organic sulfur. All three forms are essentially oxidized to SO_2 during the combustion process. The SO_2 then combines with moisture, either in the wet scrubbing system or as moisture in the atmosphere, to form sulfurous or sulfuric acids. Sulfur content of the feed cake typically varies from approximately 0.2 to 1.0%. To estimate uncontrolled SO_2 emissions, it is reasonable to assume that all the sulfur in the feed cake is oxidized and released as SO_2.

Chlorine in the feed cake can be in the form of organically bonded chlorine such as in chlorinated hydrocarbons or in inorganic salts such as NaCl. Chlorine content typically is relatively low, approximately 0.1 to 0.4%. However, road salting can cause higher chlorine levels of 1.0 to 2.0%. The use of ferrous or ferric chloride at the WWTP and, for coastal plants, intrusion of saltwater to the sewer system can also increase chlorine levels. In the incinerator, chlorine is released and combines with hydrogen to form hydrogen chloride. In addition to the severe health effects of SO_2 and HCl, both acid gases are significant because of their highly corrosive effect on downstream equipment. It is customary to fabricate all components of the incinerator wet scrubbing system and downstream ductwork and stack of 316L stainless steel.

Wastewater treatment plant solids typically contain sufficient sulfur and chlorine to result in uncontrolled SO_2 emissions of approximately 612 to 1630 mg/Nm^3 (normal cubic meters) on a dry volume basis corrected to 11% oxygen (mg/Nm^3 dv$_{11}$) or 300 to 800 ppm on a dry volume basis corrected to 7% oxygen (ppm dv$_7$) and uncontrolled HCl emissions of approximately 116 to 348 mg/Nm^3 dv$_{11}$ (100 to 300 ppm dv$_7$). Fortunately, the conventional wet scrubbing system (Venturi and impingement tray scrubbers using plant effluent as the scrubbing liquid) can control SO_2 emissions by approximately 80 to 85% and HCl emissions to an even greater degree. It should be noted that the use of plant effluent improves acid gas control because of the higher alkalinity of plant effluent in comparison with potable water. If a greater degree of SO_2 or HCl control is required, a caustic scrubbing section can be added to the tray scrubber. Typically, this is done during design by including two additional trays on the tray scrubber and circulating a caustic solution to just these two trays. With caustic scrubbing, 95% control of SO_2 can be achieved and even greater level of HCl control can be obtained. In general HCl is easier to control than SO_2. Hydrogen fluoride (HF) is also present in small quantities in the flue gases. This acid gas is readily removed by wet and dry scrubbing systems, as shown in Table 7.2.

Although wet scrubbing typically is used for acid gas control at WWTPs, dry emission control systems are possible. Such systems are extensively used in the power utility industry to control SO_2 emissions from coal-fired power plants. Table 7.2 lists alternate acid gas control technologies, their operating temperatures, and their achievable acid gas control efficiencies. With a dry injection system, a dry adsorbent such as lime is injected to the ductwork upstream of a dry particulate collector (fabric filter or ESP). The lime reacts with the acids in the flue gas forming insoluble salts (such as CaCl and $CaSO_4$), which are collected in the dry collector. With a spray dry absorber (SDA), a lime slurry is sprayed into an absorber vessel upstream of the fabric filter or ESP. The moisture from the slurry cools the flue gas and enhances the lime neutralization reaction. A spray dryer is significantly more effective in controlling SO_2 (90%) than a dry injection system (50%). Also the use of a fabric filter instead of an ESP enhances SO_2 removal. Wet scrubbing systems are also highly effective. However, the disadvantage of wet scrubbers is that they are only marginally effective in removing sulfur trioxide (SO_3) (25 to 40%), whereas an SDA with fabric filter

TABLE 7.2 Effectiveness of acid gas control systems.

Control system	Temperature*	Pollutant removal (%) HCl	HF	SO_2
Dry injection + fabric filter	160–180°C (320–356°F)	80	98	50
Spray dry absorber + ESP	140–160°C (284–320°F)	95+	99	50–70
with recycle	140–160°C (284–320°F)	95+	99	70–90
Spray dry absorber + fabric filter	140–160°C (284–320°F)	95+	99	70–90
with recycle	140–160°C (284–320°F)	95+	99	80–95
Wet scrubber	40–50°C (104–122°F)	95+	99	90+
Wet/dry scrubber	40–50°C (104–122°F)	95+	99	90+

*Temperature at the exit of the control device.

will effectively remove SO_3. At temperatures higher than approximately 149°C (300°F), SO_3 is a vapor. Once the vapor cools below its acid dewpoint, it condenses into sulfuric acid mist (H_2SO_4) which can cause a visible plume (National Lime Association, 2002). Some coal-fired power plants with wet scrubbers have had to install wet ESPs to capture the sulfuric acid mist emitted from their wet scrubbers.

1.2.2 Carbon Monoxide

Carbon monoxide is a product of incomplete combustion (PIC) from partial oxidation of carbon in the feed cake with the oxygen in the combustion air. The formation of CO is caused by one or more of the following deficiencies with the combustion system: inadequate temperatures, inadequate residence time of the combustion gases, or inadequate mixing or turbulence which is necessary to bring the combustion gases in dynamic contact with the oxygen in the air supply. In general, the combustion environments of an MHF and fluid bed incinerator are markedly different and each will be discussed separately.

In an MHF, the feed cake first dries on the upper hearths and then is combusted on the middle hearths. Although this arrangement is efficient in terms of using the heating value in the dried feed cake to dry the incoming wet feed cake, it results in the release of partially oxidized combustion gases and PICs from the upper hearths of the furnace where the feed cake is drying and just beginning to burn. The slow, stratified flow of combustion gases (i.e., lack of turbulence) in this part of the furnace results in increased emissions of CO and PICs. Carbon monoxide emissions from an MHF without an afterburner can range from 900 to 2700 mg/Nm^3 dv_{11} (1000 to 3000 ppm dv_7); a typical CO mass emission rate is 15.5 g/kg (31 lb/dry ton) of solids incinerated. High CO and VOC emissions from an MHF are one of the primary reasons that their use has declined steadily over the last several decades. High CO emissions from an MHF can be controlled by use of a high-temperature afterburner. Several MHFs have been retrofitted with top hearth (or "zero hearth") afterburners, which have significantly reduced their CO and VOC emissions. However, the additional fuel use required to operate the afterburner is a significant drawback.

In contrast to an MHF, a fluid bed incinerator is a completely mixed, highly turbulent system in which drying and combustion take place concurrently and rapidly, within a matter of seconds. The turbulent fluid bed of a fluid bed incinerator provides complete and intimate contacting of the feed cake, the volatilized gases, and the oxygen in the fluidizing air. The hot combustion gases rising from the bed then enter the freeboard of the fluid bed incinerator, which provides a long gas residence time and allows the CO and other volatilized organics to fully burn out. Carbon monoxide emissions from a fluid bed incinerator are invariably less than 45 mg/Nm^3 dv$_{11}$ (50 ppm dv$_7$) and in many cases less than 9 mg/Nm^3 dv$_{11}$ (10 ppm dv$_7$); mass emission rates are typically less than 0.5 g/kg (1.0 lb/dry ton) of solids incinerated. For a new facility, state regulatory agencies typically will require a CO emission limit of 90 mg/Nm^3 dv$_{11}$ (100 ppm dv$_7$). A fluid bed incinerator can easily meet this limit. An MHF would require an afterburner operating at a minimum of 816°C (1500°F) to meet this standard.

1.2.3 Volatile Organic Compounds

Volatile organic compounds, like CO, are PICs that result from the vaporization of organic matter in the feed cake and partial oxidation of the volatilized compounds. As previously stated, incomplete combustion is caused by inadequate combustion conditions in the incinerator, such as insufficient temperature, residence time, or mixing.

Chemically, VOCs consist of a variety of compounds, including the following:

- Straight and branched chain aliphatic hydrocarbons (methane, ethane, acetylene, etc.).
- Oxygenated hydrocarbons (acids, aldehydes, ketones, etc.).
- Chlorinated hydrocarbons (perchloroethylene, trichloroethane, etc.).
- Saturated and unsaturated ring compounds (benzene, toluene, phenols, etc.).

Volatile organic compounds are regulated under the Part 503 regulations (U.S. EPA, 1993), which require that total hydrocarbons emitted from an incinerator must be less than 100 ppm as propane on a dry volume basis corrected to 7% oxygen (i.e., 100 ppm dv$_7$ [140 mg/Nm^3 dv$_{11}$]).

Because of the different combustion conditions that exist in an MHF and fluid bed incinerator, VOC emissions from an MHF and fluid bed incinerator are different. The upper drying hearths of an MHF typically will have sufficient temperatures to volatilize organic compounds but insufficient temperatures to fully oxidize them. In general, VOC emissions from an MHF will vary significantly on a daily basis depending on the feed cake rate and combustion characteristics (percent solids, percent volatile solids, and heating value) and furnace operating conditions (hearth temperatures, excess air level, burner firing rates on different hearth levels). Some MHFs can, however, meet the 140-mg/Nm^3 dv$_{11}$ (100-ppm dv$_7$) standard without an afterburner. In many cases, maintaining a top hearth temperature of 593°C (1100°F) or greater is important for achieving the standard (Waltz, 1990; Baturay, 1990).

In contrast, because of its turbulent high-temperature conditions, a fluid bed incinerator has low total hydrocarbon emissions, typically less than 14 mg/Nm^3 dv_{11} (10 ppm dv_7) as propane.

1.2.4 *Polycyclic Organic Matter*

Polycyclic organic matter (POM) is a subset of VOCs that are of particular concern to the regulatory community because of their potentially high health effect risks. Some of the primary POMs of interest are polychlorinated biphenols (PCBs), polychlorinated dibenzo-p-dioxin (PCDD), and dibenzo furan (PCDF). U.S. EPA does not have specific emission limits for these pollutants. However, some state regulatory agencies have included emission criteria for a few of these compounds in air permits of new incinerators. Methods to achieve low POM emissions are to maximize the combustion efficiency of the incinerator. For MHFs, if additional control of POMs is necessary, a high temperature afterburner may be required. This would not be necessary for a fluid bed incinerator. Emission factors from the U.S. EPA Compilation of Emission Factors (AP-42) indicate that emissions of the above pollutants are low for both MHFs and fluid bed incinerators, with ranges as follows: 0.5×10^{-7} to 0.5×10^{-9} g/kg (1×10^{-7} to 1×10^{-9}) (U.S. EPA, 1998).

1.2.5 *Nitrogen Oxide*

Nitrogen oxide is an important pollutant because it is an ozone precursor that is activated by UV light in the upper atmosphere to produce ozone. Because many of the large metropolitan areas in the United States are designated as ozone nonattainment areas, NOx emissions are carefully scrutinized by air quality control boards.

Nitrogen oxides can take many forms, with the nitrogen atom combining with one or more oxygen atoms. Nitrogen oxide is a product of all air oxidized combustion processes, including incineration. The two mechanisms that generate NOx emissions during the combustion process are fuel NOx and, to a lesser extent, thermal NOx. Fuel NOx is produced from the oxidation of the organically bound nitrogen in the fuel. Its formation rate is strongly affected by the rate of mixing of fuel and air and the local oxygen concentrations and combustion temperatures. Residuals from WWTPs have ample quantities of nitrogen (typically 3 to 6%) to generate NOx, but fortunately only a small fraction of the nitrogen in the feed cake is converted to NOx.

Thermal NOx is generated from the thermal conversion of the nitrogen and oxygen in the combustion air to NOx at high temperatures. The rate of thermal NOx formation is highly dependent on local flame temperatures and, to a lesser extent, on oxygen concentrations. Because thermal NOx becomes significant at temperatures higher than 1093°C (2000°F), thermal NOx is not as important as fuel NOx in municipal WWTP incinerators. However, auxiliary fuel burners on an MHF can be a significant source of thermal NOx.

Multiple-hearth furnaces can have significant NOx emissions, particularly if combustion temperatures exceed 899°C (1650°F). Nitrogen oxide concentrations from an MHF will typically range from 250 to 587 mg/Nm^3 dv$_{11}$ (150 to 400 ppm dv$_7$). Niessen (1990) analyzed 154 sets of MHF data and found that NOx concentrations from an MHF averaged 417 mg/Nm^3 dv$_{11}$ (284 ppm dv$_7$), with a standard deviation of 240 mg/Nm^3 dv$_{11}$ (164 ppm dv$_7$).

In contrast, because of its lower combustion temperatures of lower than 871°C (1600°F) and lower excess air levels (lower percent oxygen), a fluid bed incinerator typically has NOx emissions of less than 147 mg/Nm^3 dv$_{11}$ (100 ppm dv$_7$). A fluid bed incinerator can have higher NOx emissions if the feed cake solids are high (greater than 28%) and the volatile content is also high (greater than 80%) (Dangtran and Butt, 2004; Sapienza et al., 1998). Particularly at plants with high-solids centrifuges, it is important to appropriately size the combustion air preheated (heat exchanger) to achieve a combustion air preheat temperature that will result in combustion temperatures of approximately 843°C (1550°F) in the freeboard of the fluid bed incinerator. If the combustion air preheat temperature is too high, the freeboard temperature can climb to higher than 871°C (1600°F) and NOx emissions will increase. Preferably, the fluid bed incinerator system will be designed for as wide a range of percent solids and percent volatile solids as is practically possible. This can be done by providing a bypass duct (with flow control valve) around the bottom portion of the combustion air preheater which will allow the temperature of the preheated combustion air to be adjusted depending on the temperature in the fluid bed incinerator.

In general, NOx emissions from a fluid bed incinerator can be controlled by limiting peak temperatures to lower than 871°C (1600°F) and minimizing excess air levels while maintaining adequate combustion efficiency. If feed cake with high solids and high volatile solids content is anticipated, flexibility should be incorporated to the system design as discussed above. Achieving low NOx emissions with an MHF is more difficult and may preclude its use in sensitive air quality areas.

1.2.6 Greenhouse Gases

Greenhouse gases are gases that absorb or trap heat in the atmosphere, contributing to global warming. Although a certain amount of these gases is necessary to make the earth habitable, the concentration of GHGs in the atmosphere has been increasing steadily over the last 100 years and is now having a profound effect on the earth's climate and ecosystems. According to the National Oceanic and Atmospheric Administration (NOAA), the earth's average surface temperature has risen by 0.67 to 0.78°C (1.2 to 1.4°F) since 1900, with the majority occurring since 1970 (0.56°C [1.0°F]). The current rate of global surface warming has increased to 0.18°C (0.32°F) per decade or 1.8°C (3.2°F) per century. The NOAA also reports that the five warmest years over the last century occurred in 2005, 2004, 2003, 2002, and 1998 (NOAA, 2005). The National Research Council (NRC) states that climatic change is occurring over most of the globe and is evident from the global retreat of glaciers, reduction in the extent of snow cover, earlier spring melting of lakes and rivers, and an increase in the ocean surface temperatures and heat content (NRC, 2001).

The most significant GHGs are carbon dioxide (CO_2); methane (CH_4); nitrous oxide (N_2O); and fluorinated gases, including hydrofluorocarbons (HFCs), perfluorocarbons (PFCs), and sulfur hexafluoride (SF_6). It should be noted that there are also chlorofluorocarbons (CFCs) and hydrochlorofluorocarbons (HCFCs) that are ozone-depleting substances (ODSs) as well as being GHGs. The use of these ODSs has been decreasing since the adoption of the Montreal Protocol and the U.S. Clean Air Act Amendments of 1990. Although the emissions of the fluorinated gases are relatively low, these gases can persist in the atmosphere for extremely long periods of time, hence are very potent GHGs.

Greenhouse gases are ranked according to their global warming potential (GWP), which is the ratio, on a time-integrated basis, of how much global warming a particular gas can exert relative to CO_2. Carbon dioxide is assigned a GWP of one; CH_4 and N_2O have GWPs of 21 and 310, respectively; and the fluorinated gases have GWPs of several thousands. Despite its low GWP, CO_2 is the predominant GHG driving global warming. The U.S. EPA compiles an annual inventory of GHG emissions in which the emissions of each gas are calculated in terms of teragrams (10^{12} grams) of CO_2 equivalents. According to the U.S. EPA, in 2004 CO_2 emissions represented 84.6% of the total GHG emissions in the United States; CH_4 and N_2O constituted 7.9% and 5.5% of the GHG emissions, respectively; and the fluorinated gases composed approximately 2.0% (U.S. EPA, 2006).

Considering the global carbon cycle, CO_2 is in a continual state of flux with billions of tons of CO_2 being absorbed by oceans and plant life and billions of tons being emitted from natural processes and from the combustion of fossil fuels by human beings. The prodigious combustion of fossil fuels in the last century has upset this balance. According to NOAA, the CO_2 concentration in the atmosphere has increased from 280 ppm in preindustrial times to 379 ppm in 2005 and the current rate of increase is 1.8 ppm by volume per year (NOAA, 2005).

Control technologies for GHGs are being developed. Technologies are available to minimize CO_2 emissions and to separate and capture CO_2 in compressed or liquid form. The CO_2 can then either be stored in geological deposits (saline formations), used in oil or gas recovery operations, or sequestered (absorbed) by plants as part of the normal photosynthesis process. Regulations to control CO_2 most likely will be directed at power and other industries that consume large quantities of fossil fuels. In the future, incinerators may have to quantify and control their CO_2 emissions just like any other regulated pollutant.

2.0 CONTROL DEVICES

A variety of control devices can be used to reduce emissions. Each has its own specific characteristics, efficiency, and effectiveness in controlling one or more types of pollutants.

2.1 Afterburners

Afterburners are typically used on MHFs to control VOCs, CO, or odorous emissions from an MHF. Afterburners can have a number of shapes and arrangements, but their common purpose is to raise the flue gas temperature sufficiently high, with excess oxygen, to combust any unburned organic matter in the flue gas. The most commonly used afterburner is a refractory lined steel chamber equipped with one or more fuel oil or natural gas burners. In Figure 7.1, an afterburner consisting of a vertical cylindrical chamber located at the breech of a furnace is shown. Alternatively, as shown in Figure 7.2, the top hearth of an MHF can be used as an afterburner chamber, a so called "zero hearth afterburner." With this arrangement, feed cake is introduced to the second or third hearth with the top one or two hearths acting as the afterburner chamber. Typically, oil or gas burners will be added to the top hearth to sufficiently increase the flue gas temperature. Providing sufficient residence time and preventing short circuiting of the flue gas is critically important in the design of a

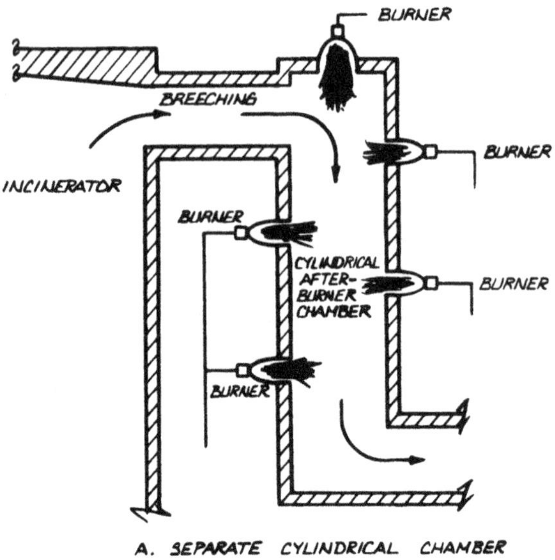

FIGURE 7.1 An afterburner chamber: separate cylindrical chamber.

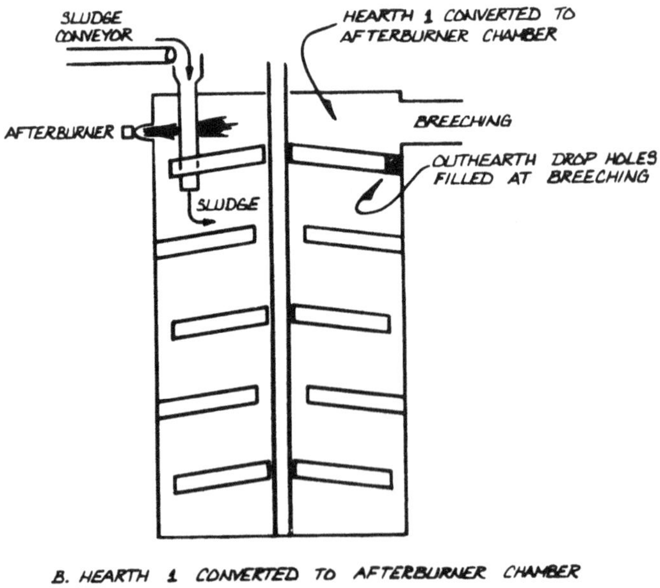

FIGURE 7.2 Hearth 1 converted to an afterburner chamber.

zero hearth afterburner. In some cases, refractory baffles are constructed on the top hearth to ensure adequate residence time.

Figure 7.3 shows a different type of afterburner known as a regenerative thermal oxidizer (RTO). These systems use inlet and outlet dampers to alternately cycle the exhaust gas through ceramic heat recovery beds. The system operates as follows. As the exhaust gas passes through a heat recovery bed in inlet mode, the exhaust gas is gradually heated up until it reaches the combustion chamber. As it enters the combustion chamber, the gas temperature will be approximately 732°C (1350°F). The gas burners in the combustion chamber will then raise the flue gas temperature to 816°C (1500°F). The hot gas resides in the combustion chamber for one to two seconds and

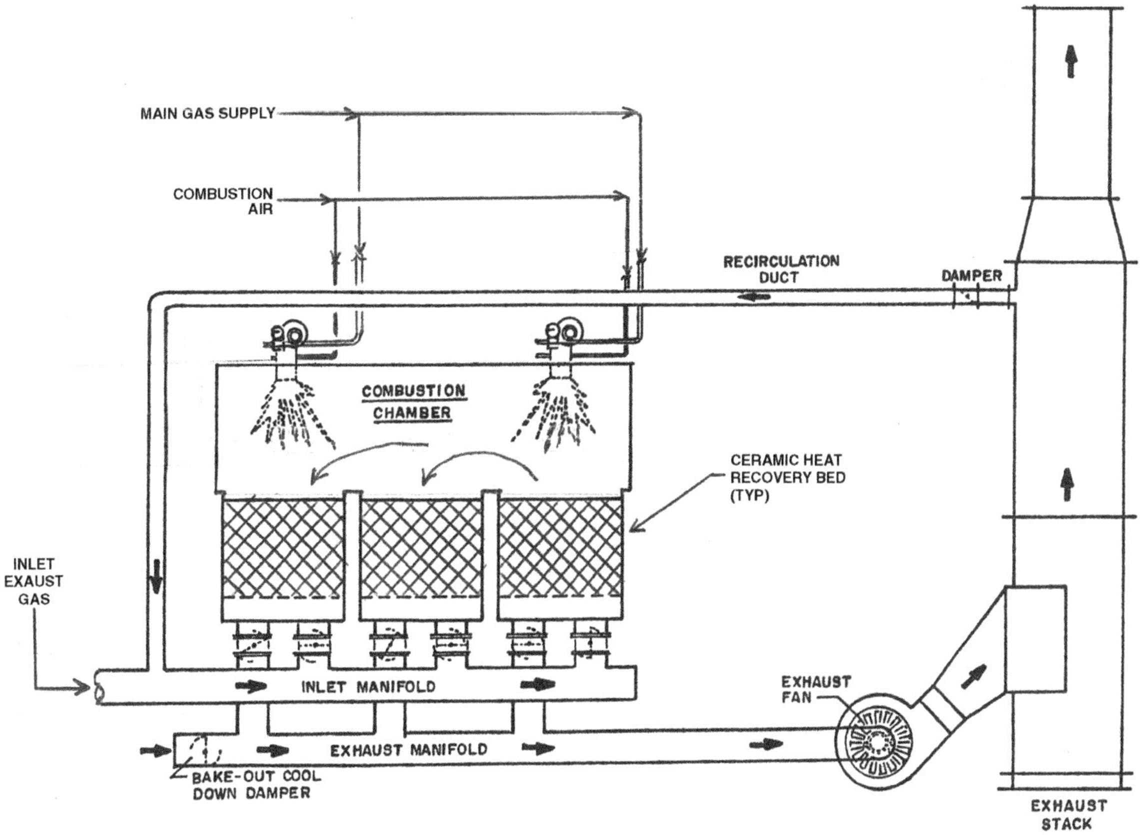

FIGURE 7.3 A regenerative thermal oxidizer.

then passes through a heat recovery bed in outlet mode. As the hot gas passes through the bed, it heats up the ceramic bed, thereby regenerating the bed. Every one to two minutes, the dampers in the inlet and outlet manifolds switch positions and the incoming exhaust gas now goes through the hot preheated bed (now in inlet mode) and out the former inlet bed now in outlet mode. By utilizing regenerative heat recovery, these systems can achieve very high thermal efficiencies of 95%, minimizing fuel use. Some RTO systems will have a third heat recovery chamber which is in transition mode while the chambers are switching. Volatile organic compound destruction efficiencies of 98 to 99% are possible with these systems.

The main drawback with the use of RTOs as afterburners is that the inlet gas must be free of particulate matter because the ceramic media will capture the particulate matter and eventually plug up the heat recovery beds. To solve this problem, an RTO on an MHF is preceded by effective particulate control device, typically a wet ESP. The addition of an RTO and upstream particulate control device is a large capital expenditure. However, the savings in fuel cost over a direct fired afterburner can justify the high capital cost.

2.2 Wet And Dry Systems

Emissions control equipment may be a combination of devices that may or may not use water for their operation. Cyclones, scrubbers, Venturi scrubbers, tray scrubbers, dry ESPs, wet ESPs, and fabric filters are common units.

2.3 Cyclones

Cyclones are now infrequently used because they have been replaced by more effective devices. However, they have been used on fluid bed incinerators to reduce the particulate loading on downstream equipment. As previously stated, 100% of the ash (and some of the bed sand) exits a fluid bed incinerator in the flue gas which results in a high particulate loading. A dry cyclone, preferably a multiple cyclone, can remove the bulk of this high particulate loading. Typical application in a system would be the use of a cyclone before a waste heat boiler to reduce the dust loading on the boiler. Another fluid bed incinerator application might be a dry cyclone before a wet scrubbing system if dry collection of the bulk of the fly ash is desired.

A dry cyclone is a vertical cylindrical vessel that takes advantage of the differences in densities of the gas stream and particulate matter. The particulate-laden exhaust gases enter the cyclone tangentially and swirl at high velocity. This high rotational speed causes centrifugal action to force the particulates to the outside of the chamber where friction with the wall of the cyclone causes them to slow down and drop vertically to the discharge at the bottom of the chamber. The sizing of a cyclone depends on the gas stream volume, the size and amount of particulate that must be removed, and the pressure drop through the cyclone. For installations where the gas volume is high, clusters of more than one cyclone can be fabricated into a single unit, as shown in Figure 7.4.

2.3.1 Performance

The performance of a cyclone depends on the dimensional relationship between the diameter and the height of the unit, gas viscosity, the number of turns that the gas takes in the cyclone, inlet and outlet gas velocities, particle and gas densities, and a dust concentration factor. The first graph in Figure 7.5 shows a typical cyclone curve relating the percent removal efficiency by weight as a function of the inlet dust particle diameter and the fractional efficiency characteristic of the cyclone, denoted by the letter "N." The fractional efficiency characteristic is a function of the parameters listed above and is specific to a particular cyclone size and configuration (i.e., single, parallel, or series arrangement). This parameter is determined by the cyclone manufacturer. The second graph in Figure 7.5 shows the percent removal efficiency of var-

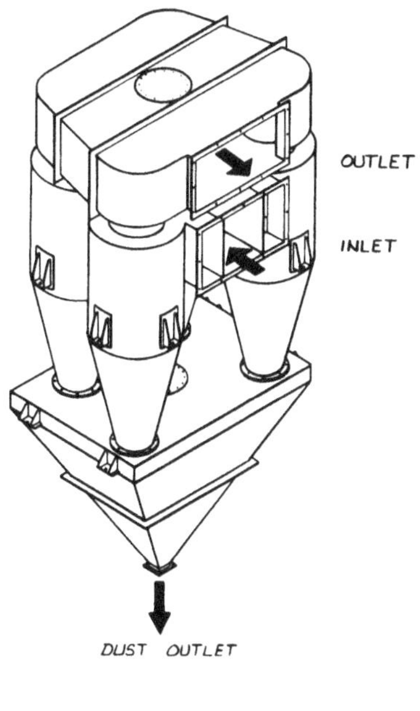

FIGURE 7.4 A multiple cyclone unit.

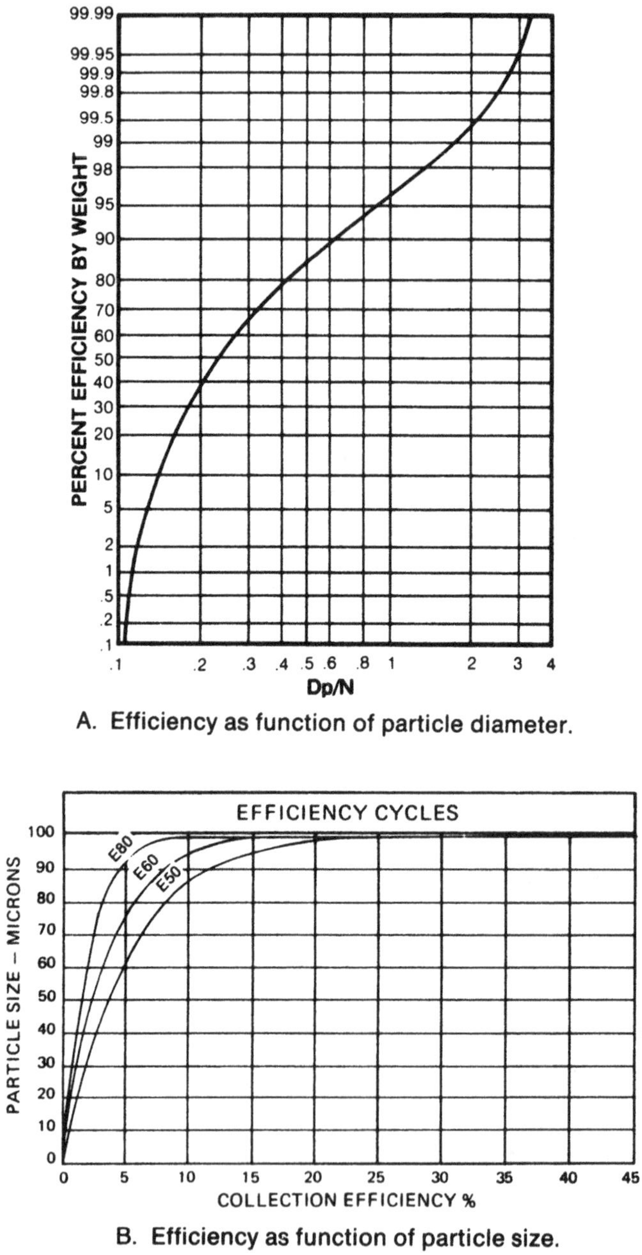

FIGURE 7.5 Cyclone efficiency graphs: (a) efficiency as a function of particle diameter and (b) efficiency as a function of particle size.

ious size particles ranging from 0 to 100 μm. The three efficiency curves reflect three different sized cyclones. Note that the curves shown in Figures 7.5 are for a low efficiency cyclone. Significantly higher efficiency cyclones are available. The predominant factor in determining collection efficiency is the inlet velocity into the cyclone and pressure drop through the unit.

2.3.2 *Advantages and Disadvantages*

A cyclone should be used where the particulate loading in the exhaust gas stream consists of large size particles. Cyclones are relatively inexpensive, and removal of particulate matter upstream of other processes will reduce loading on downstream equipment and control devices. One of the significant advantages of a dry cyclone in a fluid bed incinerator system is that it removes particulate matter in dry form, which may be a desired objective. Note that dry collection of particulate matter from a fluid bed incinerator typically will require a waste heat boiler to cool the exhaust gas before a dry ESP or fabric filter. Another significant advantage of the cyclone is that there are no operating parts so operation and maintenance requirements are minimized. It also has a relatively low pressure drop compared with wet scrubbers. A significant disadvantage of the cyclone is that it is best suited for removing larger size particles, typically greater than 10 μm.

2.3.3 *Operation and Maintenance*

Gas velocity through the unit is an important parameter regarding sizing and control efficiency of the cyclone. Unit sizing and operation should, therefore, minimize the variability of the exhaust gas flow rate. Removal efficiency of a cyclone will deteriorate significantly if there are appreciable fluctuations in the gas flow rate or if the gas velocities are significantly lower than the design velocity. Cyclones have no operational components except for a bottom air lock to allow removal of the collected particulate under negative pressure conditions. Maintenance includes periodic inspection inside the cyclone to determine if significant erosion has occurred because of high-velocity gases passing through the unit. Also, in some cases, particulate matter can collect on the walls of the cyclone and require periodic cleaning.

2.4 Venturi Scrubbers

Venturi scrubbers are the predominant particulate control devices used. In a Venturi scrubber, the exhaust gas is accelerated to a high velocity, typically 61 to 183 m/s (200 to 600 ft/sec) at the Venturi throat, where the gas comes in contact with the scrubbing liquid. The high velocity gas stream atomizes the liquid into fine droplets that collect the fine particulate. The primary collection mechanism is inertial impaction in which the micrometer and submicrometer size particles are driven into the larger size water droplets. Condensation on the submicrometer particles increases their mass and contributes to their removal by the impaction mechanism (Perry and Chilton, 1973). Downstream of the Venturi throat, the particulate-laden droplets collide and agglomerate into larger size, heavier droplets which are driven into the flooded elbow at the bottom of the Venturi. Finer size droplets are collected in the downstream tray scrubber. The pressure drop through the Venturi determines the particulate removal efficiency. The greater the pressure drop, the greater the removal of submicron size particles.

Figure 7.6 shows a vertical plume bob type Venturi scrubber which is commonly used on a fluid bed incinerator. An automatic actuator can move the plume bob up and down adjusting the size of the opening in the throat and thereby adjusting the pressure drop through the throat. Another commonly used Venturi throat, shown in Figure 7.7, has a rectangular throat and is equipped with one or two bomb-bay type dampers that allow for varying the throat opening. Other types of Venturi scrubbers operate with a fixed throat and maintain a constant pressure drop across the throat by varying the amount of water added as the gas flow changes. These types are not common on municipal incinerators.

Relatively recently, multiple Venturi scrubbers have been used. This type of scrubber is different in that condensation of the gas stream is performed before the Venturi section. A typical multiple Venturi scrubber is shown in Figure 7.8. In the first stage, the flue gas flows through a low pressure drop quenching section and then enters a vertical scrubber tower. The scrubber tower contains sub-cooling trays that cool flue gas to 38 to 49°C (100 to 120°F). The flue gas then enters the Venturi stage which consists of several Venturi throats. At the inlet of each Venturi throat, high-pressure water at 2100 kPa (300 psig) (or alternatively water and compressed air) is atomized through fine nozzles to create fine water droplets necessary to remove micrometer and submicrometer size particles.

2.4.1 Performance

Particulate removal efficiency of a Venturi scrubber is dependent on the pressure drop across the throat of the Venturi. In general, the greater the pressure drop, the greater the particulate removal. Particulate removal efficiencies as a function of particle size and Venturi pressure drop are shown in Figure 7.9 (Schiffner and Hesketh,

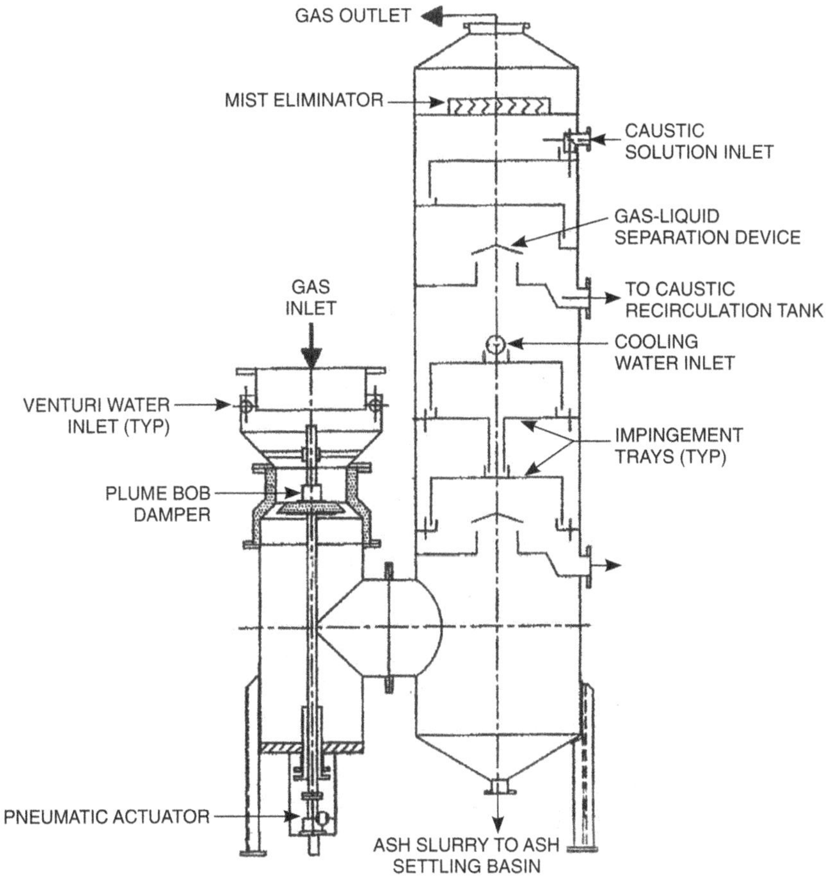

FIGURE 7.6 A Venturi scrubber with plume bob damper and tray scrubber (courtesy of Swemco International, Inc., New York, U.S.A.).

1983). The relationships shown in Figure 7.9 can be used as an initial estimation of Venturi pressure required as a function of particle size and removal efficiency. To meet the NSPS for incinerators (0.65 g/kg [1.3 lb/ton] of solids incinerated), a pressure drop of 5 to 7.5 kPa (20 to 30 in water column [w.c.]) across the Venturi, followed by an additional 2.5 kPa (10 in w.c.) across the tray scrubber, for a total of 7.5 to 10 kPa (30 to 40 in w.c.) across the complete wet scrubbing system is typically required. Some incinerator installations can meet the incentive standard (0.37 g/kg [0.75 lb/dry ton]) with the above pressure drops. Liquid flow rates of 1.4 to 1.6 L/m^3 (10 to 12 gpm/1000 acfm) are typically used.

FIGURE 7.7 A Venturi scrubber with bomb-bay dampers.

2.4.2 *Advantages and Disadvantages*

The Venturi scrubber is a relatively simple and effective particulate wetting and removal device, provided that pressure drop can be adjusted to obtain the required removal efficiency. A properly designed unit should have a variable throat section to permit a constant pressure drop while gas volume processed by the system varies. A Venturi requires a relatively clean source of water and a disposal point for the processed water. At WWTPs, effluent is typically used. One of the primary disadvantages of a Venturi scrubber is the significant power required by the incinerator induced draft fan (or fluidizing air blower for a fluid bed incinerator) to accelerate the flue through the Venturi. The higher pressures create stress for the induced draft fan, which increases maintenance requirements. With a fluid bed incinerator this is not a concern because most fluid bed incinerators in the United States do not have induced draft fans. The higher pressures required by the Venturi scrubber can easily

FIGURE 7.8 A multiple Venturi scrubber (courtesy of Envirocare International, Inc., American Canyon, CA).

be accommodated by the multistage centrifugal blower used on a fluid bed incinerator. (Many fluid bed incinerator facilities in England and Europe have waste heat boilers that necessitate use of an induced draft fan.) Another disadvantage of the Venturi scrubber is erosion of throat section and Venturi damper from high-velocity particles. Most newer Venturi scrubbers have throat sections lined with silicon carbide tiles to mitigate the erosive effect of particulate matter.

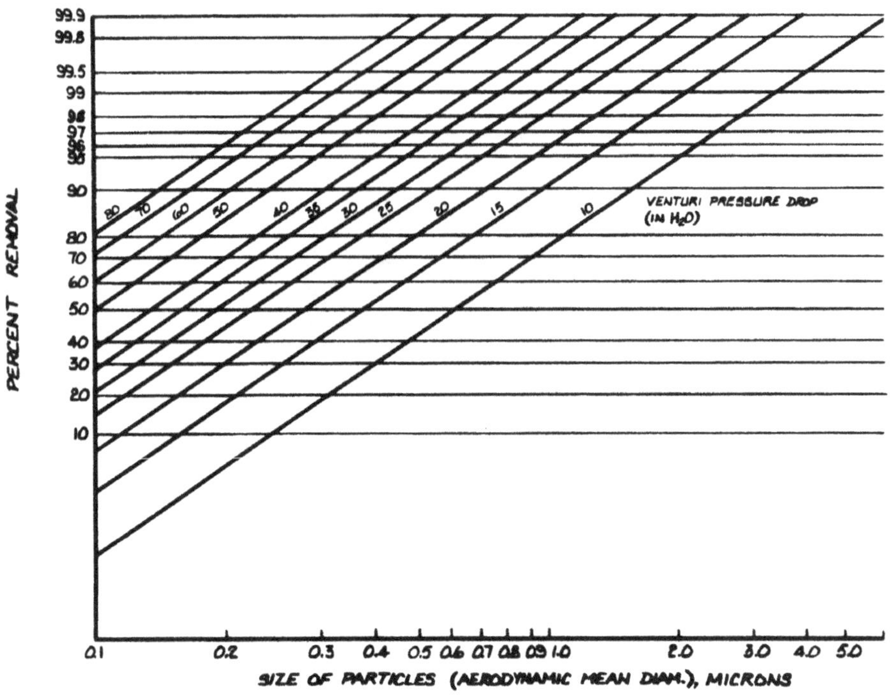

FIGURE 7.9 Venturi particle collection efficiency graph.

2.4.3 *Operation and Maintenance*

A Venturi scrubber must be separated from the incinerator breeching by either a water seal or a metallic expansion joint. This separation is required to allow differential movement between the incinerator and the relatively stable APC equipment. The incinerator will increase in height as it goes from a cold to hot operating condition. A cross-section of a Venturi scrubber with a water seal is shown in Figure 7.10. As shown in the figure, at the top of the Venturi throat is a collar with a water seal. Water is continuously supplied to allow for evaporation. The depth of the seal must be greater than the negative pressure that can be expected at the inlet to the Venturi. For fluid bed incinerators (without an induced draft fan) the pressure at the Venturi inlet is approximately 10 to 12 kPa (40 to 50 in w.c.), which precludes the use of a water seal. Thus, for fluid bed incinerators a metallic expansion joint is typically used.

On some Venturi scrubbers quench water (or throat water) is added directly above the throat to cool the gases to the adiabatic saturation temperature before the

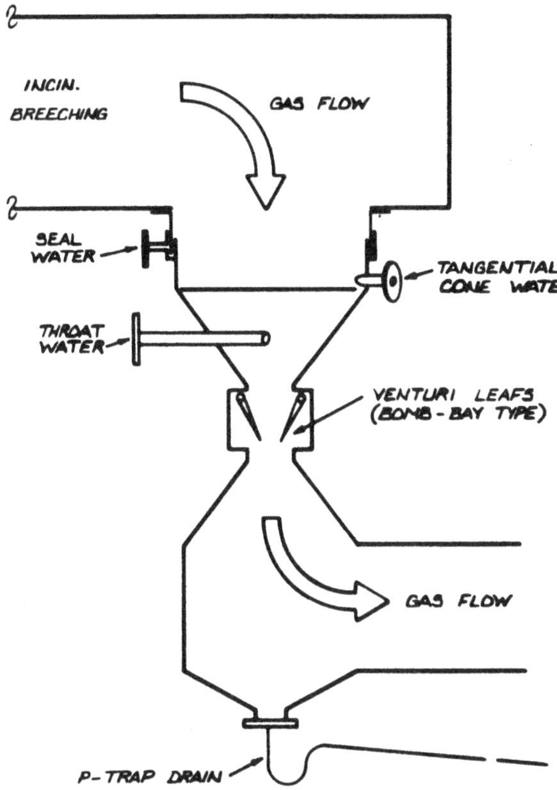

FIGURE 7.10 A Venturi scrubber cross-section view.

throat section. As the unit begins to converge toward the throat section, a series of pipes tangentially supply water that floods the throat section. The flue gas is accelerated to a high velocity through the flooded throat, which atomizes the water into fine droplets and drives particulate matter in the liquid phase. Below the throat in the diverging section of the Venturi, the fine, particulate laden droplets coalesce and agglomerate into larger droplets, forming a liquid slurry that is collected at the bottom of the scrubber. The water also continually wets the surface of the cone and thereby minimizes erosion and corrosion of the throat section. In the throat section, either a plum bob damper or single or double leaf dampers vary the throat opening to maintain a constant pressure drop to compensate for changes in the exhaust gas volume.

Maintaining adequate water flows to wet all of the metallic surfaces of the throat section is important for minimizing corrosion and erosion. Periodic inspection and repair of the Venturi damper are necessary to ensure that dampers are in proper working order. Also silicon carbide linings are subject to high velocities and stresses and can break away from the scrubber housing. If possible, operation at pressure drops of 6.2 kPa (25 in w.c.) or less will prolong the life of these linings.

2.5 Tray Scrubbers

As shown in Figure 7.6, a tray scrubber is commonly used following a Venturi scrubber to act as a complete wet scrubbing system. The tray scrubber consists of a vertical cylinder in which the exhaust gases enter at the bottom and exit through the top. The gas stream first passes through a gas-liquid separation device that is used to separate the Venturi water (heavily laden with particulate matter) from the tray water. The water contains a relatively small amount of particulate matter and typically is returned to the WWTP headworks. The above gas stream then proceeds through three or more scrubber trays that allow contact between the exhaust gas and the scrubbing water. The scrubber water is introduced on the top tray and flows countercurrent to the gas stream. Large quantities of water are used to cool the gas to temperatures of 38 to 49°C (100 to 120°F).

Two types of trays are commonly used: perforated sheet metal plates or impingement trays. With perforated sheet metal plates, each tray can have one or two layers of perforated plates. With two perforated sheets, the metal plates are separated by a small gap and positioned so that the holes are not aligned and the gas flows through a contoured pathway. An impingement tray consists of a perforated metal sheet with metal strips or "targets" located above the holes in the perforated plate. As the gas flows through the holes in the plate, it is then forced to contact the metal targets, which increases turbulence and scrubbing effectiveness. Packed scrubbers have also been used on some systems but they are more prone to fouling and plugging, particularly if they are not preceded by a dry collector such as fabric filter or ESP. A demister section is located above the trays to remove large water droplets from the gas stream. Typically a zigzag-type baffle demister is used because it is less prone to fouling.

2.5.1 *Performance*

A tray scrubber has three beneficial effects:

1. Condenses aerosols in the flue gas, hence enhances particulate capture.
2. Removes water vapor from the gas stream, hence minimizes plume formation from the stack.
3. Removes the bulk of the acid gases in the flue gas.

The additional particulate capture will contain some volatile hydrocarbons, which would be detected in the "back-half" of the method 5 particulate test. The tray scrubber shown in Figure 7.6 contains an additional gas-liquid separation device and two additional trays. With this design, a caustic solution is circulated to just the top two trays to achieve a high degree of acid gas control. Typically with this arrangement, SO_2 and HCl removals of 95 and 98%, respectively, can be achieved.

2.5.2 *Advantages and Disadvantages*

The advantages of the tray scrubber are that pressure losses in the system range from 0.5 to 0.7 kPa (2 to 3 in w.c.) per tray. Because three trays are typically used, the overall pressure drop is approximately 2.2 to 3 kPa (9 to 12 in w.c.). To achieve subcooling of the flue gas to 38 to 43°C (100 to 110°F), liquid flow rates of 3.4 to 6.7 L/m^3 of saturated flue gas (25 to 50 gpm per 1000 acfm) (i.e., following the Venturi scrubber) are required.

Because this type of scrubber is inefficient in removing small diameter particles, it cannot be used as the sole particulate control device. It also requires large quantities of water to operate, which then must be treated. This excess water is typically not a problem at a WWTP. However, this can be a significant concern at WWTPs that are required to produce a high-quality effluent and want to minimize recycle flows to the WWTP.

2.5.3 *Operation and Maintenance*

Because the performance of a tray scrubber is a function of the pressure drop across the trays, a differential pressure indicator is typically used to continuously record these figures. Removal efficiency of an impingement tray scrubber is relatively constant from 40 to 100% of the maximum design flue gas flow rate.

Even without moving parts, tray scrubbers can require significant maintenance. Water cooling of the gas stream can form condensed organics on the interior surfaces of the chamber and on the trays if combustion in the incinerator is incomplete, as is the case with an MHF. On some MHFs, this buildup of greasy slime can reach thicknesses of 6 to 13 mm (0.25 to 0.50 in) and can also occur in ductwork and in the induced draft fan housing. This type of fouling has not been a significant problem with fluid bed incinerators.

The warm, moist atmosphere in the scrubber can also promote the growth of a biological mass on the trays, requiring high-pressure sprays for removal. The use of a pressure differential indicator across the trays can identify this problem during operation. The use of chlorinated plant effluent can lessen the growth of biomass on the scrubber trays. The demister also is subject to fouling and plugging from soot or biological growths. The scrubber should be designed with an access door for each of the scrubber trays and the demister. Furthermore, the trays and demister should be constructed in sections such that they can be removed from the scrubber for periodic cleaning.

Another concern with this type of scrubber is maintaining a uniform flow of water across the trays. The flow of water across the trays is controlled by weirs at one end (and sometimes both ends) of the trays. The weir depth should be at least 80 mm (3 in) to maintain a pressure drop of 0.5 to 0.7 kPa (2 to 3 in w.c.) across each tray. Weirs are sometimes not installed perfectly level. Also slime buildup or metal warping from temperature excursions (resulting from temporary loss of scrubber water) can alter weirs and result in a non-uniform depth of water across each tray. Thus, it is important to periodically inspect the inside of a tray scrubber particularly after process upsets. A typical scrubber tray and weir combination is shown in Figure 7.11.

Another design consideration is that the scrubber drain pipe must be atmospherically separated from the scrubber. Otherwise, ambient air will be drawn into the scrubber from the drain pipe, or, if it is under positive pressure, flue gas will be blown out the drain pipe. To prevent this from happening, an S or P trap must be provided at the scrubber drain as shown in Figure 7.11. If the scrubber is under negative pressure, the height of the drain trap, dimension A, must be greater than the maximum negative pressure experienced by the scrubber. If the scrubber is under positive pressure (as is often the case with a fluid bed incinerator), the depth of the U-shaped section of the trap must be greater than the positive pressure in the system.

Formation of foam within the scrubber can also clog the unit. Some plants have experienced excessive foaming of scrubber water, which can prevent exhaust gases from moving through the trays. In such circumstances, a complete chemical analysis of the scrubber water should be performed to identify the foaming agent so that it can be removed from the wastewater stream. If this is not possible, antifoaming agents can be used.

2.6 Dry Electrostatic Precipitators

Dry ESPs have been used for years on utility power boilers that combust coal. They have not been used on MHF incinerators in the United States because of the high moisture and volatile organic content of exhaust gases. However, dry ESPs have been used on fluid bed incinerator installations in Europe, the United Kingdom, and, on at

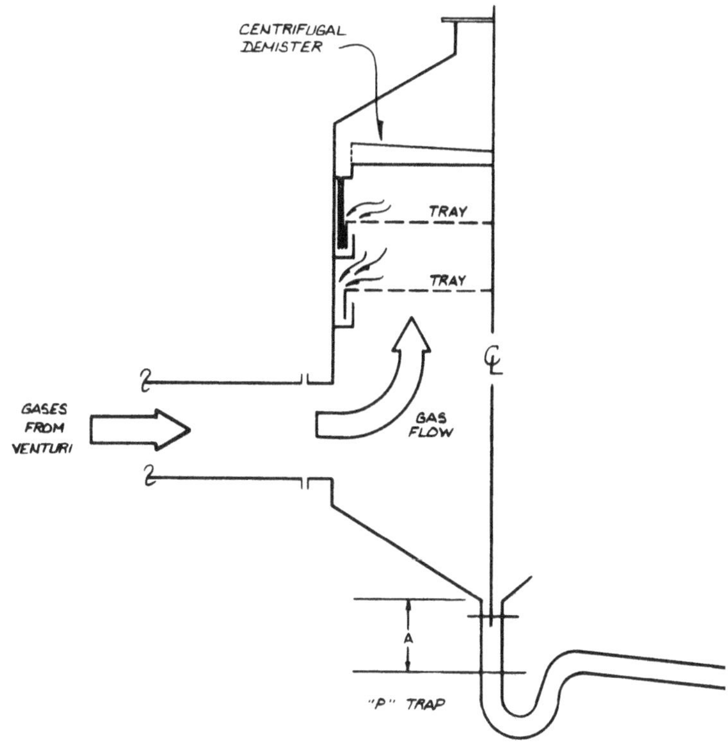

FIGURE 7.11 A tray scrubber with a P trap.

least one installation, in Canada. On these installations, the earlier fluid bed incinerators, built in the 1980s and early 1990s, had single field ESPs; whereas, plants built after 1995 have two to three field ESPs, which typically form the first stage of the flue gas treatment process. On these installations the exhaust gas is first cooled in a waste heat boiler before the dry ESP.

In a dry ESP, as shown in Figure 7.12, the exhaust gases pass through a large chamber where electrodes impart a negative charge to the particulate matter in the exhaust gas stream. The electrodes are negatively charged and are provided with a direct current from 20 to 100 kV. Plates with a positive electrical charge run parallel to the flow of gases through the chamber. The negatively charged particles are attracted by the positive electrical force to the plates. Periodically, buildup of particulate matter on the plates is removed by rapping the plates, which causes it to fall to the bottom of the chamber for removal.

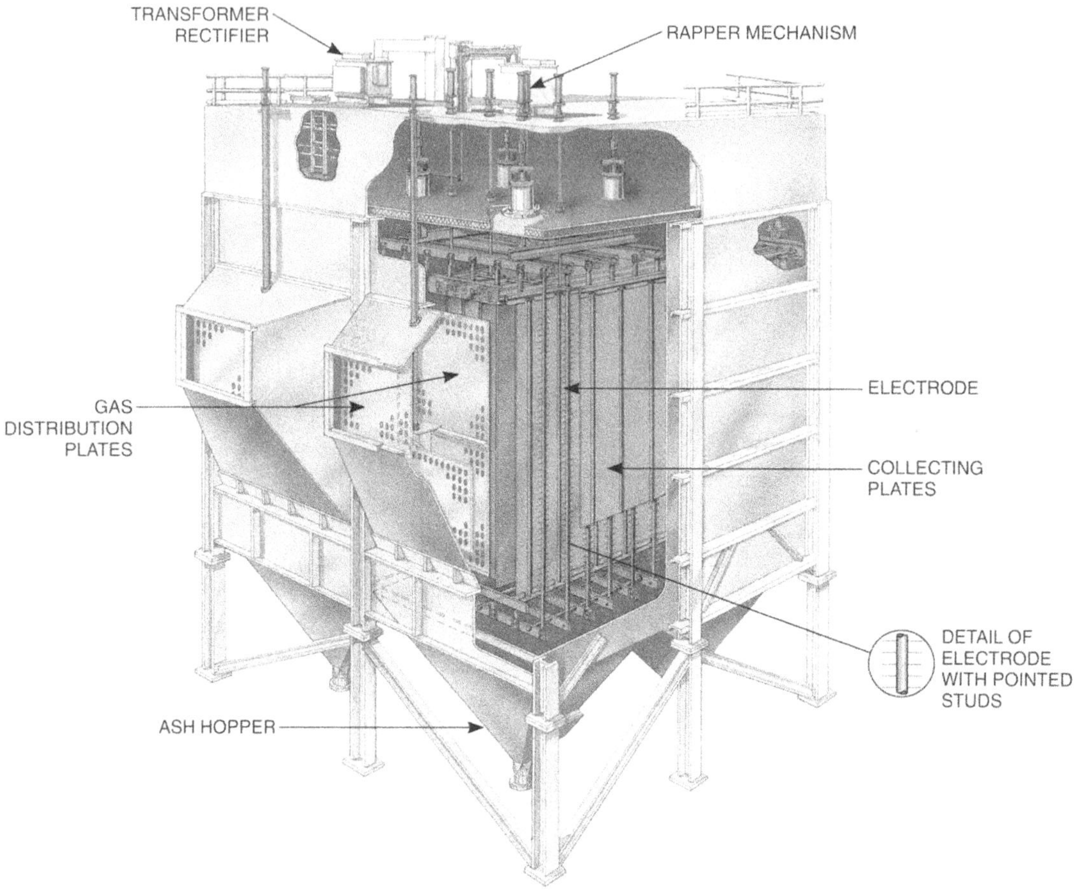

FIGURE 7.12 A dry electrostatic precipitator (courtesy of Environmental Elements Corp.).

Several factors influence the design and particulate removal efficiency of a dry ESP, including particle size distribution, gas flow rate, uniformity of gas flow in the ESP, resistivity of the fly ash, particle density, and temperature of the exhaust gas.

2.6.1 Performance

Most dry ESP installations comprise three or more fields. This allows one of the fields to be bypassed while the others remain in service. On recent installations at power plants, as many as five fields are used with a minimum of four in operation at all times. Each field removes a greater percentage of the total inlet particulate loading. Depending on resistivity of the fly ash, particle size distribution, flue gas temperature, and flow distribution through the unit, particulate removal efficiencies of 99% and greater are achievable.

2.6.2 *Advantages and Disadvantages*

Dry ESPs have certain advantages over other types of APC devices. For example, because they do not use water, there is no requirement to supply clean water or to treat processed water. Compared with a medium or high pressure drop Venturi scrubber, dry ESPs use much less electric power.

There are also several disadvantages to dry ESPs. The inlet gas to the precipitator must be cooled to less than 316°C (600°F) to prevent warping of the plates. This requires use of a heat removal device, such as a waste heat boiler. In addition, exhaust gas must be relatively free of VOCs to prevent particulate matter from sticking to the plates. This is not a problem for a fluid bed incinerator, but application on an MHF would require an afterburner. Also, a flue gas with high moisture can promote discharging of the electrical charge between electrodes and plates which will prevent a high charge buildup on the particles.

Operation of the unit depends greatly on the characteristics of the fly ash and may limit the use of certain chemicals (such as ferric chloride, alum, lime, or methanol) in liquid processes at the WWTP. The dry ESP also has no effect on gaseous pollutants and, therefore, must be followed by other APC devices. Another disadvantage of dry ESPs is their relatively high capital cost.

2.6.3 *Operation and Maintenance*

The operation of a dry ESP does not require significant effort by the operator because voltage from electrodes to plates is automatically controlled by automatic voltage controllers. Occasional observation of the voltages and measurements of various plates and fields is all that is required to ensure effective particulate removal. Correct alignment of the electrode in the center of the collection field is important to achieving a highest voltage potential across the gap from discharge electrode to collection plate. In general, the greater the voltage potential, the greater the particulate removal efficiency. In time, if the voltage and current measurements indicate a buildup of particulate and degradation of performance, manual cleaning of the plates and electrodes will be required. A dry ESP, particularly one located outdoors, should be well insulated because cold spots on the ESP housing will cause condensation of the moist gas on the inside, causing corrosion of the steel housing. This problem is particularly severe during cold startup of the unit when warm moist flue gas enters. Provisions for preheating the unit before entry of the exhaust gas can greatly mitigate this problem. Also, the ash hoppers should have heaters to keep the ash warm and in a free-flowing state. If the ash is allowed to cool and absorb moisture, it will set up and cause plugging problems. A related problem is plugging or malfunction of the rotary valves that control the discharge of ash from the ash hoppers.

Although a host of clogging and particulate build-up problems can occur when an incompletely burnt out exhaust gas is treated in a dry ESP, these problems have not occurred on such installations on fluid bed incinerators. In fact, European experience with dry ESPs on fluid bed incinerators indicates that they operate well and require minimal maintenance.

2.7 Wet Electrostatic Precipitators

A wet ESP operates similarly to a dry ESP but contains a washing mechanism to counteract the buildup of volatile or particulate matter on collector surfaces. A typical wet ESP, as shown in Figure 7.13, operates like a dry ESP in that there are electrodes which charge the incoming particles and collection plates or tubes to attract the particles. With a wet ESP, however, a thin film of water is continuously flowing down the surfaces of the collection tubes. This water film helps to remove the collected particles. The exhaust gas entering the wet ESP must be cooled and fully saturated before entering the unit. In addition, a fine mist generating system is typically used at the inlet of the wet ESP to create very fine water droplets which are collected on the surface of the collection tubes and are important in maintaining a wetted surface and continuous flow of water down the collection tubes. Also periodically, flushing sprays above the collection tubes are used to wash the particles off the tube collecting surfaces.

2.7.1 *Performance*

Performance of the wet ESP is equal to or better than the dry ESP. Particulate collection efficiencies of 99% or more can be expected.

2.7.2 *Advantages and Disadvantages*

The advantages of the wet ESP are as follows:

- The unit does not require a heat recovery device such as a waste heat boiler for cooling the incoming gases.
- The unit can be used after a conventional wet scrubbing system (Venturi and impingement tray scrubbers) to achieve additional removal of particulate matter and metals.

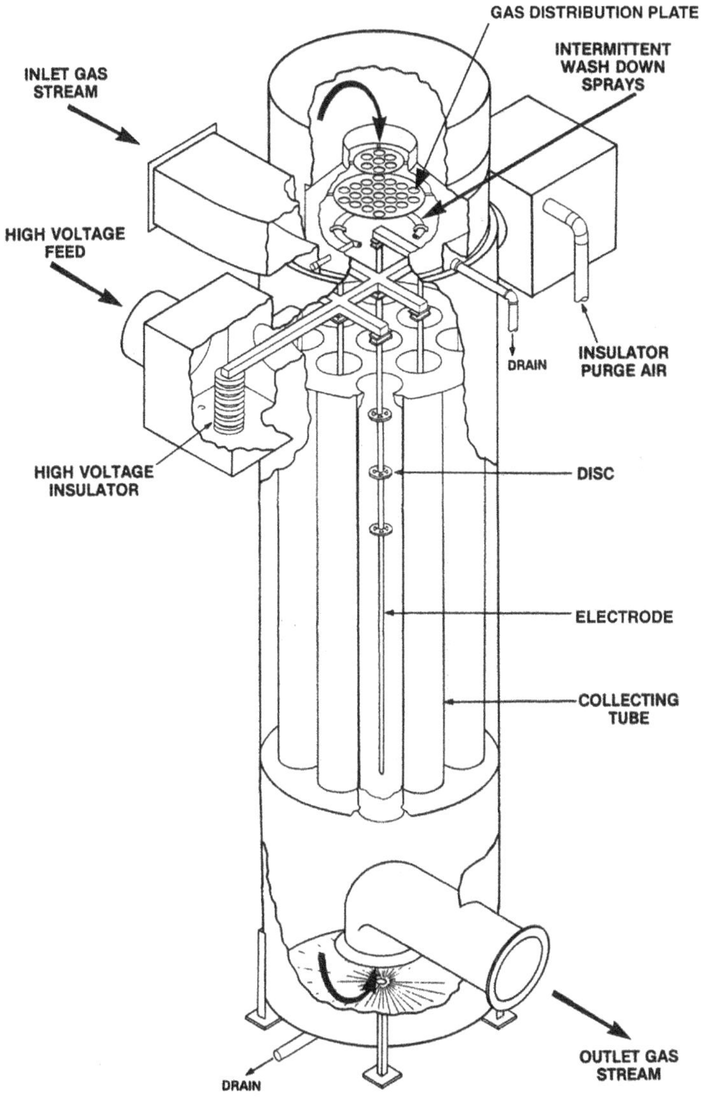

FIGURE 7.13 A wet electrostatic precipitator (courtesy of Sonic Environmental Systems).

- The unit uses relatively small quantities of water for irrigating and flushing the collection tubes.
- Power consumption by the unit is relatively modest in comparison with the fan or blower power that a Venturi scrubber would require to achieve comparable particulate removal efficiency.

Disadvantages of the wet ESP are as follows:

- The unit works best when it is lightly loaded with particulate matter, essentially acting as a final control device after a Venturi scrubber or other device has removed the bulk of the particulate matter.
- If a wet ESP is exposed to heavy particulate loadings, collection surfaces can build up particulate matter and choke the flow of exhaust gas through the unit.
- If the surface of the collecting tubes becomes dry from an upset in the condition of the flue gas or interruption of the fine mist generating system, particulate matter will adhere to the dry section of the collection tubes and can cause a buildup.
- The unit collects the particulate matter in liquid form, which may require separate clarification to remove the particles from the flushing water.

2.7.3 *Operation and Maintenance*

Similar to the dry ESP, the wet ESP requires little operator attention because the voltage and current across the electrodes and collection tubes are controlled automatically to achieve a given particulate removal efficiency. Periodic checking of voltage drops is necessary to ensure that no dry spots or particulate buildups are occurring. If these conditions do occur, more frequent flushing of the collection surfaces or an adjustment of the mist generating system may be required. If the design voltage drop between the electrodes and the collection tubes cannot be attained, realignment of the electrodes so that they are precisely positioned in the center of the collection tubes should be performed.

2.8 Fabric Filters

Fabric filters, or baghouses, have been increasingly used on power plants, refuse incinerators, and other emission sources where the absolute maximum level of particulate control is required. Fabric filters are large structures that house an array of long cylindrical cloth or fabric bags. Inside each bag is a rigid wire cage that keeps the bags from collapsing. The dirty exhaust flows through the fabric bags from outside to inside. The cleaned gas flows out of the top of each bag. Particulate matter is collected on the outside of the fabric. Periodically the bags are cleaned by either reversing the airflow, mechanically shaking the bags, or sending a pulse of compressed air down the bags which dislodges the particulate matter from the surface. Particulate matter falls into large hoppers at the bottom of the unit where it is collected for removal.

As shown in Figure 7.14, the fabric filter has three dust tight sections: a plenum at the top, a collector housing in the center, and a hopper at the bottom. The dust-laden air enters the bottom of the unit and is distributed across the chamber by either a diffuser plate or inclined baffles. The dirty gas then flows through the bags and exits through the top of each bag into the outlet plenum. A tubesheet separates the collector housing from the outlet plenum. The bags are secured to the tubesheet with clamps.

2.8.1 *Performance*

On industrial, power utility, and refuse incineration facilities, the use of fabric filters has allowed particulate removal of as great as 99.9%. These high removal efficiencies are accomplished not only by filtering of particulate by the fabric, but also by buildup of particulate on the fabric. As particulate matter builds up on the fabric, the built-up layer acts as a further filtering medium to remove even smaller size particles.

2.8.2 *Advantages and Disadvantages*

The advantages of fabric filters are as follows:

- Highest collection efficiencies of all particulate control devices available.
- Significantly less electrical power use than low and medium pressure drop Venturi scrubbers; slightly greater power use than dry ESPs.
- Fly ash is collected dry, requiring less handling than wet ash.

In the past, use of a fabric filters was very limited because of high water-vapor content in the incinerator flue gas and the overriding concern of moisture in the gas condensing out and causing muddy bags. The same phenomenon can occur with VOCs in the exhaust gas; thus incinerator exhaust must be well burnt out and essentially free of VOCs. Despite these concerns, fabric filters have been used on fluid bed incinerators in England and Europe and recently at the Metro WWTP in St. Paul, Minnesota. To use a fabric filter, the flue gas must be cooled to less than 260°C (500°F), which typically requires a waste heat boiler.

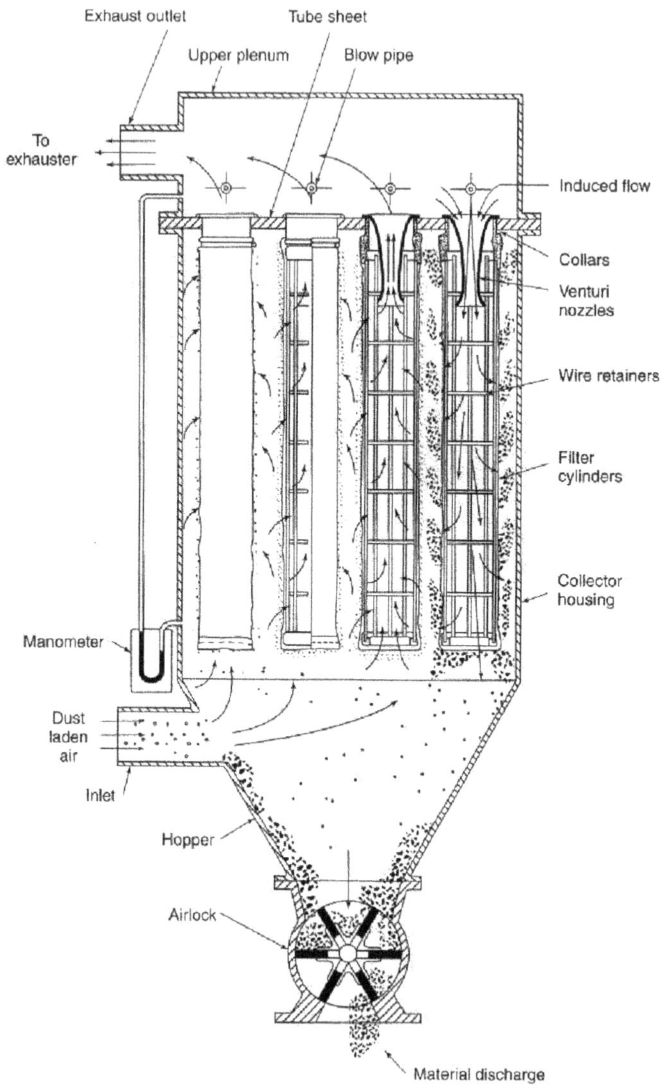

FIGURE 7.14 A fabric filter (baghouse).

Fabric filters are frequently used in spray dry absorption (SDA) systems to control acid gases and particulate matter. As shown in Figure 7.15, a lime slurry is sprayed into the flue gas stream to cool the gas and disperse the lime into the gas stream. The lime reacts with acidic pollutants in the gas phase and is collected on the surface of the bags in the baghouse. This forms an absorptive filter which provides

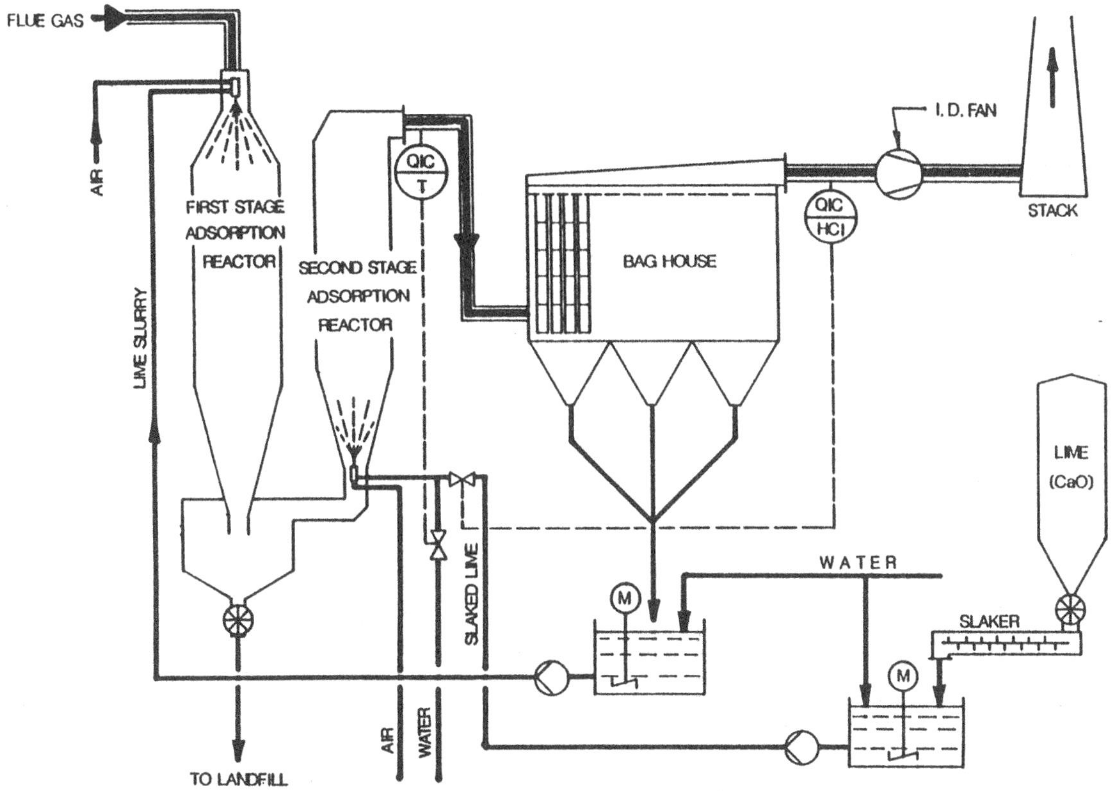

FIGURE 7.15 A spray dry absorption system (I.D. = induced draft).

further capture of acid gases. Spray dry absorption systems can be one or two stages with single-stage systems being more common. Activated carbon slurry can also be used in an SDA system for mercury control as discussed later in this section.

2.8.3 Operation and Maintenance

The primary maintenance requirement of a fabric filter is periodic replacement of the bags. An important monitoring device is the pressure drop between inside and outside of the bags. A gradual increase in pressure drop indicates particulate buildup on the outside of the bags and signals that the unit should go into a cleaning cycle. A sudden drop in pressure most likely indicates a breach in the fabric on one or more bags. A broken bag detector should be installed downstream of the bag filters to indicate any increase in particulates escaping the bag filter. Alternatively, if the incinerator has a particulate or opacity continuous emission monitoring system (CEMS), a bag failure will be noticed by an increase in the particulate emissions or flue gas opacity. Periodic maintenance includes inspections inside the units and replacement of the bags. Immediately after installation of a new set of bag filters, a bag leakage check using a fluorescent dye and dark light should be conducted to ensure there is no leakage of particulates.

3.0 CURRENT AIR POLLUTION CONTROL SYSTEMS

State departments of environmental protection and local APC boards have required more stringent control of regulated pollutants, including particulate matter, acid gases, and metals such as mercury. More elaborate APC systems have been developed and installed on newer incinerators in the United States. A noteworthy examples of such an installation is found in Ypsilanti, Michigan. The Ypsilanti plant has one fluid bed incinerator rated at 69 metric tonne/d (76 dry ton/d).

A simplified process flow diagram of the Ypsilanti, Michigan, plant is shown in Figure 7.16. The plant uses a fluid bed incinerator equipped with a primary heat exchanger to recover sufficient heat to achieve autogenous combustion. At the Ypsilanti plant, flue gas exiting the primary heat exchanger proceeds through a secondary heat exchanger, a Venturi scrubber, a tray scrubber, and a wet ESP. The cooled and saturated flue gas then proceeds through a gas conditioner which heats the gas stream and lowers its relative humidity. The heat source to the gas conditioner is hot air supplied by the secondary heat exchanger. The heated flue gas is then conveyed to an activated carbon adsorber which removes the mercury.

Pollutant emissions from the plant are well below all regulatory requirements. Typical emissions achieved at the plant are shown in Table 7.3. Note that the APC system achieves 99% and greater removal of mercury.

4.0 EMISSIONS MONITORING

Continuous emissions monitoring systems are now required as a result of the U.S. EPA Part 503 regulations. These regulations require continuous monitoring of stack gas concentration of total hydrocarbons (THCs) using a flame-ionization detector. Because concentrations must be reported on a dry basis corrected to 7% oxygen, the flue gas moisture content and percent oxygen must also be determined. As an alternate to total hydrocarbons, the U.S. EPA will accept continuous monitoring of carbon monoxide. In general, a carbon monoxide analyzer is easier to operate and maintain

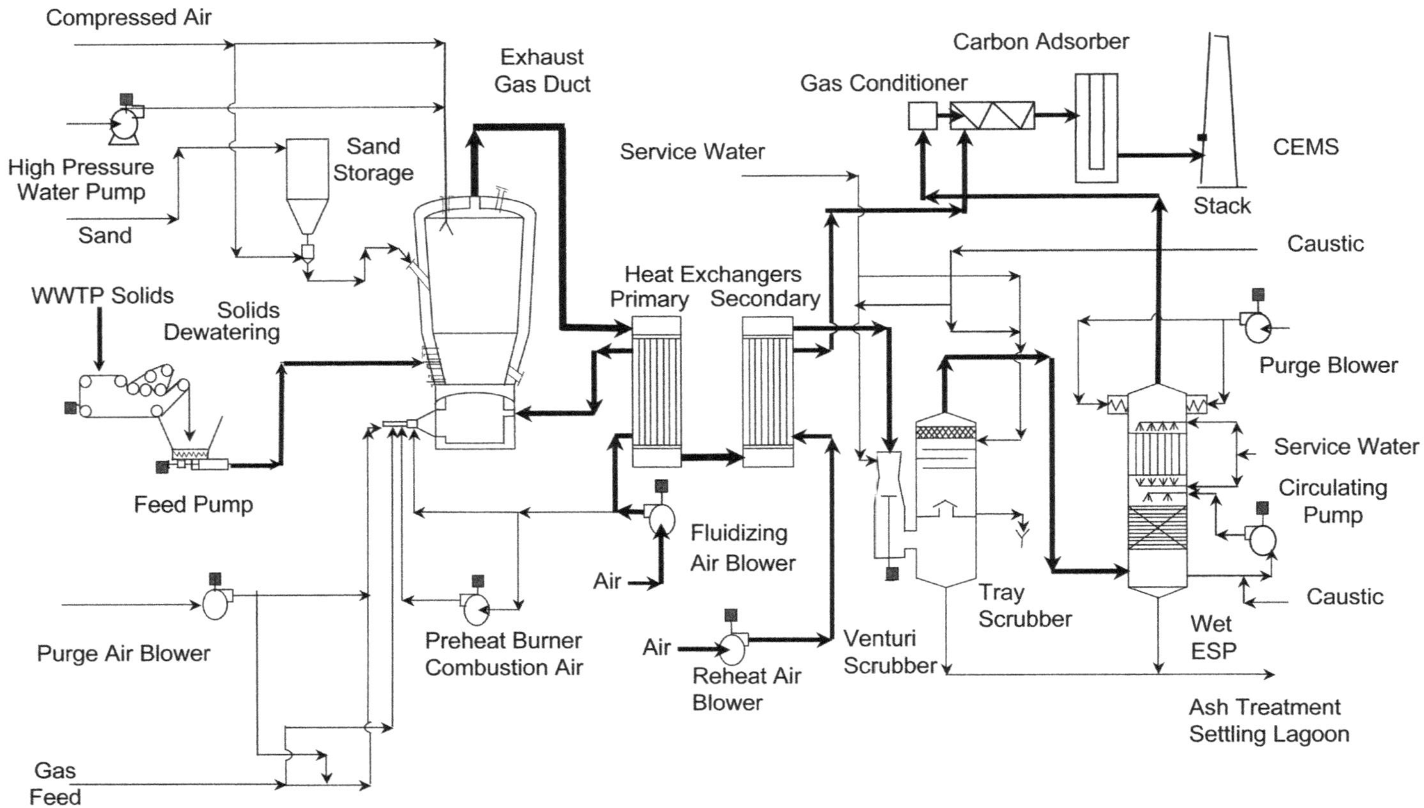

FIGURE 7.16 A process flow diagram of the Ypsilanti, Michigan, fluidized bed incineration system (courtesy of Degremont Technologies – Infilco).

TABLE 7.3 Typical emissions from a fluid bed incinerator with current APC system.

Parameter	
Particulate matter	< 0.05 kg/metric tonne (< 0.10 lb/ton)
PM10	< 0.05 kg/metric tonne (< 0.10 lb/ton)
CO	< 36 mg/Nm^3 at 11% O2 (< 40 ppm dv)
VOC	< 7 mg/Nm^3 at 11% O2 (< 5.0 ppm dv)
NOx	22–117 mg/Nm^3 at 11% O2 (15–80 ppm dv)
SO_2	< 20 mg/Nm^3 at 11% O2 (< 10 ppm dv)
HCl	< 5 mg/Nm^3 at 11% O2 (< 4.0 ppm dv)
2, 3, 7, 8 TCDD, TEQ	$< 0.7 \times 10^{-9}$ kg/metric tonne ($< 1.4 \times 10^{-9}$ lb/ton)
Total PCBs	$< 0.6 \times 10^{-6}$ kg/metric tonne ($< 1.2 \times 10^{-6}$ lb/ton)
Metals	
Hg, Ar, Be, Cd, Cr, Pb, Mg, Mn, Ni, Se, Zn	Removal efficiencies for all listed metals = 99 + %

and is preferred over a THC analyzer. Because the Part 503 regulations require that either THC or CO be continuously monitored and that stack gas concentrations of less than 100 mg/kg (100 ppm dv) on a monthly average basis be continuously demonstrated, a carbon monoxide analyzer is typically used on a fluid bed incinerator, which can achieve low emissions of CO, typically less than 20 mg/kg (20 ppm).

Some state regulatory agencies will require continuous monitoring of other pollutants depending on specific region air quality requirements.

A survey of CEMS conducted by the Water Environment Research Foundation, Alexandria, Virginia, and the National Association of Clean Water Agencies (formerly Association of Metropolitan Sewerage Agencies), Washington, D.C., found that incinerators consistently meet their THC and CO emission limits set by the Part 503 regulations. The survey indicated that the annual monthly average THC and CO concentrations in 2003 were both approximately 36 mg/Nm^3 dv_{11} (26 ppm dv_7). However, the survey also revealed that operators with THC monitors have extensive problems operating and maintaining these units. Operators with THC monitors report an average monitor life span of seven years and an average annual operating and maintenance cost of $25,000. Problems with the THC monitors include difficulty obtaining calibration gases, excessive cost of calibration gases, plugged sample lines, clogged filters, sample pump failures, problems keeping the analyzer in service, and problems with the software and data acquisition systems. In contrast, no significant operating or maintenance problems were reported for the CO monitors (AMSA/WERF, 2004).

5.0 REFERENCES

Association of Metropolitan Sewerage Agencies and Water Environment Research Foundation (2004) *AMSA/WERF 2004 Survey of Total Hydrocarbon/Carbon Monoxide Continuous Emissions Monitoring Systems;* Association of Metropolitan Sewerage Agencies: Washington, D.C. and Water Environment Research Foundation; Alexandria, Virginia.

Baturay, A. (1990) *Case Studies of Total Hydrocarbon Emissions (THC) From Multiple Hearth Sewage Sludge Incinerators and THC Reduction Strategies;* Prepared for Association of Metropolitan Sewerage Agencies Incinerator Workgroup Meetings in New Orleans, Louisiana; Association of Metropolitan Sewerage Agencies: Washington, D.C.

Dangtran, K.; Butt, J. (2004) *Minimization of CO and NOx Emissions By Optimization of Fluid Bed Design/Operating Conditions and By Chemical Additives;* Prepared for Bioenergy Workshop—Permitting, Safety, Plant Operations, Unit Process Optimization, Energy Recovery and Product Development; Cincinnati, Ohio, Aug 11–12; Water Environment Federation: Alexandria, Virginia.

Gerstle, R. W.; Albrinck, D. N. (1982) Atmospheric Emissions of Metals from Sewage Sludge Incineration. *J. Air Pollut. Control Assoc.,* **32,** 1119–1123.

National Oceanic and Atmospheric Administration (2005) U.S. Department of Commerce; State of the Climate in 2005 in *Bulletin of the American Meteorological Society,* **87;** National Oceanic and Atmospheric Administration: Washington, D.C.

National Lime Association (2002) *Dry Flue Gas Desulfurization Technology Evaluation;* Prepared by Sargent and Lundy, L.L.C.; National Lime Association: Arlington, Virginia.

National Research Council (2001) *Climate Change Science: An Analysis of Some Key Questions;* National Academy Press: Washington, D.C.

Niessen, W. R. (1990) The MHF Model: A Tool for Evaluation and Optimization of Multiple Hearth Incineration Systems; Prepared for WPCF Residuals Management Conference; New Orleans, Louisiana; Dec 2–5; Water Pollution Control Federation: Alexandria, Virginia.

Perry, R. H.; Chilton, C. H. (1973) *Chemical Engineers' Handbook,* Fifth ed.; McGraw-Hill: New York.

Sapienza, F.C.; Canham, R.; Baturay, A. (1998) NOx Emissions From PWCSA's Fluidized Bed Sludge Incinerator. Paper Presented at 12th Annual Residuals and Biosolids Management Conference; Bellevue, Washington, July 12–15; Water Environment Federation: Alexandria, Virginia.

Schiffner, K.; Hesketh, H. (1983) *Wet Scrubbers;* Ann Arbor Science: Ann Arbor, Michigan.

U.S. Environmental Protection Agency (1989) *Incineration of Sewage Sludge;* Technical Support Document, Office of Water, U.S. Environmental Protection Agency: Washington, D.C.

U.S. Environmental Protection Agency (1991a) *Standards of Performance for New Stationary Sources;* Code of Federal Regulations, 40 CFR, Part 60, Appendix A, Method 5; U.S. Environmental Protection Agency: Washington, D.C.

U.S. Environmental Protection Agency (1991b) *Standards of Performance for New Stationary Sources;* Code of Federal Regulations, 40 CFR, Part 60, Subpart O; U.S. Environmental Protection Agency: Washington, D.C.

U.S. Environmental Protection Agency (1993) *Standards for the Use or Disposal of Sewage Sludge;* Code of Federal Regulations, 40 CFR, Part 503, Subpart E; U.S. Environmental Protection Agency: Washington, D.C.

U.S. Environmental Protection Agency (1998) *Compilation of Air Pollutant Emission Factors, Volume I: Stationary Point and Area Sources,* AP-42; Section 2.2; Office of Air Quality Planning and Standards, U.S. Environmental Protection Agency: Research Triangle Park, North Carolina.

U.S. Environmental Protection Agency (2006) *Inventory of U.S. Greenhouse Gas Emissions and Sinks: 1990–2004;* Office of Atmospheric Programs, U.S. Environmental Protection Agency: Washington, D.C.

Waltz, E.W. (1990) *Technical Discussion of Proposed EPA Hydrocarbon Regulation for Sludge Incinerators—Charts and Graphs;* Prepared for Association of Metropolitan Sewerage Agencies Incinerator Workgroup Meetings in New Orleans, Louisiana; Association of Metropolitan Sewerage Agencies: Washington, D.C.

Chapter 8

Ash Handling and Recycling

1.0 INTRODUCTION

Ash is the end product of solids incineration, essentially consisting of the noncombustible portions of the feed material. There are numerous methods and equipment available for handling ash and there are various uses for recycled ash. Handling methods and the final destination or use of material is often site-specific. Ash handling equipment can be the most troublesome subsystem associated with incinerators.

Multiple-hearth furnaces (MHFs) and fluid bed incinerators use either wet or dry conveyance systems. Ash is abrasive and sometimes nonuniform, making it difficult to convey as a bulk material. Fugitive ash emissions often cause problems with the equipment; steps need to be taken to seal all leaks. Fugitive ash can also have detrimental effects on ancillary or surrounding equipment. Controlling discharges of fugitive ash is also often required by the air emissions permit for the facility. Working in an ash-laden environment can present challenges. An improperly designed or maintained system may create severe environmental conditions that require continuous use of particulate masks by operating personnel. In such circumstances, maintaining a clean facility can be problematic. A clean facility typically indicates that equipment is being properly operated and maintained, which is an important consideration for visitors.

Ash historically has been sent to landfills. In recent years more emphasis has been placed on finding beneficial uses for the ash. Ash has been used as landfill cover, a soil amendment, a substitute for fly ash in concrete, a fine aggregate in asphalt, flowable fill material, and an additive in brick manufacturing. This chapter does not cover all the possible configurations and equipment available for handling ash. It does, however, present several techniques and systems of ash conveying and its application to MHFs and fluid bed incinerators.

2.0 SOURCES OF ASH

The multiple-hearth incineration process produces two types of ash: bottom ash and fly ash. "Bottom ash" is discharged from the bottom of the incinerator and constitutes the largest portion of ash generated. "Fly ash" is discharged from the incinerator in exhaust gases and typically is captured in a wet scrubbing process and returned to the plant in a recycle stream. If a waste heat boiler follows the incinerator, the ash may accumulate and need to be removed periodically with purge air or soot blowers. The waste heat boiler will have a hopper at the bottom where the ash accumulates and is then combined with the bottom ash system. Bottom ash will sometimes contain "clinkers."

Clinkers are chunks of fused or partially fused ash or chunks of refractory. A grinder can be installed to crush the clinkers. The grinder is often referred to as a "clinker breaker" and consists of two counter-rotating, motor-driven rollers that grind clinkers into smaller particles. The clinker breaker is typically installed in the ash discharge chute at the bottom of the MHF. The grinder should be installed before mechanical conveying equipment and before sluicing, if applicable. After grinding, the ash is discharged to the conveying system.

All ash from a fluidized bed incinerator is carried off in the exhaust gas as fly ash. In addition to the ash, fluid bed sand breaks down into smaller particles, which also will be carried off in this gas stream. This ash is subsequently captured in the waste heat boiler, economizer, baghouse, wet scrubber, or electrostatic precipitator, as applicable for the given installation.

3.0 ASH HANDLING

In the wet or hydraulic ash handling system, ash is removed as a slurry. Wet systems tend to be more prevalent on fluid bed systems coupled with wet scrubbers because the majority of ash is removed from the process train in a wet form. Wet systems may also be found in MHFs, where sluicing water is used to convey bottom ash to disposal. In a fluid bed incinerator, ash slurry is drained from the bottom of the wet scrubber. In either case, the ash slurry is conveyed to a lagoon or to mechanical thickening and dewatering equipment. In the MHF system, dust can be released at grinders, access doors, and discharge chutes before sluicing. Steam can be generated from sluicing water evaporation as the hot dry ash mixes with the water. Seals and gaskets must be inspected regularly and replaced as necessary. In any wet ash handling application, the temperature of the slurry must be considered from a safety as well as a material properties viewpoint.

3.1 Wet Systems

3.1.1 Conveyance

3.1.1.1 Ash Sluiceways

Sluiceways transport ash by gravity. The sluiceway will discharge ash to a lagoon. The sluiceway is typically a rectangular concrete channel that may be lined with an abrasion-resistant material. The sluiceway should be designed using the principles of open-channel hydraulics. When a wet scrubber is used on the incinerator exhaust gases, scrubber water can be used to sluice the incinerator bottom ash.

3.1.1.2 Ash Slurry Well

When ash slurry is pumped to a disposal point, a well is typically required. The well provides some storage capacity and the proper hydraulic conditions for suction of the ash slurry pumps. A mixer may be installed in the well to provide proper mixing of water and ash. Alternatively, a self-cleaning wet well approach may be used.

In this method, ash also can be discharged to a well, sluicing water added, and the slurry discharged through a drain pipe system to a lagoon.

3.1.1.3 *Ash Pumps*

Pumps are used to pump ash slurry and thickened ash from gravity thickeners. Slurry pumps are generally centrifugal units with end suction and suitable for pumping a 2 to 6% mixture of ash and water. Pump passageways are large enough to pass the product of the ash grinder. A horizontal-shaft pump located so that ash slurry will flow by gravity into the pump suction is preferred in some installations; however, vertical pumps and submersible pumps can also be used. Frequently, ash slurry pumps are of the recessed impeller type, or the rubber-lined volute design. In some installations, "chopper"-type pumps have worked well. If the recessed impeller design is used, then the pump impellers and volutes need to be made of hard metal wear-resistant alloys.

Care must be used in design of the slurry pumping system to make sure that net positive suction head conditions are met. The high temperature of ash slurry can sometimes cause unexpected cavitation in pump suction. Also, other parts of the pump, such as the seals and bearings, need to be considered because of the abrasive nature of the slurry and the higher fluid temperature being handled.

To match pump capacity to the ash sluicing rate or to the discharge condition, varying pump speed may be necessary. When pumping ash to a lagoon, the discharge condition—including pipe elevation and length—may change as the lagoon is filled.

Ash from a gravity thickener may be pumped with a mechanical or air actuated diaphragm pump. The thickened ash is typically approximately 10% solids.

3.1.1.4 *Ash Pipelines*

Ash pipelines are designed to transport the ash slurry at a velocity as low as practical but high enough to keep the solids suspended. Good hydraulic and engineering practice suggests velocities from 0.6 to 1.5 m/s (2 to 5 ft/sec). A variety of pipe materials are available for ash slurry service. Careful consideration must be given to the pipe material and piping layout. Pipe bends should be long radius and minimized as much as possible because of the highly abrasive nature of the slurry. A variety of pipe materials are available for ash slurry, including carbon steel, cast iron, Ni-Hard, ceramic, and basalt-lined steel pipe.

If space on the plant site allows, then ash piping should be installed abovegrade. The abovegrade installation allows easier pipe inspection and repair.

3.1.1.5 *Mechanical Conveyance*

Screw conveyors may be used for conveyance of wet ash in a dewatering operation or ash conditioning application. Where used in wet ash conveyance, a ribbon screw is an effective design, allowing some minimal roll back of ash. Wet ash can easily form a hard and thickened layer at the bottom of the trough which can lead to stress and potential breakage of the conveyor shaft if it builds up sufficiently. An ultra-high molecular weight plastic liner to facilitate "slippage" along the trough can be a valuable addition.

3.1.2 *Storage*

3.1.2.1 *Ash Lagoon*

Ash lagoons provide a simple and basic means for storing ash. The ash lagoon receives the ash slurry and provides an area where the ash can separate from the water and settle to the bottom of the lagoon. Water is decanted from the lagoon and recycled to the plant for treatment. The settled ash is then excavated and transported for ultimate disposal or is recycled. These lagoons can be constructed to suit local topography; depth can range from 1 to 3 m (3 to 10 ft), with length and width variable. At least two cells should be constructed to allow clearing of one while the second receives slurry.

3.1.2.2 *Bins*

Bins or hoppers are used for storage of ash that is dewatered mechanically. Alternatively, dewatered ash is directly conveyed to a truck or dumpster.

3.1.2.3 *Mechanical Thickening and Dewatering*

If mechanical thickening and dewatering are used, then the ash slurry is typically conveyed to a gravity style thickener or decanter. Thickened ash slurry is then pumped to a dewatering device. Vacuum filters are predominantly used for mechanical ash dewatering. Belt filter presses also have been used for dewatering thickened ash. The dewatered ash is then transported to a disposal site or is recycled.

3.2 Dry Systems

Dry systems historically have been installed only on MHFs. More recently, dry ash conveyance systems have been installed on fluid bed incinerators as well. Dry ash handling systems may be mechanical or pneumatic. The pneumatic systems may be either pressure or vacuum and dilute or dense phase. Ash grinding may be required for bottom ash systems to allow effective ash transport and to protect downstream conveyance equipment. Both systems typically convey dry ash to storage bins. While ash is being discharged from the bins to disposal trucks, it is typically wetted with water to reduce fugitive dust during the loading process. This "wetting" is commonly called ash conditioning. The ash is then transported to the ultimate disposal site or is recycled. Dry conveyance systems are common on MHFs because the majority of ash is removed in dry form as bottom ash.

Dry conveyance systems are also suitable on fluid bed incinerators that are coupled with baghouses or electrostatic precipitators because the majority of the fly ash is captured in dry form.

3.2.1 *Conveyance*

3.2.1.1 Mechanical Conveyance Systems

Mechanical conveying systems frequently transport ash from the discharge point at the bottom of MHFs to ash storage bins. A vertical chute in the bottom of the incinerator will discharge ash into a clinker breaker. The clinker breaker will break up large pieces of slag, refractory, or other large items that may be discharged with the bottom ash. Not all facilities currently in operation use clinker breakers. Some installations use a simple clinker separator with a sloped bar screen device to catch clinkers and allow them to roll off into a small hopper adjacent to the ash chute.

Clinker breakers can become jammed with large items. If the clinker breaker becomes jammed, then an alarm is needed to alert the operator. Clinker breakers should be equipped with a removable inspection plate. The inspection plate can be taken off, the object causing the jamb removed, and the clinker breaker restarted. If a clinker breaker or separator is not installed, then a large object could cause a jam in downstream conveying equipment, such as screw conveyors or bucket elevators.

Following the clinker handling, the ash typically either discharges directly to a bucket elevator or to a screw conveyor. If used, a screw conveyor would transport the ash to a bucket elevator. The bucket elevator will lift the ash vertically to the top of an ash bin where it discharges to another screw conveyor. This conveyor then transports the ash to a discharge point at the top-center or directly to the ash bin. The size of the ash bin and conveying equipment should be sized for maximum loading of the incinerator. The ash bins commonly have some type of vibratory device mounted to the side or bottom of the vessel to facilitate discharging to the dump truck below. Bin vibrators or an activator may be used.

At the bottom of the ash bin is another device that can convey ash from the storage bin to a truck and simultaneously wet the ash. This device, referred to as a "conditioning conveyor," uses one or multiple screws specifically manufactured for this purpose. This conditioner is equipped with water spray nozzles in the cover. The ash is sprayed with water as it is conveyed to help eliminate fugitive dust during loading and transport. The ash typically is loaded into dump trucks or roll-off containers. Sealing the conditioning discharge chute to the cover of the transport container also will help eliminate fugitive dust. A mechanical ash handling system, shown in Figure 8.1, consists of screw conveyors, bucket elevators, ash bins, ash conditioners, and associated drive equipment.

Hanger style bearings typically are used for the screw conveyors where intermediate bearings are required. Screw conveyor covers and gaskets must be kept securely in place to minimize the amount of ash dust discharged to the immediate operating area. The covers provide safety for operators because ash dust can potentially pose a health hazard. Although not explosive, ash dust may require that operating personnel wear particulate masks in the area. Containing the dust with the covers also reduces housekeeping and maintenance needs. All covers, gasketing, and

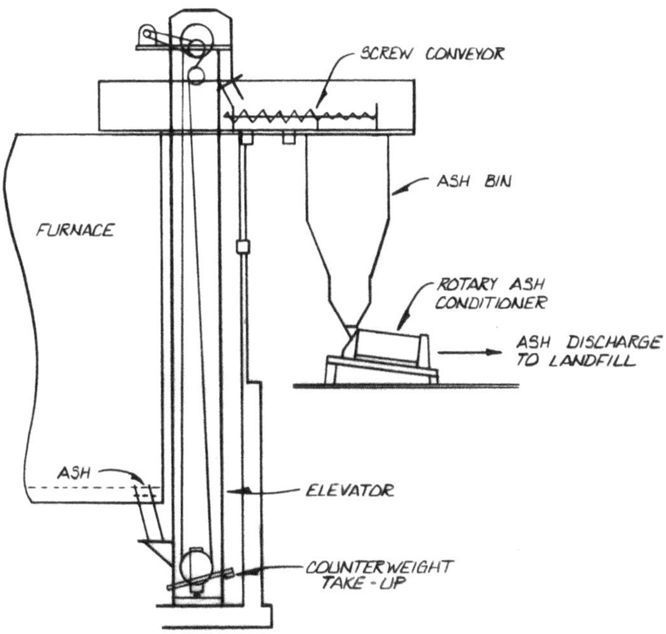

FIGURE 8.1 A typical mechanical ash handling system.

cover fasteners should be reinstalled immediately following conveyor maintenance. Gaskets should be inspected and replaced if deteriorated.

Mechanical systems may also be used to convey ash collected in waste heat boilers, economizers, and baghouses.

3.2.1.2 Pneumatic Conveyance Systems

Pneumatic conveying devices are either pressure- or vacuum-driven. They consist of airlock equipment—to prevent cross mixing of transport air and air within the incinerator on the lower hearths—and may include various combinations of ash coolers, blowers or vacuum pumps, transporters, conveyance piping, ash bins, ash conditioners, associated drive equipment, and dry dust removal equipment typically of the baghouse type.

The dilute-phase pressure system (Figure 8.2) uses a positive-displacement or centrifugal blower to move the air and ash mixture through piping to the storage bins (Dynamic Air Inc, 2006a). Another type of dilute-phase pressure system

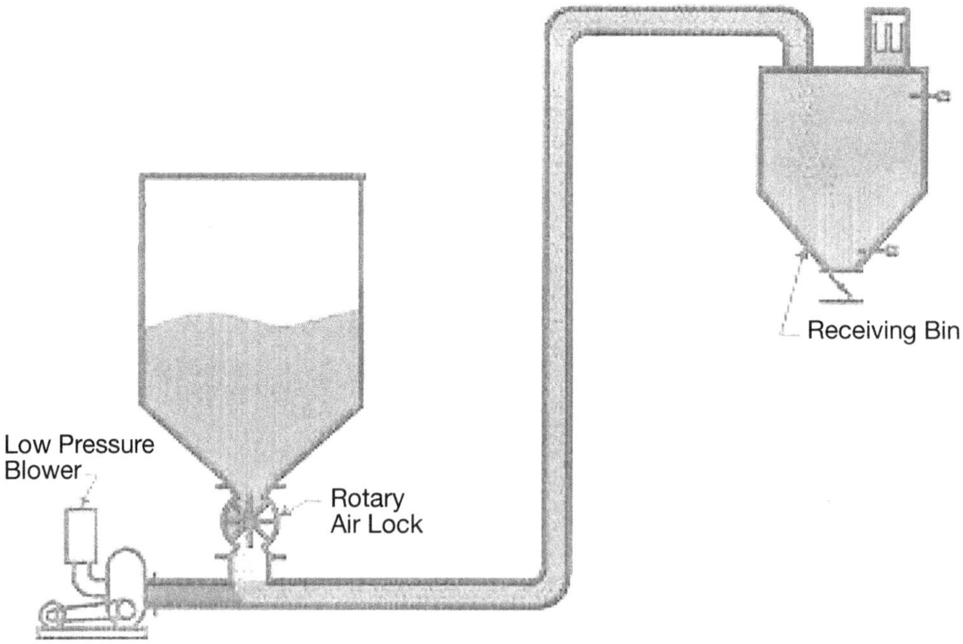

FIGURE 8.2 A typical dilute-phase pressure system for ash handling (courtesy of Dynamic Air, Inc.).

operates continuously and uses a blower to drive an eductor that pulls ash from the collection chamber and mixes it with transport air. The dilute-phase conveyance system is designed to transport ash at high velocity and low density.

Dense-phase pressure systems (Figure 8.3) have been installed on MHFs and fluid bed incinerators (Dynamic Air Inc., 2006b). These pressure systems typically include hoppers, baghouses, transporters, piping, and valves. In fluid bed incinerators the hoppers collect ash that accumulates in the waste heat boiler, economizer, and baghouse, if so equipped. In typical operation, a valve at the bottom of one of the collection hoppers opens to fill a transporter vessel with ash. The transporter inlet valve closes, the transporter is pressurized with compressed air, and then the outlet valve is opened. The material is conveyed in dense "slugs" and may be assisted by pneumatic booster stations along the discharge pipeline. The ash is conveyed through piping to a storage hopper that is typically fitted with a vent filter. The dense-phase system is designed to transport ash at low velocities and high density.

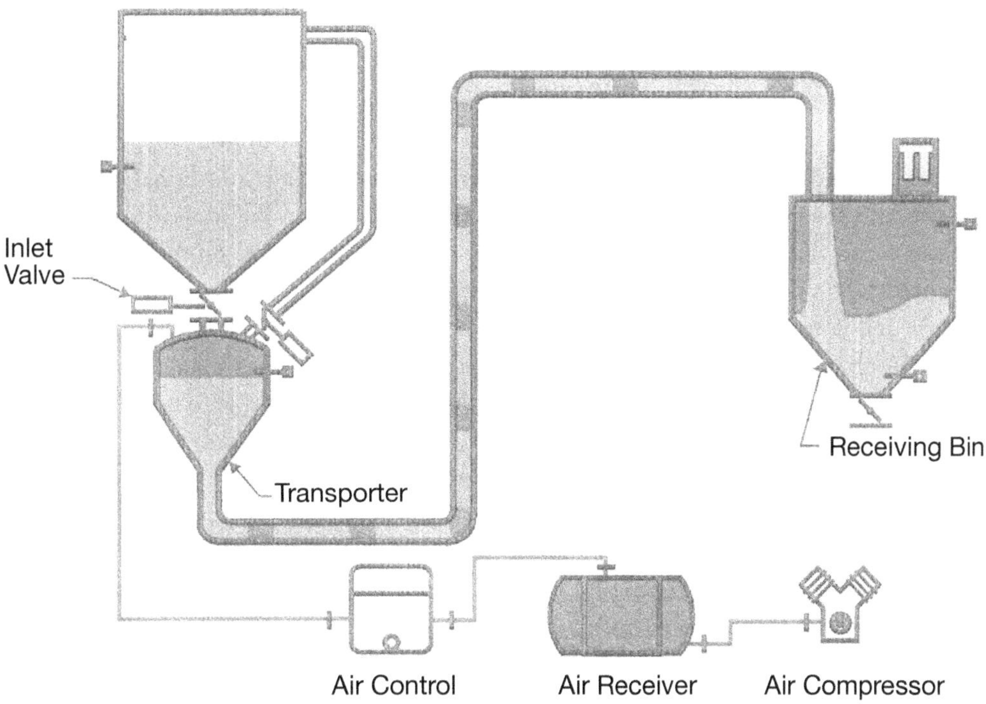

FIGURE 8.3 A typical dense-phase ash conveying system (courtesy of Dynamic Air, Inc.).

Some advantages of the dense-phase system include substantially reduced amounts of air, lower energy cost, smaller piping, and reduced abrasive wear and tear. Disadvantages include the requirement for a high-pressure air system conditioned to −40°C (−40°F) dewpoint, and the potential need for booster stations along the pipeline.

All components of pressure systems must be thoroughly sealed to reduce the amounts of fugitive ash dust released to the surrounding areas of the facility. Even small leaks can result in substantial amounts of dust. To avoid accumulation, an extensive housekeeping program must be implemented. In addition, reduced airflow caused by system leaks can reach a point where insufficient motive force exists to transport the ash load. This can result in system plugging.

Depending on facility and system design, the vacuum system operates either continuously or is sequenced between pick-up points using vacuum pumps located on top of the ash storage bins. These vacuum pumps draw ash from collection chambers at the bottom of the incinerator to the storage bins. Vacuum systems also include bag filters upstream of the pumps that must be cleaned periodically to maintain system performance. Bag house filters may be equipped with automatic cleaning controls that initiate cleaning based on differential pressure across the filters or on a time clock.

The vacuum system (Figure 8.4) has the advantage of avoiding fugitive ash dust emissions. It is subject, however, to reduced capacity from undetected leaks. This reduced capacity can result in system plugging and failure. An additional advantage of vacuum systems is that the vacuum pumps are located on the roof of the building or ash bins, well away from water and dust that can accumulate on the lower floors of incinerator facilities. A properly sized vacuum system fitted with appropriate access and connection ports can be used in housekeeping operations to remove any ash dust that accumulates in the incinerator area during maintenance activities. In contrast to vacuum systems, the pressure system is less susceptible to capacity reduction and plugging associated with system leaks. In a pressure system, leaks are quickly located because ash dust discharges and accumulates near the leak opening.

Equipment for all of these systems must be constructed of materials that can withstand the abrasion of dry ash. In fluid bed incinerators, the ash may contain sand from the fluid bed and potentially increase abrasiveness. To achieve abrasion resistance, several different techniques are used depending on manufacturer or designer. These techniques include use of hardened materials such as Ni-Hard, and special linings such as glass, ceramic, basalt, and concrete. Material selection is best accomplished through testing of ash material by the manufacturer for specific abrasion

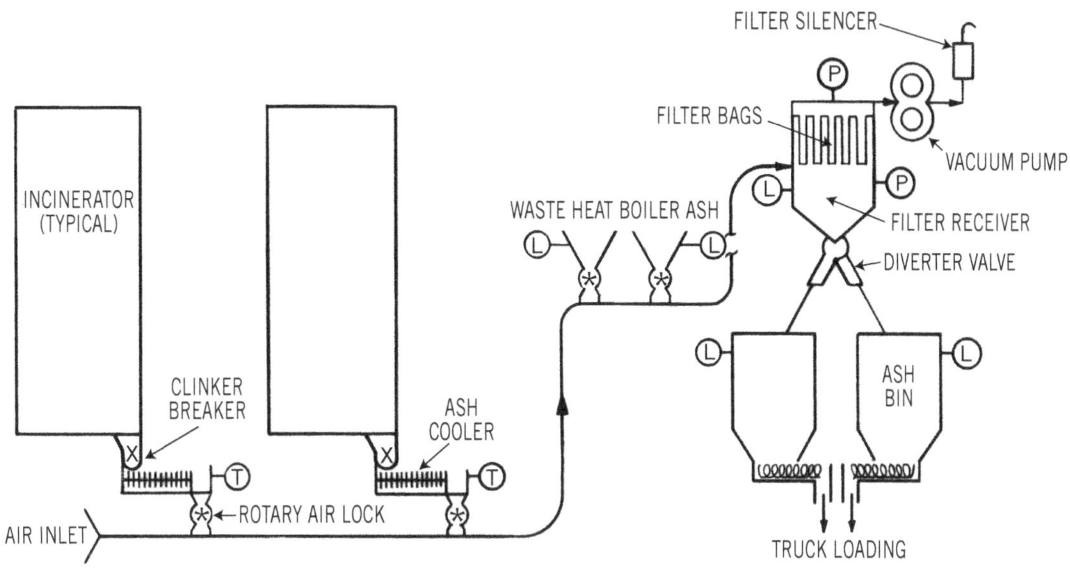

FIGURE 8.4 A typical vacuum system for ash handling.

characteristics. Linings that have worked successfully in one installation may not be adequate in another because of significant differences in abrasiveness.

Piping should be configured with a minimum of bends to reduce the probability of ash accumulation. Generally, piping bends should be manufactured with substantially thicker walls than adjacent straight sections. This extra thickness will allow for additional wear that occurs because of increased velocities at these points and impingement of particulate on the backside of the elbow. Bends should also have a radius that is as long as possible. Some manufacturers use segmented bends that permit selected removal and replacement as wear-through occurs. Another approach is the use of a "concrete form" elbow that can be removed and reformed with a new layer of concrete lining. Special pipe fittings are manufactured with replaceable sections on the outer radius of the fitting where the wear from abrasion is most prevalent.

3.2.2 Storage

3.2.2.1 Ash Storage Bins

Dry ash storage bins receive and accumulate ash from the conveyance systems. The bins are sealed vessels under which roll-off style transport containers or dump trucks are placed for periodic loading. Associated storage bin equipment includes bin vibrators or activators, ash dust filters, and material-level measurement equipment. Problems with ash bins include bridging and uneven distribution of accumulated ash. Bridging is caused when portions of the stored ash achieve enough structural strength in combination with the bin sidewalls to support ash accumulation above while allowing free flow of ash from below. Moisture in the ash, which can accumulate from diurnal temperature and humidity changes, increases the probability of bridging. Bridging reduces usable bin capacity and complicates its emptying into the transport container or vehicle. Bridging can be reduced through the use of bin vibrators or activators or pulsed air jets.

Level measurement devices that contact ash are prone to failure because of the hostile environment. Load cell sensors and noncontact ultrasonic level sensors have been used successfully where the specific density of the ash is known and is consistent. However, this instrumentation does not replace routine inspections and manual soundings of ash bin levels. The greatest operational success is achieved through programs that include the following:

- Frequent operator inspection,
- Frequent unloading of ash bins before maximum storage levels are reached, and
- Daily tracking of generated and disposed ash quantities.

The storage vessel size should be based on the quantity of ash that is to be produced and frequency of removal.

3.2.2.2 Dry Ash Conditioners

Ash conditioners are mechanical devices designed to facilitate wetting of ash before discharge to a truck or roll-off container. They may consist of a helical screw conveyor or dual screws or a rotary conditioning drum, and may include service water addition nozzles, baffling, scrapers, and associated drives and isolation gates.

Dry ash conditioners must be properly maintained to perform effectively and equipment drives must be sufficiently sized for wetted ash conditions. These loadings can be significantly higher than loadings imposed by dry ash alone. Control of service water flow requires frequent operator attention because of changing ash wetting characteristics and dry ash flow rates from the ash storage bins.

Conditioning water can contribute significant added weight to the ash to be transported. This added weight translates into increased vehicle trips and tipping fees at commercial landfills. Added water accounts for as much as 30% of the conditioned ash mass. In some instances, a wetting agent is added to the water to enhance its ability to penetrate the ash.

4.0 INSTRUMENTATION AND CONTROLS

Ash systems consist of multiple, sequential components that require integrated operation, monitoring, and control. The system components are often distributed over several levels of a facility, are located both inside and outside the building, and are always remote to manned control areas.

Regardless of the system used, key system monitoring parameters should be displayed and alarmed in the control room. Control room start-stop devices should be grouped together and located adjacent to the parameter indicators to permit operation of the equipment as a system.

Electrical interlocking should be provided to permit automatic system shutdown when a component fails. Interlock bypass controls should be available for maintenance and safety purposes. Local start-stop-jog controls should be located as close to the equipment as practical. These controls should include electrical lockout and emergency stop features for personnel protection.

5.0 RECYCLING

The end product of incineration is ash, which is an inert material. Interest in beneficial reuse of ash is increasing. Historically, the most common way to dispose of ash is to a landfill. More recently, ash has been used as landfill cover, soil amendment, fill material, in brick manufacturing, and in concrete and asphalt mixes. One more innovative approach is in a vermiculture process. Detail regarding these options follows.

- Landfill—disposal of incinerator ash in a landfill is an option available to most operators. Landfill tipping fees vary across the country. A beneficial use of ash at landfills is to use it as a landfill cover or to blend it with soil and use as a cover.
- Fill material—ash can be used as fill material for excavations. One utility, for example, has a contractor using the material to fill old sludge lagoons. The material can also be used as a flowable fill.
- Soil amendment—in some specific areas (particularly areas with high clay soils), incinerator ash may be used as a soil amendment through an additive

process, which produces a soil that handles more easily, allows better drainage and airflow, and includes some valuable minerals.

- Brick—ash has been used in brick manufacturing by various utilities quite successfully. The brick manufacturers normally require large quantities of ash at a time. Such quantities could be obtained from a lagoon that needs to be emptied.
- Concrete fly ash—ash has been used as a fly ash substitute in concrete mixes.
- Asphalt additive—the ash has been used as a mineral filler and fine aggregate in asphalt mixes.
- Others—one of the more innovative uses of incinerator ash is in vermiculture process. The ash is blended with food waste material and worms are then added. After an adequate period of time, the worms are separated from the mixture, and the remaining material is used as a soil amendment.

Reuse options for incinerator ash tend to be site-specific. Utilities should pursue all avenues available for recycling the ash. State departments of transportation should be contacted to determine requirements for use of incinerator ash as a fly ash substitute in concrete or as a mineral filler or fine aggregate substitute in asphalt mixes. If the incinerator ash is approved for use in mix designs by the state department of transportation, then a considerable market can be opened for reuse of incinerator ash. Table 8.1 contains the results of an ash survey conducted by the Northeast Ohio Regional Sewer District (Dominak et al., 2005).

6.0 REGULATIONS

Regulations vary from state to state. In some states, ash is not regulated; in others, it is treated as a waste product. Local, state, and federal regulations need to be checked before disposal or reuse of incinerator ash. It is recommended that each facility do its own research regarding local, state, and federal regulations with regard to ash disposal and reuse.

Some landfills will require a toxicity characteristic leaching potential (TCLP) test prior to accepting ash. The results of a TCLP test on MHF ash are shown in Table 8.2 (Hampton Roads Sanitation District, 2004). The pH of the ash is also a test that may be required before landfilling.

TABLE 8.1 Results of an ash survey conducted by the Northeast Ohio Regional Sewer District (Dominak et al., 2005).

Wastewater treatment agency	State	City	Plant name	Ash handling method	Moisture content	Ash generated per year	Disposal method: Landfill	Disposal method: Beneficial reuse	Disposal cost	Description of beneficial reuse and/or landfilling
Central Contra Costa Sanitary District	California	Martinez	Main	Wet	65%	5110 wet ton[b]	X	X	\$21/ton	Most ash is beneficially reused as additives in landfill cover materials and brick making. Some ash is, however, disposed of in a commercial MSWLF[f].
Albany County Sewer District	New York		Albany County	Wet	30–50%	8000 cu yd[c]		X	No cost	Ash is blended with compost in a 50/50 mix and used as final landfill cover at no cost to Albany County. In return, Albany County accepts and treats landfill leachate.
Allegheny County Sanitary Authority	Pennsyl-vania		Alcosan	Dry	5–15%	6500 dry ton	X		\$17–\$19/ton	Water is added to dry ash to control emissions when off-loaded into trucks. Costs provided by Alcosan are for tipping fees only.
Metropolitan Council Environmental Service	Minnesota	St. Paul	Metro	Dry	NA[a]	15 000 dry ton		X	\$25.75/ton	Ash is used as a raw material in the manufacture of Portland cement.
Metropolitan Council Environmental Service	Minnesota	St. Paul	Seneca	Dry	NA	1800 dry ton		X	\$25.75/ton	Ash is used as a raw material in the manufacture of Portland cement.
City of Canton Water Pollution Control Center	Ohio	Canton	City of Canton	Dry	NA	1820 dry ton	X		\$210 per roll-off box plus taxes	Ash is disposed of at a commercial MSWLF. Ash is removed from the plant in 30-cu yd roll-off boxes.

(continued on next page)

TABLE 8.1 Results of an ash survey conducted by the Northeast Ohio Regional Sewer District (Dominak et al., 2005) *(continued)*.

Wastewater treatment agency	State	City	Plant name	Ash handling method	Moisture content	Ash generated per year	Disposal method: Landfill	Disposal method: Beneficial reuse	Disposal cost	Description of beneficial reuse and/or landfilling
Kansas City Water Services	Missouri	Kansas City	Blue River	Wet	50%	3500 dry ton	X		\$20/ton	Ash is disposed of at a commercial MSWLF after it is dewatered in a storage lagoon. Kansas City may shift to a dry ash system and close the ash-filled lagoon as is.
Upper Blackstone Water Pollution	Massa-chusetts		Upper Black-stone	Dry	NA	7000–9000 dry ton	X		?	Ash is disposed of at a publicly owned treatment works facility where screenings and grit are also disposed.
Narragansett Bay Commission			Field's Point	Wet	25%	2341 wet ton		X	\$16/ton	Ash is mixed with soil and used as a cover material at a quasi-public owned monofill.
Hampton Roads Sanitation District	Virginia		Army Base	Dry	NA	1300 dry ton	X	X	\$46/ton (if BR[d]) \$55/ton (if landfilled)	Water is added to dry ash to control emissions when off-loaded into trucks. Some ash is disposed of in a commercial MSWLF; some ash is beneficially reused as select fill.
Hampton Roads Sanitation District	Virginia		Boat Harbor	Dry	NA	3100 dry ton	X	X	\$46/ton (if BR) \$55/ton (if landfilled)	Water is added to dry ash to control emissions when off-loaded into trucks. Some ash is disposed of in a commercial MSWLF; some ash is beneficially reused as select fill.

Hampton Roads Sanitation District	Virginia		Chesapeake/ Elizabeth	Dry	NA	2500 dry ton	X	X	\$46/ton (if BR) \$55/ton (if landfilled)	Water is added to dry ash to control emissions when off-loaded into trucks. Some ash is disposed of in a commercial MSWLF; some ash is beneficially reused as select fill.
Hampton Roads Sanitation District	Virginia		Virginia Initiative	Dry	NA	4500 dry ton	X	X	\$46/ton (if BR) \$55/ton (if landfilled)	Water is added to dry ash to control emissions when off-loaded into trucks. Some ash is disposed of in a commercial MSWLF; some ash is beneficially reused as select fill.
Hampton Roads Sanitation District	Virginia		Williamsburg	Dry	NA	4000 dry ton	X	X	\$46/ton (if BR) \$55/ton (if landfilled)	Water is added to dry ash to control emissions when off-loaded into trucks. Some ash is disposed of in a commercial MSWLF; some ash is beneficially reused as select fill.
City of Palo Alto	California	Palo Alto	Palo Alto	Dry	NA	1460 dry ton		X	?[e]	Water is added to dry ash to control emissions when off-loaded into trucks. 500+ tons of ash is land applied; the rest is used as landfill cover.
City of Columbus Department of Public Utilities	Ohio	Columbus	Jackson Pike	Wet	35–65%	3102 dry ton	X		\$30–\$38/ton	Ash is disposed of at a commercial MSWLF approximately 20 miles[g] from the plant.
City of Columbus Department of Public Utilities	Ohio	Columbus	Southerly	Wet	35–65%	4458 dry ton	X		\$30–\$38/ton	Ash is disposed of at a commercial MSWLF approximately 20 miles from the plant.

(continued on next page)

TABLE 8.1 Results of an ash survey conducted by the Northeast Ohio Regional Sewer District (Dominak et al., 2005) *(continued)*.

Wastewater treatment agency	State	City	Plant name	Ash handling method	Moisture content	Ash generated per year	Disposal method: Landfill	Disposal method: Beneficial reuse	Disposal cost	Description of beneficial reuse and/or landfilling
Green Bay Metropolitan Sewerage District	Wisconsin	Green Bay	Green Bay	Dry	NA	3766 wet ton	X		\$26.77/ton	Water is added to dry ash to control emissions when off-loaded into trucks. After wetting, ash has moisture content of 25–27%. Ash is disposed of at a county-owned MSWLF.
Buffalo Sewer Authority	New York	Buffalo	Buffalo	Dry	NA	6100–6500 dry ton	X		\$27.40/ton	Ash is disposed of at a commercial MSWLF.
City of Youngstown	Ohio	Youngs-town	Youngs-town	Dry	NA	1065 dry ton	X		\$15.24/ton	Water is added to dry ash to control emissions when off-loaded into roll-offs. Plant transports ash to a MSWLF (Mahoning Landfill) in New Middletown, Ohio.

[a] NA = not applicable.
[b] ton × 0.9072 = Mg.
[c] cu yd × 0.7646 = m^3.
[d] BR = beneficially reused.
[e] ? = information not available.
[f] MSWLF = municipal solid waste landfill.
[g] mile × 1.609 = km.

TABLE 8.2 Results of a TCLP test on MHF ash (Hampton Roads Sanitation District, 2004).

Analytical Report

Project:	**Hampton Roads Sanitation District TCLP Monitoring – Boat Harbor Plant**
Project code:	**BH**
Sample point:	**Ash**
Sample date:	**06/19/04**
Sample receipt date:	**06/21/04**

Analyte	SW846 method	Unit	Result	Report limit[a]	Regulatory limit	Analysis date	Analysis time
Wet chemistry							
Ignitability	1030		Negative		Yes	06/22/04	15:05
Free liquid	9095		Negative		Yes	06/22/04	15:05
Corrosivity by pH	9045C	SU	5.57		12.5<pH<2.0	06/21/04	11:43
Reactivity	7.3				Positive		
Reactive cyanide	9012	mg/kg	<0.05	0.05	250	06/24/04	13:05
Reactive sulfide	9034	mg/kg	7	5	500	06/25/04	14:05
Metals							
Arsenic	6010B	mg/L	0.092	0.020	5	06/25/04	11:50
Barium	6010B	mg/L	0.204	0.001	100	06/25/04	11:50
Cadmium	6010B	mg/L	<0.002	0.002	1	06/25/04	11:50
Chromium	6010B	mg/L	<0.005	0.005	5	06/25/04	11:50
Lead	6010B	mg/L	<0.005	0.005	5	06/25/04	11:50
Mercury	7470	mg/L	<0.0001	0.0001	0.2	06/25/04	11:01
Selenium	6010B	mg/L	<0.040	0.040	1	06/25/04	11:50
Silver	6010B	mg/L	<0.002	0.002	5	06/25/04	11:50
Volatile organics							
Benzene	8260B	mg/L	<0.010	0.010	0.5	06/23/04	18:33
2-Butanone	8260B	mg/L	<0.010	0.010	200	06/23/04	18:33
Carbon tetrachloride	8260B	mg/L	<0.010	0.010	0.5	06/23/04	18:33
Chlorobenzene	8260B	mg/L	<0.010	0.010	100	06/23/04	18:33
Chloroform	8260B	mg/L	<0.010	0.010	6.0	06/23/04	18:33
1,2-Dichloroethane	8260B	mg/L	<0.010	0.010	0.5	06/23/04	18:33
1,1-Dichloroethene	8260B	mg/L	<0.010	0.010	0.7	06/23/04	18:33
Trichloroethene	8260B	mg/L	<0.010	0.010	0.5	06/23/04	18:33
Tetrachloroethene	8260B	mg/L	<0.010	0.010	0.7	06/23/04	18:33
Vinyl chloride	8260B	mg/L	<0.010	0.010	0.2	06/23/04	18:33
Pesticides							
Lindane	8081A	mg/L	<0.00050	0.00050	0.4	07/02/04	12:45
Chlordane	8081A	mg/L	ND[b]	0.00020	0.03	07/07/04	11:13
Endrin	8081A	mg/L	<0.00050	0.00050	0.02	07/02/04	12:45
Heptachlor	8081A	mg/L	<0.00050	0.00050	0.008	07/02/04	12:45
Heptachlor epoxide	8081A	mg/L	<0.00050	0.00050	0.008	07/02/04	12:45
Methoxychlor	8081A	mg/L	<0.00050	0.00050	10	07/02/04	12:45
Toxaphene	8081A	mg/L	ND	0.00075	0.5	07/07/04	11:13

(continued on next page)

TABLE 8.2 The results of a TCLP test on MHF ash (Hampton Roads Sanitation District, 2004) *(continued)*.

Analytical Report

Project:	**Hampton Roads Sanitation District TCLP Monitoring – Boat Harbor Plant**
Project code:	**BH**
Sample point:	**Ash**
Sample date:	**06/19/04**
Sample receipt date:	**06/21/04**

Analyte	SW846 method	Unit	Result	Report limit[a]	Regulatory limit	Analysis date	Analysis time
PCB							
PCB 1016	8082	mg/L	ND	0.0010		07/07/04	11:13
PCB 1221	8082	mg/L	ND	0.0010		07/07/04	11:13
PCB 1232	8082	mg/L	ND	0.0010		07/07/04	11:13
PCB 1242	8082	mg/L	ND	0.0010		07/07/04	11:13
PCB 1248	8082	mg/L	ND	0.0010		07/07/04	11:13
PCB 1254	8082	mg/L	ND	0.0010		07/07/04	11:13
PCB 1260	8082	mg/L	ND	0.0010		07/07/04	11:13
Semivolatile organics							
1,4-Dichlorobenzene	8270C	mg/L	<0.010	0.010	7.5	06/29/04	21:53
2,4-Dinitrotoluene	8270C	mg/L	<0.010	0.010	0.13	06/29/04	21:53
Hexachlorobenzene	8270C	mg/L	<0.010	0.010	0.13	06/29/04	21:53
Hexachlorobutadiene	8270C	mg/L	<0.010	0.010	0.5	06/29/04	21:53
Hexachloroethane	8270C	mg/L	<0.010	0.010	3.0	06/29/04	21:53
o-cresol	8270C	mg/L	<0.010	0.010	200	06/29/04	21:53
p/m-cresol	8270C	mg/L	<0.020	0.020	200	06/29/04	21:53
Total cresol	8270C	mg/L	<0.030	0.030	200	06/29/04	21:53
Nitrobenzene	8270C	mg/L	<0.010	0.010	2.0	06/29/04	21:53
Pentachlorophenol	8270C	mg/L	<0.010	0.010	100	06/29/04	21:53
Pyridine	8270C	mg/L	<0.010	0.010	5.0	06/29/04	21:53
2,4,5-Trichlorophenol	8270C	mg/L	<0.010	0.010	400	06/29/04	21:53
2,4,6-Trichlorophenol	8270C	mg/L	<0.010	0.010	2.0	06/29/04	21:53
Subcontracted analysis*							
2,4-D	8151	ug/L	<2	0.2	10 mg/L	07/01/04	05:57
2,4,5-TP	8151	ug/L	<2	0.2	1.0 mg/L	07/01/04	05:57

Authorization: ______________________ **Date** 9/16/04

[a] Report limit is lowest concentration at which quantitation is demonstrated. Report limit must be less than regulatory limit.

[b] ND = not detected.

7.0 REFERENCES

Dynamic Air, Inc. (2006a) *16 Pneumatic Conveying Concepts; Dilute Phase System;* Dynamic Air, Inc.: St. Paul, Minnesota.

Dynamic Air, Inc. (2006b) *16 Pneumatic Conveying Concepts; Dense Phase System;* Dynamic Air, Inc.: St. Paul, Minnesota.

Dominak, R.P., et al. (2005) *Long Term Residuals Management Plan for the Northeast Ohio Regional Sewer District;* Northeast Ohio Regional Sewer District, Cleveland, Ohio.

Hampton Roads Sanitation District (2004) *Analytical Report; HRSD TCLP Monitoring - Boat Harbor Plant;* Hampton Roads Sanitation District: Virginia Beach, Virginia.

Chapter 9

Instrumentation and Control

1.0 MODERN SYSTEMS PROVIDE INTEGRATED FUNCTIONALITY

Modern instrument and control and supervisory control and data acquisition (SCADA) systems are more properly considered management support systems, or decision support systems. They provide for the integrated needs of many levels or functional categories of users, from the operator, manager, and accountant, to the agency director. These systems bring together various functions within the plant, helping to ensure that their use is supported as necessary capital expenditures.

Like their wet stream process counterparts, operators of wastewater treatment plant bioenergy processes must ensure effective and economical operations while complying with all permit requirements. Agencies implementing a biosolids environmental management system increasingly are aware of planning, quality control, and quality assurance (i.e., critical control points) needs for all aspects of plant and process operations—especially related to bioenergy processes.

The good news about instrument and control systems is that the distinctions are becoming blurred between computer, smart components, controllers, and other components. It is no longer necessary to begin with the design of system hardware, but rather by outlining the functionalities desired. For any need, there are now multiple means of meeting it, and accomplished system integrators can design the hardware and software to meet requirements.

2.0 DESIGNING INSTRUMENT AND CONTROL SYSTEMS

This chapter will introduce important concepts and serve as a guideline in the conceptual design or review several systems: instrument and control, data acquisition, SCADA, plant control systems, and plant monitoring systems for bioenergy processes.

In this chapter, emphasis is placed on combining fundamental operational information with an intelligent knowledge sharing and management decision support system. It should seem obvious that the best systems are those that most fully integrate and demonstrate these concepts. Even so, all systems are intended to accomplish basic functions.

In the traditional view, computers excel in repetitive tasks; operators and technicians, on the other hand, excel at those tasks that require insight, discrimination, analysis, and experience. The relatively new field of adaptive process control is bridging the gap that separates these two views.

It is important to remember that no amount of automated process control can make a poorly designed mechanical process system perform well. Automation is not a solution for misapplied or improperly sized process equipment components.

In a recent broad-industry survey, more than two-thirds of data acquisition software systems purchasers identified the following features as necessary or desirable (Harrold, 2006):

- Data logging and historian functions.
- Control and simulation.
- Statistical quality control and statistical process control.
- Developer capabilities of database connectivity through an open source structured query language (SQL), such as ANSI SQL.
- Integration and customization tools and functions such as ActiveX and OPC (object-linking and embedding for process control), which specify the communication of real-time plant data between programmable logic controllers, distributed control systems (DCS), and other control devices to human-machine interfaces and display clients from different manufacturers (http://www.pacontrol.com/OPC.html).

Fundamentally, functional design or modification of instrument and control systems for bioenergy processes must address the following categories and functions:

- Process control,
- Reporting and recordkeeping (including regulatory compliance), and
- Safety, security, and maintenance.

3.0 THE FUTURE IS NOW

Forward-looking agencies are following the lead of other industry sectors by linking process and production information into a vertically integrated system that provides accounting, budgeting, purchasing, maintenance and production planning, staffing, and even goal-setting and reporting functions.

Although it is not typically practical to demand all elements in a single system immediately, agencies often discover that they can realize significant value-added functionality in a tightly integrated system.

4.0 INSTRUMENT AND CONTROL SYSTEMS BACKGROUND AND TERMINOLOGY

4.1 Telemetry

The definition of telemetry is "to measure from a distance." A physical property, state, status, or quantity is measured, converted to an electric signal, and transferred to another (remote) location. The transmitting means can be any medium capable of serving as a connecting link: copper wire, glass fiber, electromagnetic waves in any spectrum or routing, either directly to the receiver, or perhaps bounced through space by satellite.

4.2 Data Acquisition Systems

Data acquisition (DAQ) is the process of collecting data from digital and analog measurement sources through automation. Depending on industry focus, DAQ definitions vary. A pharmaceutical researcher may need high-speed, high-accuracy temperature and pressure measurements to analyze a new experimental process. A trucking company may want to observe locations of its vehicle fleet. A dairy may want to query production capabilities, a product being processed, current inventory of products, and status of shipping of those products. The wastewater utility business is a blend of these needs and typically considers DAQs fundamental components of a SCADA system. The typical arrangement provides data gathered from remote locations via a combination of links of the type mentioned previously.

4.3 Direct Digital Control and Distributed Control System

The common definitions of direct digital control (DDC) and DCS take their birth in the generation of the "mainframe" computer, and are defined based on that classic perspective.

4.3.1 Direct Digital Control

In DDC, all process control calculations and functions were managed or calculated in a mainframe, or central computer, which communicated to field terminals or units to directly control them. Communications were through proprietary communication hardware and protocols, transmitted through direct cabling links. Depending on the generation of hardware, dedicated wire pairs sent signals directly. Later, wire pairs were still used, but modules shared (multiplexed) several signals on one wire pair. The modules were known as multiplexers (MUXs) and demultiplexers (deMUXs). Communications improvements allowed devices to randomly talk (broadcast) on a piece of wire, with an identified talker and identified target (listener) but was heard by many.

4.3.2 Distributed Control System

A DCS was similar to a DCC in that a central computer or mainframe still performed process calculations. But instead of directly controlling an element, it transmitted set-points, or parameters, for the desired action to the hardware that would operate or move a valve or other device. Therefore, it exerted "supervisory control" rather than "direct control." The "local" hardware was intelligent enough to control the actions but needed oversight for the process set points. As with the DDC systems, communications were encoded onto wire pairs through MUX/deMUX. The remote field units were known as remote terminal units (RTUs) and responded to supervisory signals from master terminal units (MTUs). The RTUs at their inception were capable of only primitive ladder control: to turn on an output (for example, a pump) and recognize a state of output and communicate that back to the MTU. As the price of electronics has plummeted, field devices have grown increasingly "smart" to the point that they rival the level of sophistication found only in a mainframe computer a few decades ago.

4.4 Supervisory Control and Data Acquisition

In the water and wastewater treatment industries, the term SCADA has been used most often to refer to a system that communicates by radio, performs low-level control such as controlling pumping stations, and communicates back liquid levels of elevated water towers and other similar control processes. A central control room interface is typically provided to display an overview of the status of the controlled system, known as the human-machine interface, or man-machine interface. Today's radio and computer systems have replaced early hard-wired systems that typically controlled a primitive "light board" to display status conditions.

Simply restated, SCADA "talks" to various remote locations from a central location and relays a command or request that an operator enters at a console. The SCADA listens for data reports from those remote locations and, typically, enters this information to a database. It may include the capabilities to "playback" this information on chart form or output data in printed tables or electronic file form.

5.0 INSTRUMENTS IN BIOENERGY PROCESSES

Useful and intelligent systems are designed from the top down but constructed from the bottom up. First, sensors and instruments, such as gauges, are applied to monitor basic conditions of temperature, flow, pressure, voltage, current, weight, level, speed, and position. It is illustrative to consider the very beginnings of instrument systems and the concepts of remote viewing and remote control, such as for a steam boiler installation.

In their most fundamental form, these instruments provide direct information to the operator but no one else. Even to the operator, their usefulness can be compromised because they cannot see more than one process condition at one location or at one time. To remedy this drawback, higher-level components were added to provide the capability to transmit or provide remote readings to one central area, or co-locate the readings of these process conditions to a more convenient location, typically the control room. This allowed the operator to be many places at once. When chart recorders were added, the operator was allowed to be other places and still have a record of what occurred. But, more importantly, the operator was able to view information, trends, and responses to previous operational changes and apply judgment to make sound decisions about operations. From a management perspective, the components increased both the quality and quantity of work an operator could undertake and complete.

The next level of sophistication entailed use of "controllers" in conjunction with the instruments. They can be open- or closed-loop and use feedback control. When process conditions vary, the controller takes corrective actions to vary (increase, slow) a process condition, further freeing the operator from repetitive tasks and allowing increased attention to higher level decision-making. For example, turning your home furnace on if the outdoor temperature is low and off when it is warm involves an open-loop system; measuring the temperature in the house (the controlled space) and tuning the furnace on or off to maintain a desired temperature based on feedback from a thermostat involves a closed-loop system.

Once these instruments, controllers, and recorders were co-located in one place, it was but a short step to recognize that they could be located nearly anyplace. Thus primitive instrument and control systems were born, incorporating remotely located control rooms, remote-control of local systems, remote information, and remote sharing of information.

6.0 PROCESS AUTOMATION

The focus of instrumentation and control systems is changing. The initial focus on labor savings has been essentially supplanted by the opportunity to increase safety and quality, minimize process deviation, and eliminate process inconsistencies.

Dating back to the industrial revolution, individual wires and instrument piping have served these systems well. Present micro- and nano-electronics and digital computer advances have provided expanded, sophisticated means to collect, analyze, store, share, and communicate data so that operators can make decisions in the plant, outside the plant, and, perhaps, from anywhere in the world.

The benefits of overall unit automation were recognized by large steam utility boiler plants at least 50 years ago. Such a focus is still appropriate. Stultz (1972) described automation's advantages as follows:

- Improved protection of personnel and equipment through more complete instrumentation, simplified information display, and more extensive and thorough supervisory control.
- Reduced outages, reduced maintenance, and longer equipment life through more uniform and complete control procedures of startups, online operations, and shutdowns.
- Better plant efficiency through continuous and automatic adjustments to the controls with the objective of optimizing plant operation.
- More efficient use of manpower during startup and online operation.

Its author continues with a discussion of installation and service needs and closes with a cautionary and critical point: "A planned program of preventative maintenance should be developed for the control system."

There is no doubt that carefully planned preventive maintenance of instruments, programmable logic controllers, computers, software backups, and similar activities of bioenergy process control systems is absolutely essential for their proper and continued functioning and the ability to provide safe, efficient, and economical operations. Such systems are no longer an optional add-on but rather are integrated parts of the entire process.

7.0 PROCESS CONTROL MEASURING AND MONITORING

The wastewater industry most often controls "elements" that typically are powered by electric motors: pumps, fans, compressors, blowers, conveyors, centrifuges, or hydraulic power packs. This includes equipment that directs the output of or modifies the power requirements of those elements, such as motor-operated valves, positioners, and dampers.

For an individual piece of equipment (motor), control is provided by field elements and sensors, often wired to the unit's local control panel or the motor control center (MCC) starter cubicle (bucket.) At this level, fundamental safety and interlocking is managed. The typical means of interconnection is by actual physical wiring to the MCC or control panel.

More and more, however, devices that were considered solely mechanical or electrical are provided with integral microprocessor control packages, blurring their prior distinct operation. Presently, control, communications, life-cycle assessment, and speed of response to a control communication may all be implemented or monitored in one device package.

An incineration instrument and control system is, as noted, an extremely complex amalgamation of many factors. Table 9.1 is a compilation of seven "control loop" categories necessary for the operation, monitoring, and control of incineration systems. These loops are feed, combustion, pressure, temperature, emission, air pollution, and utility controls. Presented in grid format, it provides easy reference to four "management area" categories:

- Process control,
- Safety and personnel protection,
- Regulatory compliance, and
- Business categories.

It highlights where such control loops relate functionally to the overall business of incineration and emphasizes the interconnectedness of the various process control, business needs, and regulatory compliance. Design and review of instrumentation systems cannot be effectively undertaken and managed without such an understanding and overview of their interplay.

Continuous emission monitoring systems (CEMs) and continuous opacity monitoring systems (COMs) are requirements of various federal, state, and local regulations, most notably the Clean Air Act Part 60 and Clean Water Act Part 503 regulations. The 40 CFR 60 spells out requirements for opacity monitoring and oxygen analyzers, and 40 CFR 503 provides requirements for a total hydrocarbon (THC) CEMS system. Although they are not technically instrument and control systems themselves, they provide needed information to control bioenergy processes, thus must be deeply integrated to control, monitoring, data acquisition, reporting, and recordkeeping functions. This chapter provides an overview but cannot fully discuss the complete and particular requirements of these systems. Table 9.2 provides

TABLE 9.1 Process control measuring and monitoring.

Control loop	Process control	Safety and personnel protection	Regulatory compliance (excess emissions control and emission compliance monitoring, recording, and recordkeeping)	Business (production, inventory, accounting, budgeting, and asset management planning)
1. Feed control • Biosolids • Auxiliary fuel	•	•	•	•
2. Combustion control • Combustible atmosphere safety purge • Incinerator flame loss control • Burner control • Oxygen control	•	•		
3. Pressure control: incinerator vessel pressure and draft	•	•		
4. Temperature control • Incinerator vessel • Hearths, bed, freeboard, chamber • Air pollution control equipment • Heat recovery equipment, ducts, related equipment	•	•	•	
5. Emission measurement and control • Total hydrocarbon, CO, NO_X, SO_X • O_2, CO_2, moisture • Opacity • Particulate matter	•		•	
6. Air pollution control systems • Wet scrubber water flow, ΔP • Dry fabric filter flow, ΔP • Electrostatic precipitator (ESP), wet ESP applied volt, ampere, kilowatt • Chemical addition, pH adjustment	•		•	
7. Utility system • Electric use, fan/blower horsepower • Water and sewer use • Compressed air • Fuel use—natural gas, oil • Chemical use • Ash handling	•			•

TABLE 9.2 Summary of regulatory requirements for incinerators.

Regulation	Continuous monitoring system description/requirement
40 CFR Part 60 Subpart O for Sections 60.150 through 60.156	Instruments for the following: • 60.153 (a): Mass/volume sludge charged • 60.153(b)(1): Pressure (differential pressure) for scrubber(s) • 60.153(b)(2): Oxygen analyzer for flue gas • 60.153(b)(3): Temperature (thermocouple) for all hearths • 60.153(b)(4): Flow auxiliary fuel
40 CFR Part 60 Section 60.152 Standard for Particulate Matter (Opacity) 60.11(b) and 60.11(e)(5) Compliance with standards and maintenance requirements, and 60.13 (Monitoring)	Analyzer for the following: • 60.152(a)(2): COMS • 60.11: COMS • 60.13 (a): COMS • Continuous opacity measurement systems are optional under 40 CFR 60 Subpart O. If used, they must meet the following standards: ▪ 40 CFR 60 App A, Method 9 Visual Determination of the Opacity of Emissions from Stationary Sources ▪ 40 CFR 60 App B, Performance Specification 1—COMS ▪ 40 CFR 60 App F—Quality Assurance Requirements for Gas Continuous Emission Monitoring Systems Used for Compliance Determination
40 CFR 503.40 Applicability, 40 CFR 503.44 Operational Standard Total Hydrocarbons and 40 CFR 503.45 Management Practices	Analyzers for the following: • 503.40: carbon monoxide (CO) • 503.44(a), 503.45(c): moisture • 503.44(b), 503.45(b): oxygen • 503.44(c), 503.45(a): total hydrocarbons Reference: 40 CFR 60 App B, Performance Specification (PS) • Performance Specification 3—O_2, CO_2 • Performance Specification 4, 4b, CO, O_2 • Performance Specification 8, 8A volatile organic compounds, THC • 40 CFR 60 App F—Quality Assurance Requirements for Gas Continuous Emission Monitoring Systems Used for Compliance Determination (Also please refer to 503.41 Special Definitions)

detailed references to Code of Federal Regulations appendices and performance specifications. It must be emphasized that these requirements are exceedingly complex and tedious. An entity must study them thoroughly and implement them carefully to achieve and maintain compliance.

Indeed, many regulations require not only notification of the failure of such a monitoring system but also shutdown of the process after a failure of emission-monitoring equipment. Failures as a result of the incompleteness or inadequacy of quality assurance/quality control (QA/QC) systems may also lead to regulatory penalties or fines. Process and workplace safety requirements preclude the facility from continuing process operations without the benefit of safety systems (and it is typically illegal to do so). Maintenance and operation of instrument and control systems and particularly continuous emissions monitoring systems are critical.

8.0 OTHER RELATED REGULATIONS

Federal regulations that govern other air pollution sources may apply to incineration systems, which often creates confusion. Table 9.3 provides a look at three regulatory references. They are mentioned here to caution the reader against designing or implementing unneeded monitoring and instrument and control systems. Often, commercial suppliers are not fully aware of incineration needs, and instead attempt to sell or substitute a system designed for another emission program. The obvious results are noncompliance with regulations and additional expense related to implementing unnecessary instrumentation monitoring, recording, and recordkeeping systems.

The Clean Air Act (CAA) Part 70 program is often applicable to incineration systems because emissions typically exceed the threshold to be declared a major source under the CAA. A specific determination must be made for facility-wide emissions and may recognize efforts undertaken by the designer to limit the facility's and source's "potential to emit" codified emissions. Recently designed incineration systems have been able to be declared a "minor" source category definition or a "synthetic minor" category definition, thus are not subject to Part 70 rules.

Another federal program, compliance assurance monitoring (CAM) systems (40 CFR Part 64) often are mentioned. It is not applicable to incineration systems. The CAM rule (see Table 9.3) requires owners and operators to oversee and evaluate the effectiveness (QA/QC) of the instruments, control devices, and systems that monitor process emissions. The rule requires reporting the status and accuracy of the monitoring system as well as reporting the facility's actual determination of compliance or

TABLE 9.3 Additional references.

Related regulation	Continuous monitoring system description/requirement
40 CFR 64 Compliance Assurance Monitoring Rule	The regulations provide procedures for coordinating these new requirements with U.S. EPA's operating permits program regulations under 40 CFR Parts 70 and 71. Revisions to the operating permits program regulations clarify the relationship between the Part 64 requirements and periodic monitoring and compliance certification requirements in Parts 70 and 71.
	The CAM rule requires owners and operators to monitor the operation and maintenance of their control equipment so that they can evaluate the performance of their control devices and report whether their facilities meet established emission standards. The CAM rule requires them to take action to correct any malfunctions and to report such instances to the appropriate enforcement agency. Monitoring will focus on emissions units that rely on pollution control device equipment to achieve compliance with applicable standards.
40 CFR 70 Title V or Permits Program	Part 70 program theoretically adds no requirements. It does amplify the language and impose considerable requirements to quantify and certify how continuous compliance with the underlying permit regulations is attained.
40 CFR 75 Continuous Emission Monitoring	Part 75 is not applicable to wastewater sludge incinerators; however, U.S. EPA region, state, and local authorities may adopt or reference these Code of Federal Regulations sections. These sections provide a detailed view of requirements and practices for continuous instrument systems installation, calibration, precision, maintenance, reliability, performance auditing, QA/QC, uptime, reporting, data verification and validations, and substitution for missing data.

noncompliance with its emission limits. Regulations clarify the relationship between the Part 64 requirements and periodic monitoring and compliance certification requirements in Parts 70 and 71. Part 75 is not applicable to wastewater sludge incinerators. U.S. Environmental Protection Agency (U.S. EPA) regional, state, and local authorities may adopt or reference these Code of Federal Regulations sections. These sections provide a detailed view of requirements and practices for continuous instrument systems installation, calibration, precision, maintenance, reliability, performance auditing, QA/QC, uptime, reporting, data verification and validations, and substitution for missing data.

Although the previously mentioned are not requirements, they provide excellent reference and reading material to more fully understand the philosophy of what is required under the regulations.

9.0 DATA ACQUISITION SYSTEMS DATA INTEGRITY

Unfortunately, the more data acquisition is automated, the more likely operators tend to ignore its validity, fail to question the "data integrity chain," or subsequently fail to correct errors in the data set. This is manifested in two distinct areas—system specifications and system utilization by the owner. First, the owner must oversee the specification, design, and selection of a data acquisition system with data integrity features at a highly detailed level. Second, the owner must implement the use of the data integrity features specified in the product. Many systems and their underlying databases have been designed and purchased that do not anticipate the need for the owner to review data and subsequently mark data as valid / invalid. (For example, a field pH probe may be challenged during calibration with unknown or standard solutions and sends that signal to the system database. While the pH reading might have been (or not been) accurate, it was transmitted properly and entered into the database. It was not a representative signal from the process and must therefore be annotated for validity and possibly correction or augmentation with appropriate data from another instrument. Because this validity management need was not anticipated, a vital functionality is absent that would allow the owner to perform examination and correction of the data, supplementing correct data gained from another instrument or source. If data are modified or augmented, it must be done through a comprehensive feature that provides an audit trail documenting the "who, what, and when" of such changes. Though this feature is typically integrated to laboratory information management systems (LIMS), it is not at all common for plant-wide data acquisition systems. The details are not trivial, however. Sample data in LIMS are entered infrequently, whereas continuous process data are typically measured multiple times per second.

Automatic data acquisition and storage must be compared with the familiar wet sample chain-of-custody questions implemented through LIMS systems:

- Where have your data been?
- Have they been collected and stored as they should?
- Have they been tampered with?

From a regulatory standpoint, this information provides the basis for demonstrating a facility's compliance with regulations—or visible evidence of noncompliance. Data quality should be carefully considered regarding whether it shows careful compliance or willful noncompliance. The difference may result in significant financial loss or even serve as the basis for criminal charges. The ability to store, access, view, evaluate, extract, and present information from the data sources and archives must be carefully considered.

Such requirements and features reside deep in the system architecture and must be addressed from the beginning of the project. They must be clearly understood and accounted for in the design and cannot be added in piecemeal at the end.

10.0 INSTRUMENT AND CONTROL SYSTEMS CODES AND STANDARDS

The design and development of bioenergy process equipment and systems are highly complex and subject to many different industry codes, standards, and requirements. Instrument and control and SCADA systems necessarily reflect that complexity. Many different types of organizations—including multinational insurers, safety boards, and industry groups—promote safe operation through both required and voluntary codes and standards (refer to Table 9.4). Again, these requirements and recommendations can be highly variable, prone to change, and sometimes contradictory.

TABLE 9.4 A summary of applicable regulatory and voluntary codes and standards.

Code and standard	Title/description
ANSI/ISA-84.00.01 2004	Functional safety: Safety instrumented systems for the process industry sector (IEC 61511-1 Mod)
ANSI/ISA-91.00.01 2001	Identification of emergency shutdown systems and controls that are critical to maintaining safety in process industries
ANSI/ISA-95.00.01 2000	Enterprise-control system integration, part 1 models and terminology
ISA-5.41991	Instrument loop diagrams
NFPA 499	Classification of combustible dusts and of hazardous (classified) locations for electrical installations in chemical process areas, 2004 edition
NFPA 54	National fuel gas code
NFPA 85	Boiler and combustion systems hazard code
NFPA 86	Ovens and furnaces
NFPA 820	Standard for fire protection in wastewater treatment and collection facilities
NFPA 654	Standard for the prevention of fire and dust explosions from the manufacturing, processing, and handling of combustible particulate solids
NFPA 8501	Standard for single burner boiler operation
NFPA 8502	Standard for prevention of furnace explosions/implosions in multiple burner boilers

11.0 FINAL NOTES

Facilities should regularly review and renew existing systems, even though such considerations are often made only during the initial design of an instrument and control system or during initiation of a capital construction project. Although an installation may have met initial requirements when installed, equipment failures and obsolescence must be considered and managed.

This chapter provides an overview of industry and U.S. EPA requirements and practices for continuous instrument systems and data quality and integrity requirements. Local, state, national, and international requirements are in a constant state of flux, however, and should be consulted.

12.0 SUGGESTED READINGS

Stultz, S. C. (1972) Controls for Steam Power Plants. In *Steam: Its Generation and Use*; 38th ed.; Kitto, J.D., Ed.; Babcock and Wilcox Company: Lynchburg, Virginia.

Harrold, D. (2006) Product Focus: Data Acquisition. *Control Engineering*. Jan 1.

Lipták, B. G. (2003) *Instrumentation Engineers' Handbook—Process Measurement and Analysis*. CRC Press: Boca Raton, Florida.

U.S. Environmental Protection Agency (1991) *Standards of Performance for New Stationary Sources;* Code of Federal Regulations, 40 CFR, Part 60; U.S. Environmental Protection Agency: Washington, D.C.

U.S. Environmental Protection Agency (1993) *Standards for the Use or Disposal of Sewage Sludge;* Code of Federal Regulations, 40 CFR, Part 503; U.S. Environmental Protection Agency: Washington, D.C.

U.S. Environmental Protection Agency (1996) *Federal Operating Permits Programs;* Code of Federal Regulations, 40 CFR, Part 71; U.S. Environmental Protection Agency: Washington, D.C.

Chapter 10

Incinerator Operations

(continued)

Safe incinerator operations are critical for plant operations. Incinerator operators should strive to meet or exceed all applicable permit requirements to protect human health and the environment, maintain a safe operating environment for plant personnel and the incineration-related equipment, and operate the incineration system in a cost-effective manner. Operators must understand how the incinerator operates, take special care in operating and maintaining the system, and be able to respond appropriately to minor changes as well as emergency situations. The operators must also be familiar with the individual components of the incinerator system. Therefore, wastewater treatment plant management and operating personnel must ensure that

- Incinerators are operated in a manner that ensures complete combustion of wastewater solids.
- An efficient air pollution control system is used to minimize particulate matter and emissions of other regulated pollutants.
- The incinerator and all component parts are properly operated and maintained in accordance with industry standards.
- Safe operating procedures are established for the entire incineration system.

- All operators are properly trained for the tasks they are assigned.
- There is compliance with all regulatory, recordkeeping, and reporting requirements.

This chapter addresses key issues for operators to consider and address when operating either a multiple-hearth furnace (MHF) or a fluid bed incinerator.

1.0 COMPLIANCE WITH RULES AND REGULATIONS

The incineration unit process manager and key incineration operating staff members must be aware of and understand all federal, state, and local rules and regulations relating to the operation of the installed incinerators. Up-to-date copies of all required permits should be located within the incineration complex and be easily accessible by operating personnel. Several wastewater treatment agencies have developed electronic permit tracking applications that enable incinerator operators to quickly obtain electronic copies of all permits.

Compliance with all regulatory requirements and terms and conditions of all applicable permits is necessary to protect employees, the general public, and the environment. Noncompliance can result in penalties, fines, and the potential loss of an operating permit. Permitting and emissions regulations are discussed in detail in Chapter 3, Permitting and Emissions Regulations.

2.0 MULTIPLE-HEARTH FURNACE OPERATIONS AND PROCESS CONTROL

2.1 Pre-startup Inspection

Before an MHF can be placed into service, trained personnel must conduct a thorough inspection of the incinerator's interior, exterior, and ancillary equipment (e.g., pumps, blowers, conveyance and feed system, and air pollution control system). Proper lockout/tagout procedures must be implemented and followed before an incinerator is inspected or serviced.

2.1.1 Internal Inspection

If the incinerator has been out of service for an extended period of time, all ash, clinkers, and slag should be removed from the incinerator and the ductwork should be cleaned before an internal inspection. Chapter 11, Incinerator Maintenance, contains a discussion of cleaning issues.

A thorough internal inspection can only be completed when the incinerator is out of service and at ambient temperature to allow inspection from inside the incinerator. The inspection area is typically a confined space entry, and all applicable confined space entry procedures must be followed. However, some of the equipment can be inspected by observation from outside the incinerator when the incinerator is in the hot standby mode. This type of inspection is described later in this section.

The internal inspection of the MHF includes, but is not limited to, inspecting the following components:

- Refractory, including hearths, ceiling, sidewalls, and center shaft.
- Lute caps.
- Ash outlet.
- Out-hearth drop holes.
- Emergency bypass stacks and dampers.
- Exhaust gas ductwork.
- Rabble arms and rabble arm teeth (cracked or warped arms and teeth should be replaced).
- Burners
  - Burner ports and tiles should be inspected for cracks or slag buildup,
  - Target the burner flame to ensure that it is centered in the burner port, and
  - Verify that the burner flame does not impinge on the rabble arms or the thermocouples.
- Thermocouples.
- Air pollution control system internal components such as the scrubber trays, Venturi throat, spray nozzles, drains, and other components.

2.1.2 *External Inspection*

The external inspection of the MHF system includes, but is not limited to, the following components:

- Air dampers, blowers, cooling air system, and induced draft fans.
- Upper and lower sand seals.
- Conveyance and feed system.

- Auxiliary fuel system.
- Heat recovery system.
- Instrumentation and control system.
- Monitoring and recording systems.
- Air pollution control system.
- Center shaft drive, gear, and shear pin.
- Clinker grinder.
- Process water system.
- Lance ports.
- Ash conveyance system.
- Valves and actuators.
- Outer shell.

In the outer shell, operators should be looking for evidence of "hot spots" or corrosion. Infrared thermography, similar to that used for electrical switchgear monitoring, is useful for identifying and quantifying hot spots. Metal thickness testing may be required to determine if there is interior deterioration of the shell.

2.2 Multiple-Hearth Furnace Startup

There are two types of startup: cold startup and hot standby startup. A cold startup raises the internal temperature of the incinerator from ambient air temperature to at least 593°C (1100°F) at the hottest location within the incinerator. The temperature reached may be higher than 593°C (1100°F), depending on the percentage of solids and the heat content of the feed. When properly performed, a cold startup will cause evaporation of all moisture in the refractory without causing any damage.

A hot standby startup raises the temperature of the incinerator from a hot standby temperature of between 316 and 427°C (600 and 800°F) to at least 593°C (1100°F) at the hottest location within the incinerator.

For both cold and hot standby startups, the required incinerator temperatures will vary depending on the quality and quantity of the feed. At the Green Bay Metropolitan Sewer District, the incinerator temperature typically is set to 704°C (1300°F) for three hours before feed cake introduction. Once the incinerator has been brought

TABLE 10.1 Typical steps before warming up an MHF.

Step	Function	Action
1.	Lockout/tagout	Remove
2.	Main electrical power	On
3.	Instrument air supply	On
4.	Plant water supply	Open valve
5.	Water supply to air pollution control system	Open valve
6.	Fuel supply to incinerator	Open valve
7.	Control panel power	On
8.	Annunciator alarm panel	
	Paper and ink in all recorders, printers	Check
	Recorder functioning	Check
9.	Emergency bypass damper	Open
10.	Combustion air blower	On
11.	Center shaft cooling air blower	On
12.	Water supply to precooler	On
13.	Water supply to exhaust gas scrubber	On
14.	Induced draft fan (set draft controller to maintain a draft of −0.1 in water*)	On
15.	Emergency bypass damper	Closed
16.	Ash conveyance system	On

* in water × 0.2488 = kPa.

to the appropriate temperature higher than 593°C (1100°F), the operator introduces the feed cake to the incinerator.

The following section contains information relating to the startup of an MHF from both cold and hot standby conditions. This information is provided as a guide; the manufacturer's instructions should always be followed. Table 10.1 contains a list of general equipment that must be started and procedures that must be performed before incinerator heat-up is initiated.

2.2.1 Cold Startup

Cold startup procedures vary between incinerator installations depending on various types of equipment and system interlocks used. To purge any combustible gases that

may have built up within the incinerator, the induced draft fan should be operated to maintain negative draft. The time required to purge the incinerator of combustible gases varies based on the size of the incinerator. Thirty minutes is generally sufficient; however, in some cases a facility's insurance agency may require a specific purge time. Before purge, ensure that all safety limits are met and all equipment is operable. In addition, the center shaft cooling air should be on before burner startup.

To begin heating up the incinerator, the operator should place all of the burner pilots online. If the burners do not have pilots, the operator will proceed to the next step and light only one burner on low fire. After approximately two hours, the operator should light the burner located on the lowest hearth that has burners on low fire. Using the burner controller, the operator will increase the temperature at a rate of approximately 10°C (50°F) per hour until the burner controller is at 50%.

After one hour with the burner controller at 50%, the operator should reduce the burner controller to low fire and start another burner located on the same hearth. The operator should maintain both burner controllers at low fire. When the temperature stabilizes, the operator will start one burner located on the next higher hearth on low fire. The two burners and pilots located on the lower hearth should then be shut off.

When the temperature in any location within the incinerator reaches 260°C (500°F), the operator should start rotating the center shaft at 0.5 rpm. The operator should continue to increase the temperature on the higher hearths as done on the lower hearths. The operator will follow these procedures upward through the incinerator from hearth to hearth and burner to burner until all of the burners located on the highest hearth are on low fire. The operator will use the burner controllers to increase the temperature by 10°C (50° F) per hour.

When the temperature in any location within the incinerator reaches 538°C (1000°F), the operator should increase the rate of internal incinerator heating to 38°C (100°F) per hour until the internal temperature reaches 593°C (1100°F). The operator should allow the temperature to stabilize for three to four hours at 593°C (1100°F). This temperature stabilization is necessary because when the temperature within the incinerator first reaches 593°C (1100°F), only the surface of the refractory is at 593°C (1100°F). By maintaining the temperature at 593°C (1100°F) for several hours, the core temperature of the refractory will also be raised to 593°C (1100°F). This ensures evaporation of all moisture from within the refractory and provides the heat necessary to start combustion of the feed cake. As previously noted, an internal temperature higher than 593°C (1100°F) may be required before combustion can occur. The cold

startup process may take approximately 24 hours to raise the incinerator temperature from ambient temperatures to operating conditions.

Feeding of wastewater solids to the incinerator can begin after the air pollution control system is started and the emergency bypass stack (also known as the emergency relief stack) is closed. See section 2.3, Steady State Process Control, for additional details.

2.2.2 *Hot Standby*

Hot standby occurs when the incinerator temperature is decreased from the operating temperature by decreasing the burner controllers to low fire and reducing the number of online burners to one. The incinerator internal temperature would then decrease and stabilize between approximately 316 and 427°C (600 and 800°F) until the feeding process recommences.

Hot standby maintains sufficient heat within the refractory to allow for a more rapid startup than a cold startup and prevent the refractory from absorbing moisture. As indicated in section 2.2.1, Cold Startup, it takes approximately 24 hours to bring incinerator temperatures up to operating conditions from ambient temperatures. Therefore, a hot standby condition is preferred when there are temporarily no solids to burn, when minor maintenance needs to be performed on the incinerator or the incinerator subsystems, or when the incinerator is getting ready to be placed into service.

2.2.3 *Startup from Hot Standby*

Under hot standby conditions, the incinerator and its ancillary equipment are ready to be placed into service. To be put into service, the temperature must be raised to 593°C (1100°F) or higher to facilitate proper combustion, the air pollution control system must be started, and the emergency bypass stack must be closed before initiating cake feed to the incinerator.

To raise the temperature from hot standby to operating temperature, the operator should increase an upper hearth burner controller to raise the incinerator temperature by no more than 10°C (50°F) per hour. When the temperature in any location within the incinerator reaches 538°C (1000°F), the rate of temperature increase may be raised to 38°C (100°F) per hour until the interior temperature reaches 593°C (1100°F) or higher. As noted in the cold startup procedures, this temperature should be maintained for three to four hours before feed cake is introduced to the incinerator.

2.3 Steady-State Process Control

After startup procedures have been completed, the incinerator is ready for feed cake. As the incinerator is fed, the operator must make decisions based on characteristics of the site-specific wastewater solids (e.g., percent solids, percent volatile solids, heat content), equipment design, and operational experiences. As wastewater solids are introduced to the incinerator, its internal temperature will decrease. If the temperature decreases too much, the cake will not ignite. Proper startup temperatures ensure sufficient heat to commence the combustion process. If the temperature within the incinerator drops below the combustion temperature and the upper hearth burners are on, feed cake should be stopped and a burner on a lower hearth started. This process will assist with combustion of the cake solids that are already in the incinerator.

The optimal startup feed rate varies depending on the moisture content of the feed cake. As a general rule, the feed rate at startup should be 25% or less of the designed maximum feed rate.

The moisture content of the feed cake also dictates center shaft speed. When the cake has a high moisture content and the incinerator is near maximum feed capacity, the operator should increase the center shaft speed. A faster center shaft speed spreads the cake more quickly on the top hearths, thereby increasing the evaporation rate. However, if the center shaft speed is too high, unburned cake can be forced past the burn zone. Conversely, a slower center shaft speed can help feed with a higher moisture content burn when the incinerator is not at maximum capacity.

If the feed cake has a high solids content (28% or greater), it may be desirable to directly feed the cake to a lower hearth within the incinerator.

The combustion process is a chemical reaction; as the feed starts to burn and produce heat, the evaporation of moisture in the cake on the upper hearths is accelerated. Chapter 4, Combustion Theory, contains additional details. If the fire is allowed to work its way up through the incinerator until all of the cake solids have been burned, the reaction may increase exponentially, which is referred to as an uncontrolled burnout. An uncontrolled burnout produces high temperatures within the incinerator and should be avoided as it can result in refractory damage.

Understanding the effects of the various operating practices can provide more cost effective and enhanced operations. The various factors which affect operations are

- Autogenous operations,
- Nonautogenous operations,

- Excess air reduction,
- Combustion control,
- Draft control, and
- Air pollution control.

2.4 Autogenous versus Nonautogenous Operations

Depending on the site-specific percent total solids, percent volatile solids, and heat content of the cake solids, an MHF may operate either autogenously or nonautogenously. There are several factors to take into account when calculating the autogenous condition: incinerator exhaust gas temperature, amount of excess combustion air, and temperature of combustion air introduced to the incinerator.

2.4.1 *Autogenous Operation*

Autogenous incineration maintains stable combustion without use of auxiliary fuels. The operator calculates the percent dry solids in the feed cake at which autogenous combustion can occur.

Autogenous combustion typically occurs when the heat released by combustion of the solids is equal to the heat required to evaporate the water in the feed cake. For detailed calculations, see Chapter 4, Combustion Theory. A rapid calculation can be made using the following formula:

$$(P) \times (Q) = (100 - P) \times (W) \quad (10.1)$$

Where

P = minimum percent dry solids in feed cake required for autogenous combustion,

Q = fuel value of wastewater solids (kJ/kg [Btu/lb] of dry solids), and

W = heat required to evaporate and raise the temperature of one kilogram of water in a multiple-hearth furnace. The value of W is approximately 8141 kJ/kg (3500 Btu/lb); this value includes radiation losses and heating the gas streams and the wastewater solids.

Equation 10.1 can be solved for P as follows:

$$P = [(W)/(Q + W)] \times 100 \quad (10.2)$$

Note that kJ = Btu × 1.055, kg = lb × 0.4536, and kJ/kg = Btu/lb × 2.326.

An example follows:

- Feed rate: 4536 wet kg/h (10 000 lb/hr) or 108 864 wet kg/d (240 000 lb/d)
- Dry solids: 33%
  = (0.33) × (108 864)
  = 35 925 dry kg/d (79 200 lb/d)
- Volatile solids: 70%
  = (0.70) × (35 925 dry kg/d)
  = 25 147 kg/d (55 440 lb/d) volatile solids
- Inerts (ash) = 35 925 − 25 147 kg/d = 10 778 kg/d (23 760 lb/d)
- Heating value = 25 575 kJ/kg (11 000 Btu/lb) combustibles (volatile solids)
- Q = (Btu × volatile solids/d)/(dry solids/d)
  = (25 575 kJ/kg × 25 147 kg volatile solids/d)/35 925 dry solids/d
  = 17 902 kJ/kg (7700 Btu/lb) of dry solids
- P = (8141)/(17 902 + 8141)) × 100 = 31% dry solids

An incinerator that is operated far below the point of autogenous combustion will result in excessive use of auxiliary fuel. Operating far above the point of autogenous combustion will result in the use of additional cooling air to prevent high incinerator temperatures and clinker formation. Optimum combustion and incinerator process control occur when the percentage of solids in the feed cake are at or slightly greater than the autogenous combustion point. The theoretical point of autogenous combustion should be compared to the actual operational point of autogenous combustion. To establish autogenous conditions, the operator must frequently observe conditions inside the incinerator through a burner view port to watch for signs of combustion when auxiliary burners are shut down.

Optimum conditions at startup occur when the auxiliary burners are fired to match the heat energy required for excess moisture evaporation. When the operator observes that feed cake solid ignition has occurred, the auxiliary burners should be reduced. The extent of the auxiliary burner reduction depends on the heat energy produced from the burning feed. To ensure that the combustion reaction continues, total heat available from the auxiliary burners and the feed solids combustion must be equal to or greater than the heat required for evaporation. Therefore, as more of the feed solids mass begins to burn, the auxiliary burners should be decreased accordingly.

When the combustion process can continue without auxiliary burners the burners should be shut down; the combustion process is now autogenous. Returning hot center shaft cooling air to a lower hearth may assist in obtaining autogenous conditions. However, some insurance companies will not allow the incinerator to operate without at least one burner being on. Other insurance companies may allow the burners to be shut off but require that temperatures remain higher than 760°C (1400°F) somewhere in the incinerator.

2.4.2 Nonautogenous Operation

The goal of nonautogenous incinerator operation is to maintain a stable combustion process with the addition of a minimal amount of auxiliary fuel. Auxiliary fuels are traditionally natural gas or no. 2 fuel oil. In limited instances, coal or municipal solid waste have been used as auxiliary fuel. However, the amount of municipal solid waste used cannot exceed 30% by weight (U.S. EPA, 1993).

As in autogenous operation, the feed rate to the MHF under nonautogenous operation depends on the moisture, solids, volatility, and heat content of the feed cake. As a general rule, the feed rate at startup should be less than the optimum steady-state feed rate, as determined by the manufacturer and operators. The initial rate of feed takes into account auxiliary burner capacity as described in the previous section regarding startup for autogenous incineration.

As the moisture content of the feed cake evaporates and ignition occurs, the feed rate may be increased. Once the feed rate reaches the optimum steady-state rate, the combustion reaction will continue if the total heat available from the auxiliary burners and the feed solids combustion is equal to or greater than the heat required for evaporation.

If the feed cake solids pass through the hearth equipped with ignited burners without reaching combustion, they likely will pass through the entire incinerator unburned. If ignition does not occur before cake solids drop down from the upper hearth with ignited burners, the operator should immediately start the burners on the next burner hearth to ensure ignition. Through careful monitoring, the operator can prevent passage of unburned cake solids.

The feed cake may dry out too quickly if auxiliary burners are allowed to stay on too long. This may lead to hot spot creation, leading to clinker formation. This can result from a small burnout in a limited area of the incinerator. However, if the auxiliary burners are shut down too soon the feed cake in the immediate area can burn itself out, resulting in too little heat to evaporate the moisture and ignite the fresh feed cake entering the incinerator.

During nonautogenous operation, incinerator operators should ask the dewatering equipment operator to increase the solids content of the feed cake, if possible. The goal is to use as little auxiliary fuel as possible to maintain steady-state operations, thus reducing operating costs.

2.5 Excess Air Reduction Techniques

The air introduced to the incinerator in excess of the theoretical amount of air (oxygen) required for complete combustion of the cake solids is excess air. The amount of excess air in the incinerator can be calculated using the oxygen reading in the incinerator after combustion has occurred. A simplified equation for the percentage of excess air is

$$[(\%\ O_2\ \text{measured})/(20.9\% - \%\ O_2\ \text{measured})] \times 100 \qquad (10.3)$$

The oxygen concentration in the incinerator exhaust gases before the scrubber or any dilution air is found to be 6%.

$$\begin{aligned} \%\ \text{excess air} &= [6/(20.9 - 6)] \times 100 \\ &= 40\%\ \text{excess air} \end{aligned}$$

The reduction of excess air substantially affects the operating economics and has a positive effect on air emissions from MHFs. By reducing the excess airflowing through the incinerator, less heat energy is required to heat the makeup air and thus, flue gas energy losses are reduced. Substantial auxiliary fuel savings occur when the level of excess air is properly controlled in both autogenous and nonautogenous operations. The observed incinerator excess air rate should be tracked and compared against the design excess air rate. To determine the actual energy saved by reducing the excess air in the incinerator, a mass balance must be calculated. In the mass-balance calculation, analytical data from incinerator operations relative to the site-specific feed cake are needed.

2.6 Combustion and Temperature Control

After the initial startup, the combustion process must be stabilized. Because the combustion process is a chemical reaction which increases with higher temperatures, the reaction rate must be controlled to allow the process to reach equilibrium. To achieve equilibrium, the operator must control the feed cake rate, the incinerator temperature, and the turbulence between the hot gases and the fuel. Numerous incinerator controls may regulate these variables, which are discussed more fully below.

2.6.1 Center Shaft Speed

Some of the controls, such as center shaft speed, affect more than one of the parameters required to achieve equilibrium. The center shaft rotation controls the speed at which the cake solids are conveyed through the incinerator.

An increase in the center shaft speed will turn over the cake solids more rapidly resulting in a shorter retention time in the incinerator, higher moisture evaporation rate within the cake, quicker ignition of the cake solids, and a higher temperature within the incinerator. Combustion then moves upward in the incinerator.

If the center shaft speed is decreased, the cake solids furls (i.e., rolls of cake solids between the center rabble arm teeth) become larger, thereby reducing the frequency and surface area of the solids exposed to the hot gases. This decreases the moisture evaporation rate, slowing combustion. Combustion then moves downward in the incinerator.

In general, varying the center shaft speed can only provide a short-term correction, and without changing another variable, the fire will reestablish itself in approximately the same location, frequently only moving half a hearth. Because changes in center shaft speed take time for the results to be observed, the operator should make changes in 1/10 rpm increments and then wait at least 10 to 15 minutes to see the result of the change. However, the operator should not wait too long to avoid losing control. Most operators indicate that the center shaft speed is used only to determine required retention time in the incinerator. A limited number of MHF operators never change the center shaft speed.

2.6.2 Combustion Air

The combustion air fans and ambient air drawn in by the induced draft fan provide the oxygen required for the combustion process. This air also provides the cooling necessary to keep the combustion reaction at equilibrium. Auxiliary burner combustion air is provided to the auxiliary burners. The combustion fans are typically shut down when the burners are shut down. To prevent destruction of the burner tip, however, some burner manufactures recommend that a limited amount of air continue to pass through the tip of the burner even if the burner is shut down.

If the incinerator experiences high temperature because of an uncontrolled burnout, the operator can start the combustion air fans to lower the incinerator temperature. Additionally, ambient air may be drawn into the incinerator by the induced draft fan. The operator can control the amount of air drawn into the incinerator by using the dampers located around the incinerator. These dampers provide the quickest response to control the combustion reaction and may be operated either manually or automatically.

The dampers located on the same hearth all allow equal volumes of air into the incinerator. Air introduced at the lower hearth dampers will be heated more than air introduced at the dampers located on the upper hearths. If oxygen is needed to increase the combustion process, air should be drawn from the dampers located on the lower hearths. However, if the temperature in the incinerator must be decreased rapidly, air should be drawn from the dampers located on the upper hearths. During stable operations, the percent of each damper open generally increases through the lower hearths of the incinerator.

2.6.3 Burnouts

Uncontrolled burnouts occur when temperatures within the incinerator rise rapidly and resist returning to normal by ordinary operational controls. The operator should record the cause of the high temperature in the log and any corrective action taken. An uncontrolled burnout can be brought under control by using one or more of the following techniques:

- Open air sources. This increases the air passing through the incinerator and provides convective cooling to the combustion process. However, adding too much air to the incinerator can cause overpressurization and result in the emergency bypass stack damper opening.
- Reduce the center shaft speed. This reduces the speed at which the rabble arms turn over the burning cake solids and expose new fuel to oxygen, thus reducing the rate of combustion.
- If the incinerator temperatures continue to increase, the center shaft may be stopped for short periods of time. Shutting off the center shaft for 30 seconds out of every one minute slows down the combustion by reducing the feed cake surface area exposed to the hot gases.
- Add water to the feed by decreasing the solids content in the dewatered cake.

However, this requires changes in the operation of the dewatering units.

The operator should note that all of these actions result in a larger volume of moist feed cake on the top hearth, additional heat for evaporation, and a lower temperature within the incinerator. Of particular concern, however, is the increased volume of wet feed cake on the top hearth. If too great a volume of feed cake builds up on the top hearth, the center shaft motor can overload and trip out.

2.6.4 Draft Control

The induced draft fan must be used during normal operation. Induced draft serves three functions:

- It provides a negative pressure within the incinerator preventing gases from escaping through incinerator openings,
- It draws oxygen through the air ports into the incinerator to support and control the combustion process, and
- It provides a vehicle for exhausting the combustion gases.

Draft is measured by a pressure sensor located within the incinerator. Chapter 9, Instrumentation and Control, contains additional details. The draft control damper maintains a specific negative pressure in response to the sensor setpoint.

As the incinerator dampers function, the pressure within the incinerator changes. Under normal conditions, the negative pressure set point should be set between −0.07 and 0 kPa (−0.30 and 0 in water). The negative pressure setpoint should be set as high as possible while maintaining a negative pressure throughout the incinerator. If the pressure setpoint is too close to zero the pressure within the incinerator can increase to positive pressure conditions when the inlet air dampers open. This causes the emergency bypass stack damper to open, creating natural draft. If the pressure set point is too low, excessive fly ash can be carried out of the incinerator with the exhaust gases.

If the induced draft fan fails, interlocks must immediately stop the feed cake from entering the incinerator and open the emergency bypass stack damper. The incinerator emergency bypass stack vents incinerator gases to the atmosphere during an emergency, allowing the material in the incinerator to properly burnout. In addition, it provides draft when the incinerator is in hot standby.

The emergency bypass stack must be adequate in size and cannot be used during normal operation as it allows the incinerator exhaust gases to bypass the air pollution control system. Opening of the emergency bypass stack damper should be tied into the shutdown of the feed cake system to the incinerator. Operators have had to open their emergency bypass stack on a regular basis to relieve pressure buildup in their incinerators. However, U.S. Environmental Protection Agency personnel have indicated that this method of operation is not acceptable.

2.7 Air Pollution Control Systems

As discussed in Chapter 7, Emission Control and Monitoring, particulate matter, metals, and other pollutants are removed from the incinerator's exhaust gases by the air pollution control system. The vast majority of MHFs are equipped with Venturi / impingement tray scrubbing systems.

The key to proper operation of the offgas scrubber is optimization of the differential pressure across the impingers. Typically, if the differential pressure is too low, particulate removal is minimized. If the differential pressure is too great, excessive negative pressure may occur in the incinerator, which results in an increase in gas velocity through the system and excessive fly ash being carried out in the exhaust gas. Although an increased velocity and resulting increase in differential pressure across the impingers are methods to enhance particulate removal, an increased exhaust gas solids load may cause a higher particulate discharge. Scrubber differential optimization must be done in a manner to maintain the desired incinerator pressure and stay within permit operating conditions. For additional details see Chapter 7, Emission Control and Monitoring.

2.8 Emergency Operations

2.8.1 *Power Failure*

During complete electrical power failure, the greatest immediate danger is heat damage to the center shaft and rabble arms. In addition, a temporary unsafe condition may result because the induced draft fan shuts down and the pressure within the incinerator becomes positive. The emergency bypass stack damper should automatically open to restore a negative pressure condition.

If the power failure lasts only a short time, the incineration equipment may be restarted and the combustion process continued. If the power failure lasts for an extended period of time, the incinerator may rapidly cool down, resulting in severe damage to the refractory.

To prevent rapid cool down, the emergency bypass stack damper should be modulated to maintain the natural draft at a level that minimizes offgas heat losses. The center shaft will stop rotating during a complete power failure. Without rabbling, the combustion process is reduced to smoldering cake. The smoldering cake produces combustible gases which must be purged by the natural draft. When the power is restored, the refractory should be purged to ensure no combustible gases remain. The induced draft fan can then be started and the emergency bypass stack damper closed.

Depending on how long the power outage lasts, the moisture content of the smoldering cake can be low. One quick turn of the center shaft could release a large volume of volatile gases from the cake and create a potentially explosive condition. Therefore, before starting the center shaft after a power outage, the operator should reduce its revolution per minute output control to zero. The center shaft should be started slowly and gradually increased to its normal revolutions per minute. Some agencies have found that starting and stopping of the center shaft helps the bed to start burning slowly and uniformly. After the cake contained in the incinerator has burned out, the incinerator can be restarted using the procedures described in section 2.2.3, Startup from Hot Standby.

2.8.2 High Incinerator Temperature

High temperature within the incinerator results when the combustion process is allowed to go out of equilibrium. To bring the process back into equilibrium, follow the procedures described in section 2.6, Combustion and Temperature Control. If these procedures fail to lower the incinerator temperature, wastewater solids feed cake should be stopped and the combustion process should be controlled to burn out the cake in the incinerator. During the burnout, the temperature within the incinerator may become high; therefore, it is important to provide as much cooling air as possible. After the combustion process stops, feed cake can be restarted as described previously.

2.8.3 High Offgas Temperature

High offgas temperature can cause damage to the air pollution control system. If high temperature is a persistent problem, the operator should increase the water flow to the precooler. If the high offgas temperature is a result of high incinerator operating temperature, the following remedies are suggested:

- Increase the water flow to the precooler,
- Stop the burners,
- Decrease the feed to the incinerator,
- Go to emergency bypass stack conditions, and
- Follow the procedures in section 2.6, Combustion and Temperature Control.

2.8.4 Center Shaft Stoppage

If the center shaft stops and the remainder of the incinerator equipment continues to operate, the operator should stop the feed. This prevents a large volume of cake from building up on the top hearth. The operator should follow the procedure outlined in section 2.8.1, Power Failure. However, the induced draft fan will continue to operate for pressure control.

2.9 Multiple-Hearth Furnace Shutdown

The shutdown process for MHFs is basically the process of stopping feed to the incinerator. However, to avoid an immediate burnout the incinerator should remain in operating mode until all of the cake in the incinerator is combusted. The burners and rabble arms should remain in service with the air pollution control system until combustion of the cake within the incinerator is complete. Depending on the characteristics and quantity of cake within the incinerator, the incinerator can typically be removed from service between one and two hours after the feed is stopped. Upon shutdown of the incinerator the emergency bypass stack should be opened.

2.10 Typical Operator Duties

To ensure trouble-free operation of the MHF, the operator has a variety of duties that must be performed. Following is a list of typical duties:

- Grease the center shaft bearings.
- Maintain the level of sand or steel shot in the upper and lower sand seals.
- Check the operation of the emergency bypass stack damper.
- Check the condition of the scraper tooth on the outer hearth rabble arm.
- Check the air filters for the combustion and center shaft cooling fans.
- Check and calibrate the total hydrocarbon and carbon monoxide monitors.
- Clean and calibrate the oxygen analyzer.
- Check for clinker formation and plugging of the drop holes.
- Check precooler and scrubber sprays and flow.
- Check operation of fans and drives.
- Coordinate with the solids dewatering system operators.

2.10.1 *Data Collection*

To maintain a stable combustion process, the operator must collect and keep data on items such as the following:

- Percentages of feed cake solids and volatile solids.
- Feed cake rate.
- Temperatures throughout the incinerator.
- Center shaft speed.
- Percent oxygen.
- Draft set points.
- Auxiliary fuel use.
- Pressure drops.

Table 10.2 contains a typical spreadsheet that can be used for collecting data from an MHF. In addition to standard condition data, operators should record any incinerator malfunction, including emergency bypass stack damper opening, its cause, and any corrective action taken.

2.10.2 *Other Responsibilities*

In addition to data collection, operators have other responsibilities to ensure smooth operation of an MHF. Below is a list of some of these periodic responsibilities.

Routinely,

- Monitor all indicators on the control board, including temperature, pressure, flow, and oxygen levels.
- Respond to alarm conditions by making necessary adjustments.
- Make entries in logbook when important process changes are made, when alarms occur, or when shutdowns occur.

Every other hour, record process readings on the log sheet.

Every shift,

- At the beginning of each shift, review highlights of the previous shift, including any instructions from the supervisor, with the outgoing operator.
- Review the logbook for any instructions from the supervisor.

TABLE 10.2 Typical data collection spreadsheet for MHFs (deg F = °F and °F × 0.5556 (°F × 32) = °C; In W.C. = in H_2O and in H_2O × 0.2488 = kPa; WT T/hr = wet tons per hour and ton × 0.9072 = Mg; ID = induced draft; SL Comb Damper = sludge combustion air damper; CF = cu ft and cu ft × 0.02832 = m^3; gpm × 0.06308 = L/s; psi × 6.895 = kPa; Min 21 = this incinerator is subject to a minimum pressure differential across its Venturi/impingement tray scrubber of 21 in water column on a daily average basis [in × 25.40 = mm] [this note it to remind operators of the pressure differential limit]; A – damper in automated mode; and M = damper in manual mode).

Incinerator #______ Westerly Wastewater Treatment Center Date__________

Time	Venturi Inlet Temp Deg F	Venturi Out Temp Deg F	(Min 21) Venturi Dif Pres In. W.C.	Hearth #1 In. W.C.	Scrubber Out Temp Deg F	Scrubber Out Pres In. W.C.	ID Inlet Damper % open	ID Fan Speed RPM	Oxygen %	SL Comb Damper % Open	Center Shaft Ext Temp Deg F	Center Shaft Speed RPM	Feed Rate WT T/Hr	Natural Gas CF
0000			A M				A M	A M		A M				
0100			A M				A M	A M		A M				
0200			A M				A M	A M		A M				
0300			A M				A M	A M		A M				
0400			A M				A M	A M		A M				
0500			A M				A M	A M		A M				
0600			A M				A M	A M		A M				
0700			A M				A M	A M		A M				
0800			A M				A M	A M		A M				
0900			A M				A M	A M		A M				
1000			A M				A M	A M		A M				
1100			A M				A M	A M		A M				
1200			A M				A M	A M		A M				
1300			A M				A M	A M		A M				
1400			A M				A M	A M		A M				
1500			A M				A M	A M		A M				
1600			A M				A M	A M		A M				
1700			A M				A M	A M		A M				
1800			A M				A M	A M		A M				
1900			A M				A M	A M		A M				
2000			A M				A M	A M		A M				
2100			A M				A M	A M		A M				
2200			A M				A M	A M		A M				
2300			A M				A M	A M		A M				

Operator 11PM-7AM__________ 7AM-3PM__________ 3PM-11PM__________

(continued on next page)

TABLE 10.2 *(Continued)*

Incinerator #________ Westerly Wastewater Treatment Center Date____________

Time	Hearth 1 Temp	Hearth 2 Temp	Hearth 3 Temp	Hearth 4 Temp	Hearth 5 Temp	Hearth 6 Temp	Hearth 7 Temp	Hearth 8 Temp	Hearth 9 Temp	Hearth 10 Temp	Hearth 11 Temp	Hearths in auto	Burners in Manual On	Off
0000														
0100														
0200														
0300														
0400														
0500														
0600														
0700														
0800														
0900														
1000														
1100														
1200														
1300														
1400														
1500														
1600														
1700														
1800														
1900														
2000														
2100														
2200														
2300														

Operator 11PM-7AM______________ 7AM-3PM______________ 3PM-11PM______________

- Tour the operating equipment. Observe, listen to, and touch each piece of equipment. The sound and feel of the equipment is a ready indicator of possible trouble such as impending bearing failure, defective drive belts, or an induced draft fan out of balance because of ash buildup on the wheel.
- Record all appropriate information, including instructions from the supervisor, in the logbook.
- At the end of each shift, review highlights of the previous shift, including any instructions from the supervisor, with the incoming operator.

Table 10.2 *(Continued)*

Incinerator #_______ Westerly Wastewater Treatment Center Date____________

Time	Main Header	Main AFTR REG	Tray Scrubber		Venturi		Damper Shaft Seal Water		Expansion Seal		Quencher Spray		Total Flow	Ash System
	PSI	PSI	GPM	PSI	GPM	PSI	GPM	PSI	GPM	PSI	GPM	PSI	GPM	
0000														
0100														
0200														
0300														
0400														
0500														
0600														
0700														
0800														
0900														
1000														
1100														
1200														
1300														
1400														
1500														
1600														
1700														
1800														
1900														
2000														
2100														
2200														
2300														

Operator 11PM-7AM____________ 7AM-3PM____________ 3PM-11PM____________

Remarks ____

3.0 FLUID BED INCINERATOR OPERATIONS AND PROCESS CONTROL

Fluid bed incinerators may be either a hot wind box or a cold wind box design. Figure 10.1 shows a composite cross-section of a typical fluid bed incinerator. The left half of the cross-section shows the construction features of the hot wind box design. The right side of the cross-section shows the features of the cold wind box design. In the hot wind box design, the wind box is refractory lined and the grid support (arch) is typically constructed of refractory. The startup preheat burner is located

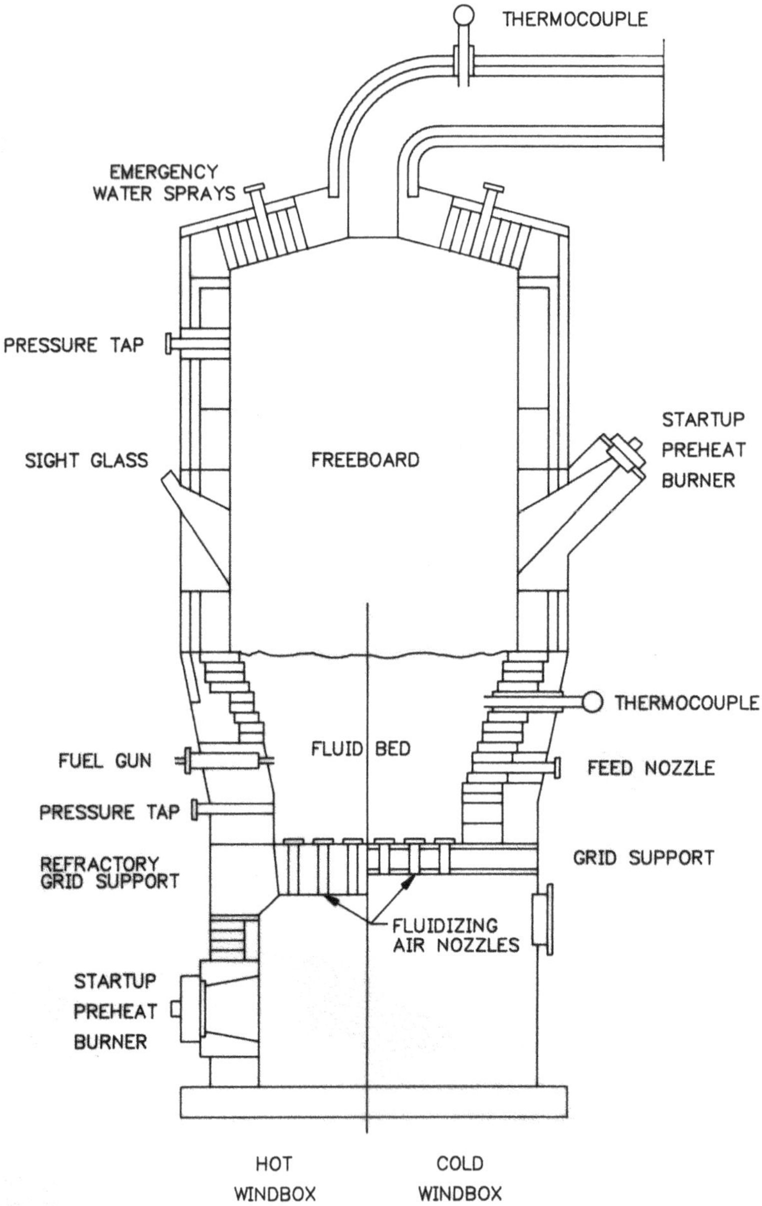

FIGURE 10.1 A typical fluid bed incinerator cross-section.

in the wind box. However, a stainless steel arch, with an abovebed preheat burner, has been successfully used in St. Paul, Minnesota, and Lynn, Massachusetts.

In the cold wind box design, the wind box is uninsulated and the arch is fabricated from steel and stainless steel. The startup preheat burner is located in the freeboard, as shown, or in the bed. Other components such as thermocouples, pressure taps, fuel guns, feed nozzles, emergency water sprays, and sight glasses are the same for both designs.

3.1 Fluid Bed Incinerator Startup

3.1.1 Cold Startup

During cold startup, the incinerator is heated from room temperature to operating temperature before cake feeding and stabilizing operations begin. When an incinerator is started for the first time, or when new refractory has been installed, the refractory must be cured or dried out. To do so, the operator should

- Raise the bed temperature at a rate of 10°C (50°F) per hour to 121°C (250°F).
- Hold the temperature at 121°C (250°F) for six hours.
- Raise the bed temperature at the same rate to 260°C (500°F).
- Hold the temperature at 260°C (500°F) for 12 hours.
- Raise the bed temperature at the same rate to 704°C (1300°F).
- Hold the temperature at 704°C (1300°F) for 12 hours.
- Raise the bed temperature at the same rate to 871°C (1600°F).
- Hold the temperature at 871°C (1600°F) until no steam is observed at the roof vents.

Before starting the refractory curing the operator must remove the pipe caps from the roof vents on the incinerator. These are replaced after the refractory is cured. The fluidized bed incinerator does not need to be cooled.

Before the incinerator is started up, the operator must check utilities and equipment and prepare the system. The operator should check that

- Electric power and emergency power, if provided, are available at the main circuit breaker.
- Fuels such as natural gas or fuel oil are available.

- The water supply for freeboard sprays is turned on.
- Air for fluidizing, purging, and preheating the burner is available.
- All necessary electrical breaker switches are closed.
- All manholes, nozzles, and sight glasses on the incinerator are closed.
- All equipment has received required preventative maintenance.
- All valves are in the required positions for startup.
- Blowers and fans are started with closed inlet valves and dampers to prevent driver overload.
- All controllers are set in manual position.
- The alarm system is functioning.

In addition, during the system preparation process, the operator should

- Test run all blowers.
- Add sand to the incinerator bed as required.
- Set valves on the preheat burner system in the required position.
- Ensure that the feed cake system is ready for operation.

The incinerator system is now ready for startup. To start up the fluid bed incinerator, the operator will

- Open the scrubbing system water valves.
- Start the purge air blowers.
- Start the fluidizing air blowers.
- Start the preheat burners.
- Begin gradual heating of the incinerator.
  - The pilot flame or the main flame at low fire will be used to accomplish the initial stages of heat up. The latter stages require increasing the preheat burner firing rate.
  - Heat the incinerator at a rate of 38°C (100°F) per hour to 104°C (220°F).
  - Hold at 104°C (220°F) for two hours.
- Operate the fluidizing air blower for 30 seconds every 30 minutes to bump the sand bed. This evenly heats the sand.

- Begin injecting fuel oil to the bed, if applicable, when the bed temperature reaches 621°C (1150°F).
  - The fluidizing air blowers and induced draft fan, if installed, must be operating before fuel oil can be injected to the bed.
  - As the operator increases the rate of fuel oil firing, the operator should decrease the rate of preheat burner firing.
- Start the feed cake when the bed temperature is higher than 649°C (1200°F); air pollution regulations may require a higher temperature before the operator injects the feed cake.
- Gradually increase the feed cake rate to the operating rate.
- Adjust the airflow rate to maintain sufficient excess air, typically 4 to 10% O_2.
- Adjust the fuel firing rate to maintain a constant bed temperature.
  - If the feed cake is autogenous the operator should decrease the fuel firing rate.
  - If the feed cake is nonautogenous, the operator should increase the fuel firing rate.
- Set all controllers to the automatic position when the operating feed cake rate has been established and the incinerator operation has stabilized.

Once these steps are completed, the incinerator should be operating normally.

3.1.2 Standby

A fluid bed incinerator may be placed in hot or cold standby modes. During hot standby, the incinerator is kept hot and is ready to accept feed cake. During cold standby, the incinerator is kept at room temperature and must go through a cold startup procedure before accepting feed cake.

Because the sand bed has a large heat capacity and thus retains heat well, the fluidized bed incinerator is well-suited for hot standby. Typically, the heat loss from the incinerator is approximately 5 to 10°C (10 to 20°F) per hour. Fluid bed incinerators may be placed in a hot standby mode if the incinerator is shutdown with no warning and without interrupting the dewatering process. Alternatively, incinerator operations may be intermittent, such as in a facility where there are only two shifts per day, five days per week.

To place a fluid bed incinerator in hot standby, the operator will

- Stop the cake feed to the incinerator.
- Heat the incinerator to 871°C (1600°F) using auxiliary fuel and a reduced fluidizing air rate.

- Stop the auxiliary fuel feed and blowers when the temperature has reached 871°C (1600°F).
- Blow out the fuel injection guns, if equipped
- Remove the fuel injection guns, if equipped, from the bed
- Check the block and bleed valves, if the fuel oil system is so equipped.
- Remove freeboard water spray nozzles.

If the incinerator is shutdown overnight, it can be restarted using the warm startup procedure discussed in section 3.1.3. If the incinerator is on hot standby for a longer period, and the bed temperature drops to 649°C (1200°F), the operator should reheat the incinerator to 871°C (1600°F) using the warm startup procedure.

3.1.3 Warm Startup

Before a warm startup of a fluid bed incinerator, the operator should check utilities and equipment and prepare the system using the procedures described in section 3.1.1, Cold Startup.

If the incinerator bed temperature is lower than 649°C (1200°F) preheat burners must be used. If the incinerator bed temperature is higher than 649°C (1200°F), the operator can immediately begin feeding cake and bed fuel into the incinerator. If the bed temperature is lower than 649°C (1200°F), the operator should perform the warm startup procedures as follows:

- Open the scrubbing system water valves.
- Start the purge air blowers.
- Start the preheat burners.
- Heat the bed at a rate no more than 38°C (100°F) per hour. Operate the fluidizing air blower for 30 seconds every 30 minutes to bump the sand bed.
- Begin injecting fuel oil to the bed, if applicable, when the bed temperature reaches 621°C (1150°F); the fluidizing air blowers and induced draft fan, if installed, must be operating before fuel oil can be injected to the bed.
- Start the feed cake when the bed temperature is higher than 649°C (1200°F); air pollution regulations may require a higher temperature before the operator injects the feed cake.
- Gradually increase the feed cake rate to the operating rate.

- Adjust the airflow rate to maintain sufficient excess air, typically 4 to 10% O_2.
- Adjust the fuel firing rate to maintain a constant bed temperature.
  - If the feed cake is autogenous the operator should decrease the fuel firing rate.
  - If the feed cake is nonautogenous the operator should increase the fuel firing rate.
- Set all controllers to the automatic position when the operating feed cake rate has been established and the incinerator operation has stabilized.

3.2 Autogenous versus Nonautogenous Operations

Autogenous operations in a fluid bed incinerator may be assisted by recovering heat from the incinerated offgases and heating the fluidizing air. A hot wind box design accomplishes this heat recovery by passing the fluidizing air and offgases through a gas-to-air heat exchanger. Thus, offgases are cooled and the fluidizing air is heated.

In general, autogenous incineration can be achieved without preheating the fluidizing air if the feed cake contains 70% volatile solids with a heating value of 23 260 kJ/kg (10 000 Btu/lb) and 35% or more dry solids by weight. This varies with both the solids' volatile content and heating value. However, if similar feed cake contains at least 26% dry solids by weight, autogenous incineration can generally be achieved if the fluidizing air is preheated using the hot wind box design. If the feed cake contains lower than 26% solids, fuel addition is required and autogenous incineration is not possible.

Because fuel addition can significantly increase the operating costs of the incineration system, operators should minimize fuel addition while still maintaining good combustion control. Methods that operators can use to minimize fuel addition include

- Optimizing the dewatering operation to get as dry a cake as possible,
- Operating at the design excess air rate while maintaining the minimum required fluidizing air rate for feed cake rates below design, and
- Using the maximum preheat of fluidizing air for hot wind box designs.

Fluid bed incinerators are designed to provide the best thermal efficiency at design loading. Therefore, autogenous operations, nonautogenous operations, and fuel addition are affected by feed cake loading to the incinerator. Heat inputs are generated from feed cake, fluidizing air, and fuel, if required. Matching heat losses are generated from offgases, ash, and radiation and convection from the incinerator shell and ducting. Heat loss because of radiation and convection is nearly constant regardless of cake feed load rates. If feed cake loading is reduced, the corresponding heat input is decreased although the total heat losses remain the same. In this case, incinerator heat loss is greater than the heat gained through incineration, decreasing thermal efficiency.

To offset this decrease in thermal efficiency, additional heat can be obtained from additional preheating of fluidizing air or from adding auxiliary fuel. The operator can reduce the fluidizing air rate to maintain the same level of excess air, but a minimum quantity of fluidizing air is required to keep the bed fluidized. If the operator instead introduces excess fluidizing air into the incinerator, additional auxiliary fuel will be required.

3.3 Combustion Control

Good combustion is achieved when all of the combustible elements in the feed cake react with the oxygen in the fluidizing air to form carbon dioxide and water vapor. Combustion occurs only if there is sufficient heat, oxygen (air), turbulence, and retention time (i.e., sufficient time to complete the combustion reaction).

In a fluid bed incinerator, the bed temperature must remain higher than the autoignition temperature of the cake solids, approximately 649°C (1200°F). The freeboard temperature must be higher than 760°C (1400°F) to destroy hydrocarbons and burn out carbon monoxide. Oxygen levels in the exhaust gases should be between 4 and 10% and the gases should remain in the freeboard for at least three to five seconds. Turbulence is provided by the fluidizing motion of the sand bed.

Turbulence in the bed and retention time in the freeboard both vary with fluidizing airflow and temperature. An increase in fluidizing airflow at constant temperature increases bed turbulence but decreases freeboard retention time. Conversely, a decrease in fluidizing airflow reduces bed turbulence but increases freeboard retention time. An increase in temperature at constant airflow has little effect on bed turbulence but decreases freeboard retention time.

The bed temperature can vary according to the cake feed rate, moisture content, combustible content, and heat content. Minor variations in these variables are masked by the fluidized sand bed, which provides a heat sink. Large variations in any of these variables will manifest as changes in bed temperature.

Variations in oxygen content in the exhaust gases also result from changes in the cake feed rate, moisture content, combustible content, and heat content. Minor variations in these variables do not typically cause the oxygen content to vary so significantly that adjustments are required. If the incinerator has an oxygen trim package installed, the controls will automatically adjust the fluidizing airflow rate. An override prevents the fluidizing air rate from being reduced below the minimum allowable rate. However, this type of automated adjustment is rare because the oxygen instrumentation may require up to five minutes to sense changes. Therefore, manual adjustments are typically sufficient.

To maintain the incinerator at the highest thermal efficiency point, the feed rate should be steady but at the maximum incineration rate to enhance burning. Cake feed rate depends on the dewatering operation; a smooth dewatering operation provides a relatively constant cake feed rate. A decrease in feed cake load only slightly lowers the electrical energy consumption for the fluidizing air blowers. Therefore, overall incinerator efficiency is also at its highest point when the feed cake is fed at the maximum incineration rate.

3.3.1 Temperature Control

The temperature control system for a fluid bed incinerator ensures safe, steady incinerator operation. The temperature control system consists of bed temperature supervision and freeboard temperature control. The bed temperature supervision system controls the auxiliary fuel feed rate, prevents fuel and cake from being fed to the incinerator if the bed temperature is too low or too high, and generates alarms for operator notification.

The auxiliary fuel feed rate is controlled by a temperature controller that receives signals from one or more bed temperature thermocouples. If the bed temperature increases, the controller reduces the auxiliary fuel rate by modulating a fuel valve or varying the speed or stroke of a fuel pump. If the bed temperature decreases, the controller increases the auxiliary fuel rate in a similar manner. If the feed cake is autogenous, no auxiliary fuel is required.

Temperature interlocks allow auxiliary fuel to be fed to the bed only if the temperature is higher than 621°C (1150°F) for fuel oil or higher than 760°C (1400°F) for natural gas. The interlocks also allow cake solids to be fed if the bed temperature is higher than 704°C (1300°F), although this temperature varies depending on the type and consistency of cake solids. Finally, the interlocks shutdown both auxiliary fuel and cake feeds if the bed temperature rises higher than the range of 871 to 982°C

(1600 to 1800°F). This temperature range can vary depending on the ash fusion temperature of the feed cake solids and design of the incinerator and refractory specifications. The bed temperature system generates low and high abnormal bed temperature alarms when the temperature drops below the cake solids enable temperature or rises above the cake solids and auxiliary fuel shutdown temperature.

The freeboard temperature system controls the temperature of the incinerator offgases to maintain a safe temperature. This temperature depends on the ash fusion temperature. If the temperature of offgases rises to the ash fusion temperature, ash builds up on the exhaust duct and eventually plugs it, thereby preventing gases from leaving the incinerator and creating a safety hazard.

The freeboard temperature control system typically consists of multiple water sprays mounted in the incinerator roof. These sprays come on in sequence as the freeboard temperature rises. The water supply to these sprays is critical; if water pressure varies, a booster pump is required. If freeboard temperature rises beyond acceptable levels, cake and auxiliary feeds will be stopped.

A secondary freeboard temperature control system that has been successfully used in some facilities consists of a separate blower that injects air to the freeboard through multiple nozzles. The air valve can be modulated automatically or manually. The freeboard air injection system maintains the freeboard temperature below the temperature that initiates the water sprays. The freeboard air injection system results in a more uniform temperature distribution to the incinerator. Some of the air that would typically be injected as fluidizing air is injected as freeboard air. This compensates for the usual temperature difference between the bed and freeboard.

Although many of the fluid bed incinerator components are automated, control of the incinerator temperature is not entirely automatic. The operator must monitor and adjust the feed cake rate and the fluidizing airflow. The feed cake rate varies according to the production of cake by the dewatering operation and the auxiliary fuel system automatically compensates for small changes in feed cake rate. However, the feed cake equipment may require adjustments in speed to compensate for changes in the cake production rate. Adjustments to the fluidizing airflow may be required if the oxygen in the offgases moves outside of the acceptable operating range. If the incinerator is a hot wind box type, the operator may also be required to periodically adjust the quantity of offgases passing through the gas-to-air heat exchanger. The bed and freeboard temperature are recorded on the incinerator temperature recorder to assist the operator in monitoring and controlling the incinerator temperatures.

3.3.2 *Draft Control*

Two types of draft systems are used for fluid bed incinerators: pressurized and balanced. In a pressurized system, the fluidizing air blowers provide the motive force for the fluidizing air and the incinerator exhaust gases. No draft control is required for this system, although a minimum quantity of fluidizing air must be injected to the bed to keep it fluidized. If the fluidizing airflow rate falls below the required level, cake and auxiliary fuel feeds are stopped.

In the balanced draft, or push-pull system, an induced draft fan is used by the freeboard controller to maintain a slightly negative pressure in the freeboard. The freeboard pressure signal is transmitted to the freeboard pressure controller which either modulates the induced draft fan inlet dampers or varies the fan speed. The fan must be running before cake and auxiliary fuel can be fed to the bed. The cakes and auxiliary fuel feeds are stopped if the fan trips or the freeboard pressure is outside the acceptable operating range. The acceptable freeboard pressure operating range is typically between −2.5 to 2.5 kPa (−10 to 10 in water).

If the incinerator is equipped with a Venturi scrubber, the pressure drop must remain constant to provide optimum particulate control. Pressure drops may be changed by adjusting the throat opening or the water flow rate, although adjustments to the water flow rate will result in only minor alterations to the pressure drop. Throat adjustment can be accomplished either automatically or manually. If the throat adjustment is performed automatically, the actuator on the throat damper will modulate to maintain a constant pressure drop. If performed manually, the operator must make periodic adjustments to the throat damper opening.

3.4 Emergency Operations

3.4.1 *Power Failure*

A power failure to a fluid bed incinerator will shut down the incinerator system, including all blowers and controls, but cake solids will continue to burn as long as oxygen is available. Although cake solids will continue to burn for only a short time, the cake's volatile solids may continue to boil off for longer. When power is restored, the operator must purge the incinerator of all combustible gases before attempting to light a burner. A cold or warm startup procedure, depending on the incinerator temperature, must be followed. The operator must perform a complete system check prior to restarting the fluid bed incinerator.

3.4.2 High Incinerator Temperature

If the incinerator temperature rises above the set temperature, the bed temperature supervision system stops the cake and auxiliary fuel feeds. Too much fuel (cake solids and auxiliary fuel) and too little air are the primary causes of high incinerator temperatures. The operator should monitor and adjust the cake feed rate, fluidizing airflow rate, and auxiliary fuel rate before the incinerator bed temperature rises to the set temperature.

If bed temperatures begin to rise the operator should increase the fluidizing air rate. If the air rate is at a maximum, the operator should reduce the fuel rate. The auxiliary fuel feed is automatically controlled, but if the auxiliary fuel feed has not responded to the temperature controller the operator should set the controller on manual and adjust it manually. Adjustments to both the fluidizing airflow and auxiliary fuel rates typically take up to five minutes before a change is indicated on the temperature recorder. The operator must wait before making additional adjustments.

Adjustment of the cake feed rate is more difficult. The dewatering equipment that produces the cake must be adjusted. However, the effects of dewatering equipment adjustment may take as long as 30 minutes to be visible. If the bed temperature cannot be controlled by adjusting the fluidizing airflow or auxiliary fuel rates the cake feed should be stopped. If the bed temperature supervision system (electrical interlocks) does not stop the cake and auxiliary fuel feeds the operator must stop these feeds immediately. This can be quickly accomplished by stopping the fluidizing air blowers.

3.4.3 High Offgas Temperature

If the incinerator offgas or exhaust temperature rises above the set temperature, the reactor freeboard temperature control system initiates water sprays and eventually stops the cake and auxiliary fuel feed as the temperature rises. If the temperature control system does not respond to increases in incinerator offgas temperature the operator must immediately stop the cake and auxiliary fuel feeds.

3.4.4 Operating Problems

Table 10.3 provides a list of typical operating problems and possible solutions.

3.5 Typical Operator Duties

During normal fluid bed incinerator operations the operator must perform numerous routine duties. These duties can be divided into categories based on frequency of action. An important part of an operator's duty is to record process information to

TABLE 10.3 Typical fluid bed incinerator operating problems.

Problem	Cause	Solution
1. Preheat burner pilot will not ignite	1a No fuel to pilot	1a Open all manual valves. Check operation of pilot solenoid.
	1b No spark	1b Remove spark plug; check, clean, or replace. Check by closing valves, pressing ignite button, and observing whether there is a spark. Check ignition transformer and power supply. Reconnect loose threads or replace transformer.
	1c Not enough air	1c Check airflow.
	1d Defective air/fuel regulator	1d Repair regulator.
	1e Pilot lights but shuts down after short time	1e Remove and clean flame scanner; replace if defective. Check flame safeguard relay or controller; and replace if defective.
2. Preheat burner main flame will not ignite	2a No fuel to main flame	2a Open all manual valves. Check operation of fuel solenoids. Check fuel pressure.
	2b Pilot flame not stable	2b Check pilot system.
	2c Not enough air	2c Check airflow. Check that air valve is in low position.
	2d Main flame lights but shuts down after a short time	2d Remove and clean flame scanner; replace if defective. Check flame safeguard relay or controller; replace if defective.
	2e Main flame lights but is unstable	2e Tune burner using manufacturer's instructions.
3. Bed high temperature	3a Auxiliary fuel rate too high	3a Reduce auxiliary fuel flow rate.
	3b Cake feed rate too high	3b Reduce cake feed rate.
	3c Cake heating value increased or moisture content decreased	3c Increase fluidizing airflow rate. Decrease cake feed rate.
	3d Bed temperature thermocouple defective	3d Bed temperature indicator and recorder read at top of scale. Replace thermocouple and well.

(continued on next page)

TABLE 10.3 Typical fluid bed incinerator operating problems *(continued)*.

Problem	Cause	Solution
4. Bed low temperature	4a Cake heating value decreased	4a Increase cake feed rate. Lower fluidizing airflow rate.
	4b Overfeeding of cake or increase in moisture content	4b Decrease cake feed rate.
	4c Fluidizing airflow rate too low and feed stopped	4c Increase airflow rate and restart cake feed system.
	4d Freeboard spray system defective and water pouring into bed	4d Check that solenoid valves are shut. Check whether spray nozzles are worn; and replace.
5. Freeboard high temperature	5a Preheat burner firing at too high a rate (if located in freeboard)	5a Lower firing rate. Increase fluidizing airflow rate.
	5b Freeboard spray system defective	5b Check whether water supply is adequate and booster pump is operating; open all manual valves. Check whether solenoid valves are actuated. Check failed or plugged spray nozzle by removing, connecting to external water; replace if defective.
	5c Insufficient excess air	5c Decrease cake feed rate or increase airflow rate.
6. Abnormal high bed pressure drop	6a Fluidizing air nozzles plugged	6a Shut down incinerator and cool; remove sand and check bed nozzles.
	6b Pressure taps clogged	6b Remove pressure taps and clean.
	6c Too much sand	6c Open bed solids overflow pipe and drain sand. Be extremely careful because sand will be hot.
	6d Clinkers formed in bed	6d Shut down incinerator and cool; remove sand and clinkers.
7. Abnormal low bed pressure drop	7a Pressure taps clogged	7a Remove pressure taps and clean.
	7b Too little sand	7b Add sand to bed.
	7c Too low fluidizing airflow rate	7c Increase fluidizing airflow rate
8. Abnormal high freeboard pressure	8 Exhaust gas duct plugged	8 Shut down incinerator and cool; check ducting. Clean.
9. Abnormal low oxygen in gases	9 Heat content of cake increased	9 Increase airflow rate or decrease cake feed rate.

assist operations as well as to provide an operating history. Log sheets and a logbook should be kept for records. These, together with charts from recorders, should be stored for future use. Table 10.4 provides a typical log sheet for fluid bed incinerator operators. Below is outlined a list of typical operating duties.

Routinely,

- Monitor all indicators on the control board, including temperature, pressure, flow, and oxygen levels.
- Respond to alarm conditions by making necessary adjustments.
- Make entries in logbook when important process changes are made, when alarms occur, or when shutdowns occur.

Every other hour, record process readings on the log sheet.

Every shift,

- At the beginning of each shift, review highlights of the previous shift, including any instructions from the supervisor, with the outgoing operator.
- Review the logbook for any instructions from the supervisor.
- Tour the operating equipment. Observe, listen to, and touch each piece of equipment. The sound and feel of the equipment is a ready indicator of possible trouble such as impending bearing failure, defective drive belts or an induced draft fan out of balance because of ash buildup on the wheel.
- Record all appropriate information, including instructions from the supervisor, in the logbook.
- At the end of each shift, review highlights of the previous shift, including any instructions from the supervisor, with the incoming operator.

Twice per week, monitor sand levels in the sand bed. Add or remove sand to the sand bed as needed.

Every year,

- Shut down and cool the incinerator for annual inspection and maintenance.
- When the incinerator is cool, remove the auxiliary fuel injection nozzles, pressure taps, and freeboard spray nozzles. In addition, close the fuel valves.
- Open the inspection doors to inspect the incinerator and all associated equipment.
- Consider conducting an annual thermographic inspection of the exterior of the incinerator and associated ductwork.

Table 10.4 A sample logsheet for fluidized bed incinerator operation (lb × 0.4536 = kg; psi × 6.895 = kPa; cfm × 0.4719 = L/s).

Incinerator No: ______________ **Date:** ______________

	FEED LBS	BED TEMP	WNDBOX PSI	BED DIFF.	FREE-BOARD PSI	SCRUBBER DIFF. PSI	FLUIDIZED BED AIR CFM	OVERBED AIR CFM	FLUIDIZED AIR-AMPS	OXYGEN %	REACTOR OUT TEMP	SCRUBBER INLET TEMP	STACK TEMP.	PURGE AIR PSI	QUENCH WATER
1:00															
2:00															
3:00															
4:00															
5:00															
6:00															
7:00															
8:00															
9:00															
10:00															
11:00															
12:00															
13:00															
14:00															
15:00															
16:00															
17:00															
18:00															
19:00															
20:00															
21:00															
22:00															
23:00															
0:00															

4.0 REFERENCE

U.S. Environmental Protection Agency (1993) *Standards for the Use or Disposal of Sewage Sludge;* Code of Federal Regulations, 40 CFR, Part 503; U.S. Environmental Protection Agency: Washington, D.C.

5.0 SUGGESTED READINGS

Lue-Hing, C.; Zenz, D. R.; Tata, P.; Kuchenrither, R.; Malina, J. F., Jr.; Sawyer, B., Eds. (1998) *Municipal Sewage Sludge Management: A Reference Test on Processing, Utilization, and Disposal, 2nd ed.;* Volume 4; Technomic Publishing Company, Inc.: Lancaster, Pennsylvania.

Metcalf and Eddy, Inc. (1985) *Multiple Hearth and Fluid Bed Sludge Incinerations: Design and Operational Considerations;* EPA-430/9-85-002; U.S. Environmental Protection Agency: Washington, D.C.

U.S. Environmental Protection Agency (1979) *Process Design Manual for Sludge Treatment and Disposal;* EPA-625/1-79-011; U.S. Environmental Protection Agency: Washington, D.C.

U.S. Environmental Protection Agency (1985) *Municipal Wastewater Sludge Combustion Technology;* EPA-625/4-85-015; U.S. Environmental Protection Agency: Cincinnati, Ohio.

U.S. Environmental Protection Agency (1994) *A Plain English Guide to the EPA Part 503 Biosolids Rule;* EPA-832/R-93-003; U.S. Environmental Protection Agency: Washington, D.C.

Verdouw, A. J.; Waltz, E. W. (1984) *Sewage Sludge Incinerator Fuel Reduction, Hartford, Connecticut;* EPA-600/S2-84-146; U.S. Environmental Protection Agency: Cincinnati, Ohio.

Chapter 11

Incinerator Maintenance

(continued)

Proper maintenance of incinerators and their ancillary equipment is an essential way to reduce operating costs and downtime, extend equipment life, protect employees, and ensure compliance with all applicable rules and regulations. This chapter addresses maintenance issues that incinerator operators and managers need to monitor.

1.0 MULTIPLE-HEARTH FURNACES

1.1 Slag Formation

One of the significant maintenance issues for incinerator operators is formation of slag. Slag is the fusion of inorganic (noncombustible) material within the feed material being burned. It is primarily the result of excessive operating temperatures, flame impingement, and hot spots within the incinerator.

1.1.1 Excessive Operating Temperatures

The actual "excessive operating temperature" at which slag will form is site specific because the inorganic fusion temperature depends on the chemistry and composition of the feed cake being burned.

Sodium, potassium, and phosphorus in the presence of sulfur, chlorides, or silica form compounds within the feed cake that have low (fusion) or melting temperatures. A switch in dewatering chemicals from ferric chloride or lime to polymers can also result in lower fusion temperatures.

In addition, skimmings such as scum, grease, and floatable materials removed from wastewater—which, in many cases, are mixed with the feed cake before incineration or are pumped separately to the incinerator—can cause excessive combustion zone temperatures that also result in slag development.

1.1.2 *Flame Impingement*

Flame temperature of auxiliary fuel burners used to sustain the combustion process also can cause slag. The style of the burner being used can increase or decrease the possibility of slag formation depending on its operating flame temperature. The actual flame temperatures within a burner are typically greater than the fusion temperature of the inorganic material in the feed cake being burned. When a burner flame impinges on any surface there is a high probability of slag buildup. This buildup can occur on the sides of a burner tile, on the ceiling immediately above a burner, and on rabble arms.

1.1.3 *Hot Spots*

Hot spots can occur in the MHF's drop holes. For example, as hot gases are pulled through a drop hole by the induced draft fan, their velocity increases, thereby increasing the temperature. This "blow torch" effect is the primary reason why slag tends to build up and plug drop holes.

1.2 Potential Methods to Minimize Slag Formation

There are several methods to minimize slag development. It is always advisable to operate the incinerator with a combustion zone temperature as low as possible and to limit operations of burners within the combustion zone. Using multiple burners and avoiding use of a single burner operating at greater than 35 to 40% of its output can minimize slag development.

Burner use above and below the combustion zone also affects slag development. Greater use of burners above the combustion zone hearth to maintain a consistent drying zone temperature and drying rate will reduce slag development. In addition, slag is reduced by using a burner or burners below the combustion zone to maintain a temperature differential between the drop holes in the combustion zone and the hearth below of no more than 93°C (200°F). By maintaining that maximum temperature differential the operator can help prevent the ash from cooling too quickly and bridging the drop holes, which could cause the incinerator to go into a positive draft condition below that point. However, the exact temperature differential required can vary depending on the composition of the feed cake and the size and configuration of the drop holes.

The addition of auxiliary ambient air through air ports or the visual access doors on or just below the combustion hearth can help reduce operating temperatures and slag formation. The addition of forced air through the combustion air lines of inactive burners on or just below the combustion hearth also will reduce operating temperatures. However, the forced air is not as effective for temperature control because it is warmer than ambient air, a result of passing through a pressurizing fan or blower. In addition, the air can also place an additional load on the induced draft fan motor.

Reducing the center-shaft speed to a minimum level to limit the combustion rate also can reduce operating temperatures. However, caution must be exercised when slowing down the center-shaft speed. If the center-shaft speed is too slow, the cake will make contact with the rabble arms in the hearth where the feed cake is introduced to the incinerator. This contact can cause an overload on the center-shaft drive or destabilize the combustion process.

A long-term defensive measure to help minimize the problem of blockages in out-hearth drop holes is to inspect the cleaner teeth on the ends of the rabble arms and replace those that are worn or missing. These teeth often are taller than most teeth by 50 to 100 mm (2 to 4 in). The extra depth provides the reach necessary to compensate for the self-supporting arch of the brick work in the hearth.

As the incinerator ages and hearths are replaced, converting the original, small rectangular or trapezoid drop holes with arch-shaped holes will create larger openings that are less likely to catch clinkers or allow slag bridging and blockage (refer to information regarding drop holes elsewhere in this chapter).

If the flame pattern is flared out, slag buildup around the burner is likely. The burner should be readjusted to reduce the flare angle. It may even be necessary to replace the burner with a "pencil" flame type (see Figure 11.1).

Open register burners have flared flame patterns. They also have high flame temperatures (higher than 1650°C [3000°F]) because of inadequate mixing of secondary air. The burners could generate excessive oxides of nitrogen (NOx). This type of burner has often has severe slag buildup problems.

Closed register burners can be adjusted to have narrow flame patterns. The secondary air mixing is excellent and results in actual flame temperatures of 1093°C

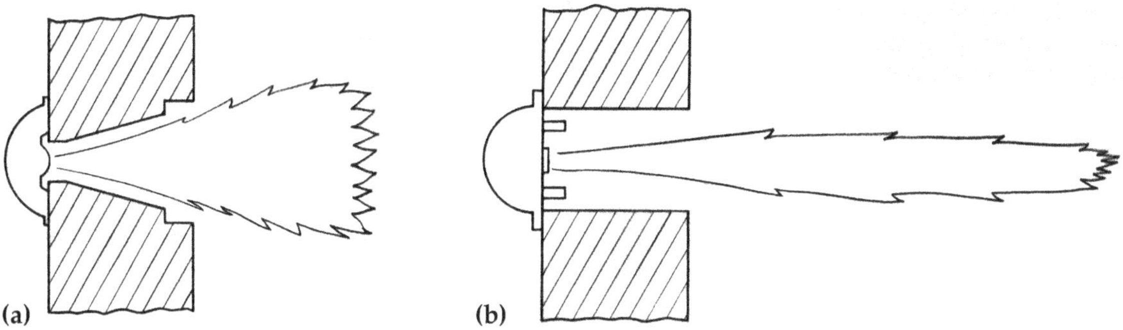

FIGURE 11.1 Burner flame shapes: (a) open register—bushy flame—and (b) closed register—pencil flame.

(2000°F). Additional dilution of secondary air is possible, which will further reduce flame temperatures.

Periodic inspection of burners is critical. If a burner tile becomes partially plugged with slag, the flame will melt out the burner, adjacent refractory, incinerator shell, and sidewalls. This can be a costly and time-consuming repair. An inline inspection port located directly across the incinerator from each burner allows a quick check of their condition.

The bottom line is that without chemical addition or a change in the chemistry of the solids being processed, the only operational defense against slag formation is reduction of operating temperatures, flame impingement, and hot spots within the incinerator.

1.3 Differing Types of Slag

Several different types of slag form in an incinerator as the temperature increases. The first slag, a lightweight sintered type, forms at the "softening" temperature. This slag builds up by agglomeration of hot, sticky fly ash particles.

Another slag, a glass-like material, occurs at the highest temperatures known as "fusion" temperatures. At this temperature ash is molten and the slag is typically a black glass-like material.

Between sinter and glass stages, several other types of slag may form. Any type of slag (from soft, lightweight to dense) can continue to build up over time. The slag that forms can be a combination of soft, medium, and hard slag depending on actual temperatures and how close the slag gets to a flame (either from a burner or burning biosolids).

It is suggested that ash at each wastewater treatment plant (WWTP) practicing incineration be thoroughly analyzed to determine the temperatures at which the different types of slag are formed. The standard test method used to determine slag formation temperature is ASTM test procedure D-1857.

1.3.1 Slag-Related Problems

Excessive slag buildup can cause several operating problems by plugging burners and drop holes, shielding thermocouples and pressure-sensing openings, plugging rabble teeth, and plugging the ash removal system. Each hearth should be visually inspected periodically for slag buildup. When excessive buildup is detected, it should be removed immediately.

1.3.2 Drop Hole Plugging

Drop holes on combustion hearths are prone to slag plugging. The soft slag that forms at the spherical and hemisphere temperatures bridges over drop holes. This bridging restricts upward movement of secondary air and downward movement of biosolids. Part of the incinerator can develop positive air pressure when drop holes plug; this must be prevented. Puffs of fly ash and smoke can be seen from door openings when there is plugging and positive air pressure.

The operator should periodically check for positive pressure and drop hole plugging and immediately correct any problems by rodding-out the drop holes. Modifications to drop holes that reduce plug ups will be discussed elsewhere in this chapter.

1.3.3 Slag Removal

Care should be taken when removing slag from the refractory to minimize damage to the refractory and maintain the integrity of the hearths, sidewalls, and burner tiles. As a result, guidelines should be prepared for the operators based on the refractory manufacturer's recommendations.

1.4 Other Maintenance-Related Problems

1.4.1 Burner Flame Impingement

Flame impingement on a rabble arm can result in high rabble arm temperatures. If the arm is properly insulated with refractory, the effect will be minimal. Arms without any insulation or broken insulation are prone to thermal stress, cracking, and sagging. Slag buildup also occurs at high-temperature locations. When large chunks of slag fall off of the arm they cause plugging of the spaces between the rabble teeth.

1.4.2 *Hearth Sagging*

All hearths are constructed in the shape of an arch to be self-supporting. The circular arch is restrained by the incinerator shell and the first row of "skew-back" bricks. The hearth assembly is critical. The temporary supports placed under each hearth during construction must have the correct arch angle. Each ring of bricks must be installed tightly and the amount of mortar on the adjacent brick surfaces must be kept to a minimum thickness.

As an incinerator heats up, the shape of the arch can change for several reasons: expansion of skew-back, hearth brick expansion, loads on top of the hearth, slag buildup, deterioration of the refractory, and bulging of the shell. The correct arch of each hearth must be maintained at all times. If the arch flattens out, the self-support will be lost and the hearth will collapse onto the hearth below. If a second hearth is overloaded, a domino effect can occur, and additional hearths will collapse.

The operators should periodically check the arch of each hearth within the incinerator. Observations from three different locations should be taken because it is possible for a portion of a hearth to sag while the remainder keeps the original arch (see Figure 11.2).

1.4.3 *Rabble Arms and Teeth*

Online inspection of rabble arms and teeth can alert operators to several issues. The sagging arm may be the result of excessive heat, a crack in the arm, or a center shaft rabble arm socket failure. By observing the shape of the arm and the clearance between teeth and hearth, an operator can tell if the arm is sagging.

Some sagging is tolerable, but excessive sagging can allow teeth to drag on the hearth, which will result in chattering of the center shaft and potentially center shaft drive failure. This should be corrected as soon as possible.

The center shaft must never be stopped where a rabble arm is located directly in front of an active burner. This will warp the arm regardless of whether it is insulated. The burner flame can also be deflected into the hearth refractory located above or below the burner and cause severe refractory damage by impingement of the burner flame on the refractory.

Refractory insulation failures on the arms can cause problems. Small cracks in the refractory surface are normal. Large cracks or missing sections of refractory that expose the metal surface of the arm are cause for concern.

Cracked rabble arm refractory can also plug drop holes. Damage or worn refractory on rabble arms should also be repaired or replaced as soon as possible.

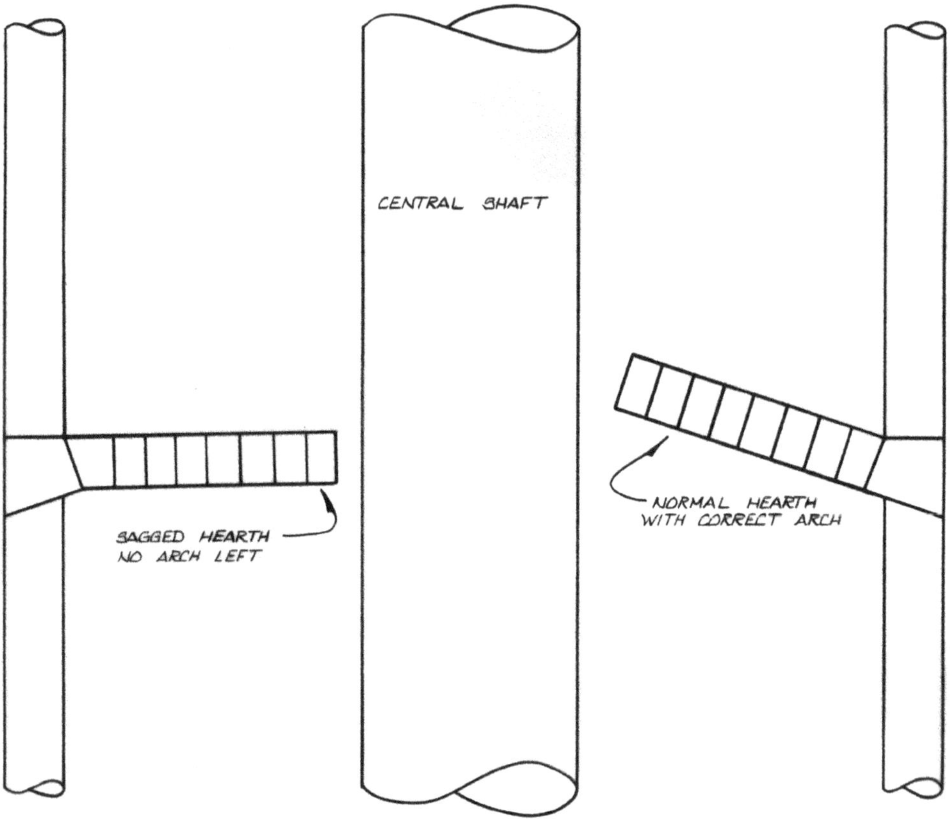

FIGURE 11.2 Hearth sagging.

Rabble teeth located within the combustion zone will eventually deform and lose their effective plowing action. In some cases the cake depth on the hearth will be thicker than normal and the rabble arm with "bulldoze" the solids from side to side.

Plugged rabble teeth result from either soft slag that has sloughed off or a brick that has become jammed between rabble teeth. This causes a local "bulldozing" condition that can damage teeth and central shaft drive. The blockage should be removed by stopping feed to the incinerator and stopping the central shaft.

Jogging the central shaft backward typically frees up the blockage. Remove the slag/brick from the incinerator. If backwards jogging does not free up the blockage, use long handled chisels and hoes to release and remove the blockage.

Rabble arm socket failure results in the arm falling out of the socket and jamming against the next rabble arm. This failure requires an incinerator shutdown to remove the broken arm/socket and repair of the socket (see section 1.8.8, Rabble Arm Replacement, for shutdown repair action).

Ash and/or slag buildup on top of the rabble arms may add excessive weight to the arm and can damage the arms and central shaft. This buildup should be removed on a regular basis.

1.4.4 *Upper and Lower Center Shaft Seals*

The upper and lower center shaft seals are typically filled with sand. Additional sand must be added to the seals on a regular basis. Recently, a number of WWTPs have replaced sand with steel shot. Because of the density of the steel shot, it does not have to be replaced as often as sand.

1.4.5 *Thermocouple Repair and Replacement*

Thermocouples within the incinerator must be cleaned, calibrated, and repaired on a regular basis. Thermocouples should be replaced on an as needed basis.

1.4.6 *Calibration and Repair of Analyzers*

Oxygen analyzers, total hydrocarbon (THC) analyzers, and carbon monoxide analyzers have to be calibrated on a daily basis and repaired on an as-needed basis. An adequate supply of spare parts should be retained on site and the equipment should be replaced as necessary.

1.5 Hot Work Cleaning and Repairs

Some cleaning and repair work can be performed without a complete shutdown (cool down to ambient temperature). To perform "hot work," the feed must be stopped and the cake within the incinerator must be allowed to burn off. After this is accomplished, the central shaft is stopped and the burner flames turned off. The entire incinerator should be under a negative draft to prevent hot gases from flaring out of inspection doors and lance tubes.

Maintenance workers should be equipped with high-temperature clothing, gloves, and face protection and be properly trained. A backup person with proper gear should also be in attendance. The tools required are typically custom-made long handled chisels, pry bars, lances, and hoes.

The following types of "hot work" jobs can be performed without shutting down (however, special care must be taken to protect workers):

- Breaking off soft slag from rabble arms, drop holes, hearth ceilings, and thermocouples.
- Removal of soft slag and loose brick that jams rabble teeth.
- Calibration, repair, and replacement of thermocouples, analyzers, etc.

1.6 Shutdown Maintenance

Shutdown maintenance can be divided into preventive and emergency maintenance. A preventive maintenance shutdown is one that is scheduled on a regular interval (for example, once per year) and includes a predetermined list of work, such as

- Rabble arm teeth replacement.
- Thorough inspection and repair of all internal refractory.
- Repair of sagging hearths.
- Replacement or repairs of rabble arms.
- Center shaft repairs.
- Modifications to improve operations.
- Burner tile cleaning and replacement (with precured burner tile, if available).

To accomplish tile cleaning and replacement, fuel, air, and electrical fittings can be shut off and disconnected from the burner. The burner can be removed from the incinerator shell and the exposed burner tile cut out from the adjacent refractory bricks.

The new burner tile should be slightly undersized from the original dimensions; sheet Kaowool® or similar fiber refractory material can be used to fill any gaps between the new burner tile and the existing tile opening.

It should be noted that emergency shutdown is required for a collapsed hearth, broken rabble arm, and cracked center shaft.

1.7 External Shell Maintenance

On incinerators, which have external insulation, there is only a small amount of exposed shell surface that must be inspected. The exposed surface will be in the immediate vicinity of all operating burners. An operator can detect hot spots from the burning of the painted surface and by the glowing of the metal. Should a hot spot be found, the cause must be immediately determined and corrected. Typically, this condition is caused by a slagged-up burner tile. If left uncorrected the shell can be seriously damaged.

On incinerators without shell insulation the operator should check the entire surface for hot spots. The operator should also check for shell cracks, particularly in the skew-back bands at each hearth. Should a skew-back band crack completely through, the arch of that hearth will settle down and quickly fail. Immediate repairs should be made.

The operator should also check the shell for corrosion in the vicinity of all hearths. Serious corrosion can occur from the inside of the shell because the shell temperature is at the dewpoint. Operators should check for all of these conditions during every shift.

1.8 Refractory Repair and Replacement

The quality of all repairs and modifications inside an incinerator affects equipment life. If high-quality material is used and installation is performed in a professional manner, service life of the refractory work will be substantially extended.

Installation of refractory brick hearths is a complex process and should not be compared with laying bricks in a building wall. The bricklayer or contractor must be trained in the art of incinerator refractory and should have the required experience. This person should have a thorough knowledge of the many refractory materials, temperature limits, strengths, curing procedures, construction procedures, and emergency hot work techniques.

1.8.1 Brick Shapes

A hearth is constructed with many individual bricks arranged in concentric rings starting from the outside and progressing inside the hearth. There are two types of bricks used for hearth construction: special shaped and standard sized. The advantages to using standard-sized brick are

- Consistent modulus of rupture throughout a lot. Standard-sized bricks are machine made and are packed at higher pressure, which produces a high-strength brick. Special-shaped bricks are traditionally hand-filled and packed. As a result, modulus of rupture values have varied so much that failures have occurred after very little running time.
- Reduction in spare brick stocking to only one size when using a standard-sized brick. With special-shaped bricks, each ring of bricks is a different size.

- Delivery time of standard-sized bricks is considerably less than for special-shaped bricks.
- Cost of standard-sized bricks is approximately one-fifth that of special-shaped bricks. The labor costs for building a complete hearth are nearly the same, though there are more standard-sized bricks to handle. The overall cost of a hearth with standard-sized bricks is approximately 40% that of a special-shaped hearth.

1.8.2 *Refractory Materials*

The significant properties of a refractory, including high-temperature strength, depend on its mineral makeup and the way these materials react to high temperatures and incinerator environments. After refractory material (bricks or castable) has been fired, fine and coarse materials form a ceramic bond.

The most significant properties of refractories are those that allow them to withstand conditions found in an operating incinerator. Refractories must withstand the maximum service temperature in the incinerator at the combustion hearths and in the vicinity of the burners.

Chemical reactions within an incinerator can contribute to ultimate destruction of refractory materials. An experienced refractory supplier must know the chemicals and gases that will be released in an incinerator before recommending the correct refractory. The experiences of past incinerator operators are invaluable in determining the long-range history of a particular refractory material.

1.8.3 *Hearth Repair and Replacement*

A hearth should be replaced only when there is danger of collapse. As long as the hearth has the proper arch, it can remain in use even if there is some localized damage (loose or missing bricks or erosion from burner flame impingement).

A section of loose or missing bricks or even a melt-through by a burner flame can be patched using castable refractory with stainless steel needle reinforcing. This is accomplished by placing plywood supports against the underside of the damaged hearth to carry the load to the next lower hearth. Loose and damaged bricks must be removed before filling the void with the castable-stainless steel needle mix.

An in-hearth may lose one or more rows of bricks, even though the remainder of the hearth has sufficient arch. To replace missing rows, workers need to install temporary supports under the hearth and carry the load to the hearth below. The exposed surface of the remaining inner row of bricks then needs to be cleaned. Finally, each missing row needs to be reconstructed using bricks that match the original (either special-shaped or standard-sized bricks). The bricks must be mortared to the remaining inner row and to each other. Once a row is started, it must be completed without interruption before the mortar sets up.

When a hearth has flattened out enough to require replacement, it must be removed very carefully. Plywood and wooden shoring must support the hearth immediately below the flat hearth to take the additional load of bricks. The hearth is removed one row of bricks at a time, starting at the row closest to the central shaft.

Typically, bricks are bonded ceramically to the adjacent row and will not release the entire row when the first brick is knocked out. After all bricks are removed back to the skew-back row, the skew-back bricks should be examined for soundness and cracks and repaired as necessary. The inner surface should be ground smooth to receive the new bricks.

Reconstruction of the new hearth begins with installation of support forms under the hearth, which establish the correct arch. The load of the new hearth is carried to the hearth below temporarily. Installation of the new bricks starts at the outer skew-back row and proceeds toward the center. Mortar is applied to the outer row of bricks and between the new bricks. Once a row is started, work must continue uninterrupted until that row is completed, before the mortar sets up.

After all bricks are in a row, they must be "tightened up" to take up any air space between bricks. A hearth that is constructed loosely will settle quickly and fail. The temporary support system can be removed as soon as the last (inner) row of bricks is completed. If special-shaped bricks are used, an adequate supply for future repairs should be stocked. Otherwise, there can be a long downtime waiting for supplies.

1.8.4 Drop Holes

Drop holes being plugged with soft slag is one of the most prevalent problems with incinerator operation. The typical drop-hole configuration of a 7-m (23-ft) diameter incinerator has 30 equally spaced rectangular holes with a cross-sectional area of 0.11 m^2 (1.14 sq ft) each. Typically, plugging starts at the four corners of a drop hole and then progresses throughout the remaining opening. By increasing the drop-hole size to 0.34 m^2 (3.7 sq ft) and using an arch construction to eliminate two corners, plugging can be reduced dramatically.

The modified larger hole actually uses the space of the two original holes and the special-shaped bricks between them. The arch can be constructed with 100% castable and standard-sized arch bricks (see Figure 11.3). "Poke" holes for cleaning out plugged holes while the incinerator remains online can be added above each drop hole.

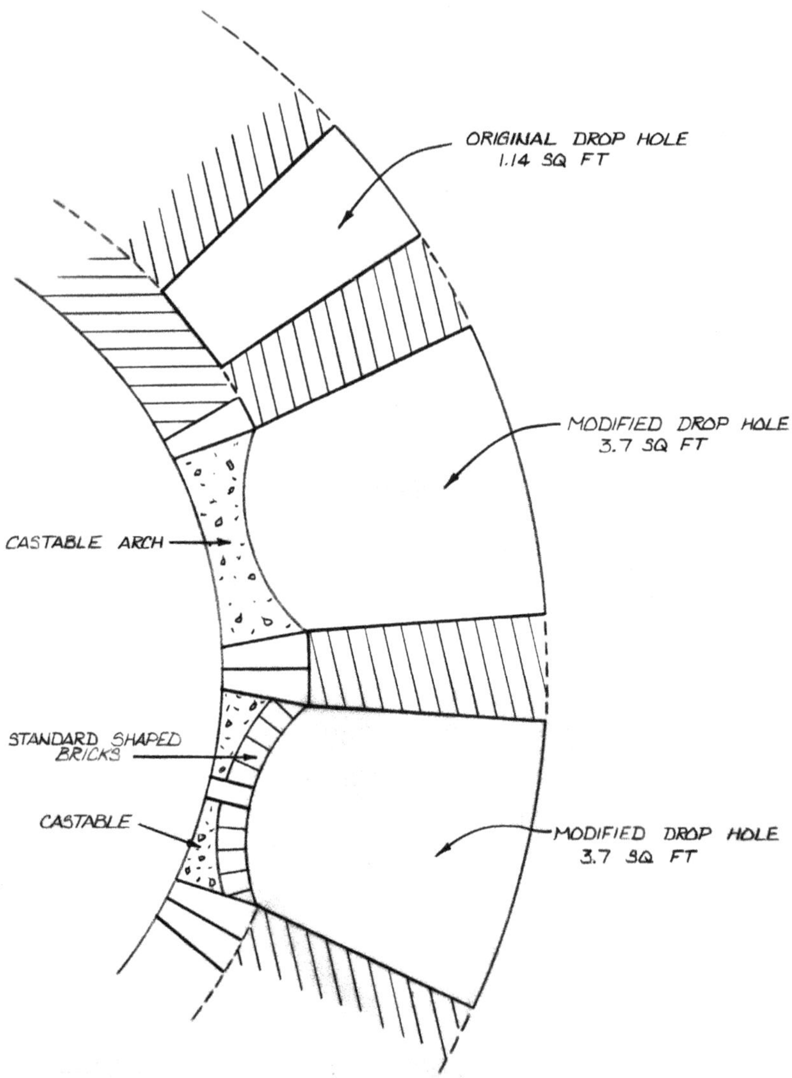

FIGURE 11.3 Larger drop hole design (sq ft × 0.0929 = m^2).

1.8.5 Center Shaft

Cracking of the center shaft can occur in two different locations on the shaft. One may be a failure of the sockets that hold rabble arms and the other may be cracking of the hollow metal shaft itself.

Repairs of these cracks can be accomplished by two different procedures. The damaged section of the shaft can be replaced by disassembling all shafting and arms above the failure, adding a new section, and assembling the remaining parts. An alternate procedure is to repair in place by "cold metal stitching." This is a procedure of lacing together a crack by means of rectangular metal blocks (see Figure 11.4). The metal blocks, which have a high nickel content, are peened tightly into the drilled-out cavity of the original casting.

1.8.6 Center Shaft Shear Pin

The center shaft shear pin should be checked on a regular basis and replaced as needed.

1.8.7 Drive Gear

Wear by the central shaft thrust bearing allows the bull gear to press against the pinion gear. Chattering and teeth wear may result. The back side of the bull gear teeth should be checked to ensure adequate clearance.

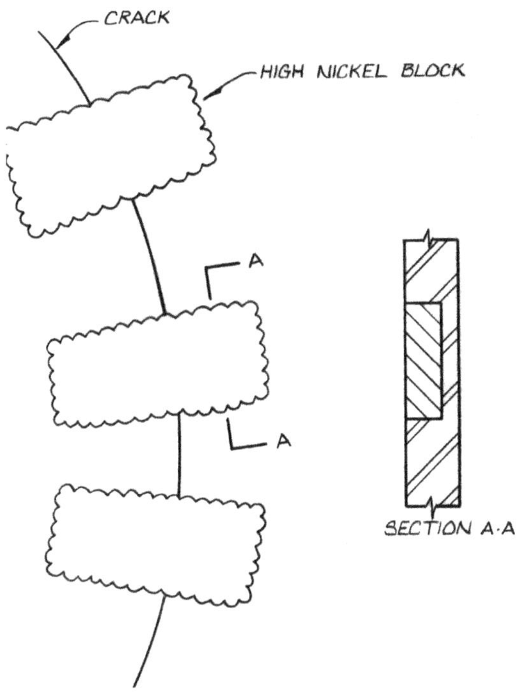

FIGURE 11.4 Cold-metal stitching.

1.8.8 *Rabble Arm Replacement*

Rabble arms may need to be replaced for several reasons: excessive arm sagging, cracked arm (broken off), or rabble teeth that no longer dovetail in the arm because the exposed metal surfaces of the dovetail have eroded over a long period of time.

Removal of the arm may require removing the teeth. The arm with teeth attached will not pass through a standard access door. Replacing the standard door with a larger one permits the arm to be removed with the teeth intact.

Teeth used for a long time are typically difficult to remove from the dovetail slot. Use of pneumatic hammers is recommended. The pin that anchors the arm in the central shaft socket may also be hard to remove; a pneumatic hammer can be used or it can be drilled out.

A new rabble arm must be insulated. It can be insulated either before installation or afterwards in the incinerator with a castable refractory. If preinsulated, care must be taken not to damage the arm while transporting and installing it. After the arm is in place, the socket-hub area should be insulated with castable refractory. No bare metal surfaces should be exposed to high temperatures.

1.8.9 *Rabble Teeth Replacement*

Rabble teeth should be replaced when they become deformed, broken, or missing. Teeth are held on the rabble arm in a dovetail slot. A pin holds the last (outermost) tooth in place, preventing the others from sliding out during use. To remove the teeth from an arm, the pin on the outer tooth needs to be pulled and the teeth can slide out of the dovetail.

Typically, teeth used for an extended period of time are difficult to slide out. Operators can use a pneumatic hammer to assist removal. Before installing new teeth, the tooth assembly drawing provided by the incinerator supplier should be consulted. The correct tooth must be installed in the correct location in the correct arm.

New teeth should be laid out on an assembly area to duplicate the assembly drawing. Each tooth needs to be checked for angle, size, and proper location. Spacer blocks should be inserted where required. High-temperature material containing 24 to 28% chromium and 18 to 22% nickel (ASTM A-297, Grade HK, Class 25-20) is typically used for the rabble teeth. However, a few incinerator installations use lower heat resistant material for rabble teeth located in the ash cooling hearths.

When inserting the new teeth to the rabble arm, workers should start with the inner-most tooth and work toward the outside. The first arm should be completed before proceeding to the second, and so on. A small (6 mm [0.25 in]) gap between the last two teeth (or spacer block) should be created to allow for thermal expansion. If the gap is insufficient, some metal can be ground off the last tooth (or spacer).

1.8.10 *Maintenance of Outside Ancillary Equipment*

While both the MHF and fluid bed incinerator portions of this chapter focus exclusively on the inner workings of the incinerator, there are several ancillary items that must also be properly maintained. These external components consist of, but are not limited to, the following:

- Burner trains and management systems.
- Wastewater solids conveyance, feed, and weighing systems.
- Fans (inducted drafts, forced air, cooling air, etc.).
- Monitoring/analyzing equipment (THC, carbon monoxide, oxygen, moisture, etc.).
- Ductwork, piping, stacks, and instrumentation and controls.

1.8.11 *Recordkeeping*

The following information should be collected when an incinerator is first assembled:

- Construction drawings showing typical in-hearths and out-hearths; the location of burners, piping, and access doors, etc.; the emergency relief stack and damper; center shaft, rabble arms and rabble teeth; etc.
- Manufacturer drawings.
- Information concerning the manufacturer, type, or trade name of castable and firebrick and brick part numbers (standard-sized or special-shaped).
- Information concerning the rabble arms and teeth, with assembly drawings.
- A listing of all spare parts.
- Manufacturer's operating and maintenance manuals.
- Changes from the original design, problems during construction or startup, and information regarding the company that performed the work.

When repairs or modifications are made after the incinerator is placed into service, the following information should be recorded.

- The name of the company that performed the work;
- A detailed description of work and the dates the work was performed;

- Information concerning type, quantities, and costs of materials used;
- Information concerning labor requirements and costs; and
- Any problems encountered during repairs.

If any changes from the original configuration are made, a record should be kept. These changes might include

- Drop holes enlarged (locations, details, and date).
- Burner view ports added (locations, details, and date).
- Rabble teeth revised (locations, details, and date).
- Hearth bricks changed (from special-shaped to standard-sized or vice versa).
- Row of bricks removed from the "in" drop zone (locations, details, and so on).

Preventive maintenance is needed once a year when the incinerator is shut down and cleaned. A thorough inspection of all internal parts should be made. In particular, check the following:

- Rabble teeth (warping and cracking).
- Rabble arms (sagging and refractory cracking).
- Slag buildup.
- Burners (tile cracking and slag).
- Hearths (sagging, loose/missing bricks, and cracks).
- Side walls (cracking and slag).
- Drop holes (particularly the bottom surfaces for deterioration of refractory).
- Center shaft (refractory cracking).
- Center shaft and rabble arm sockets (cracking).
- Center shaft drive motor and gears.
- Ceilings (cracking and slag).

2.0 FLUID BED INCINERATORS

Maintenance of fluid bed incinerators is substantially less than that associated with MHFs because of the limited number of moving parts in a fluid bed incinerator. However, maintenance is still required to extend the life of the equipment.

2.1 Slag Formation

Areas where slag buildup occurs include, but are not limited to, the sand bed below auxiliary fuel lines, burner tiles on freeboard burners, thermocouples, and offgas ducting. A detailed discussion on the formation and removal of slag and determination of fusion temperature is contained in the MHF section of this chapter.

If slag formation is observed during normal operation the incinerator should be shut down and cooled and the slag removed.

2.2 SLAG REMOVAL

Care should be taken when removing slag from the refractory to minimize damage to the refractory. As a result, guidelines should be prepared for operators based on the refractory manufacturer's recommendations.

2.3 MAINTENANCE ISSUES

2.3.1 Thermocouple Repair and Replacement

Thermocouples within the incinerator must be cleaned, calibrated, and repaired on a regular basis. The thermocouples should be replaced on an as-needed basis.

2.3.2 Calibration, Repair, and Replacement of Analyzers

Oxygen analyzers, THC analyzers, and carbon monoxide analyzers have to be calibrated on a daily basis and repaired as necessary.

2.3.3 Tuyere Inspection, Cleaning, and Replacement

Tuyeres allow air to exit the wind box and fluidize the sand bed. Tuyeres need to be inspected and cleaned on an annual basis and replaced as needed.

2.3.4 Arch Repair and Replacement

The arch separating the wind box from the sand bed needs to be checked on a regular basis. If the arch is made of refractory, both the arch and the refractory need to be checked regularly. If the arch is made of stainless steel, it needs to be checked for cracks and signs of stress and corrosion regularly.

2.3.5 Removal of Sand Bed and Sand in the Wind Box

It is good practice to remove the bed sand every two or three years. This removal is done by operating the fluidizing air blower and opening the bed drain valve. All inspection doors must be kept closed to prevent filling the building with sand.

A suitable container or conveying system is required to remove and store sand. In addition, makeup sand needs to be added to the bed regularly. Operators also need to check for sand in the wind box. This sand also needs to be removed at least annually.

2.4 External Shell Maintenance

The operator should regularly check the external shell visually for hot spots or corrosion. If hot spots are observed, the incinerator should be shut down as soon as possible and cooled. An internal inspection should be carried out to determine the cause. Hot spots typically are caused by cracks in the refractory (typically around nozzles in the bed area). Bed material and gases that are under pressure migrate to the shell and overheat it. If hot spots are found in the refractory they should be repaired.

2.5 Shutdown Maintenance

An annual inspection should be carried out on fluid bed incinerators. This requires shutting down and cooling the incinerator. The incinerator should be cooled at a rate not exceeding 37°C (100°F) per hour. This can be done by using the fluidizing air blower. Extreme care is needed to prevent the refractory from being overstressed during cool down.

When the incinerator is cooled to room temperature all inspection doors should be opened and inspection can begin (refer to Table 11.1). All fuel guns should be disconnected from the fuel source and removed. The fuel supply to the preheat burners must be isolated.

2.6 Refractory Repair

Refractory repairs in the fluid bed incinerator consist of patching spalled surfaces with plastic or castable refractory and filling cracks with refractory mortar. Occasionally bricks need to be replaced. The majority of refractory work occurs within the bed area where the fluidizing sand may cause some erosion. Most plants do not have staff qualified to repair refractory or have an inventory of refractory. A contractor must be hired to perform the work. Contractor qualifications, brick shapes, and refractory materials are discussed in the MHF section of this chapter.

2.7 Recordkeeping

The following information should be recorded when an incinerator is first assembled:

- Manufacturer and construction drawings.

TABLE 11.1 Items for inspection of fluid bed incinerators.

Refractory	Check all refractory for cracks or broken pieces. This includes all openings for nozzles. Repair or replace.
Bed support	If hot wind box, inspect arch for cracks and loose or broken fluidizing air tuyeres (nozzles). Repair refractory, replace loose or broken nozzles, and regrout. If cold wind box, inspect refractory on construction plate for cracks and loose or broken fluidizing tuyeres (nozzles). Repair refractory and replace loose or broken nozzles.
Outer shell	Buckling or bulging of shell and / or peeling paint indicate refractory failure. Locate refractory failure and repair.
Preheat burner	Inspect burner tile. Remove any clinker. If tile is damaged, patch.
Wind box	Inspect fluidizing air nozzles from below for plugging. Remove any sand that has accumulated on the floor of the wind box. This sand will have leaked from the bed through the fluidizing air nozzles. Check that no fuel has leaked from the bed into the wind box. Sand removal may be required more often. Excessive accumulation of sand indicates that one or more fluidizing air nozzles are defective.
Bed fuel guns	Inspect each bed fuel gun. Deposits of carbon may form near the tip of the gun if fuel oil is used. Rod out this obstruction. Bed fuel guns can be inspected and cleaned during operation. However, compressed air should be connected to the gun while reinserting to the bed. Otherwise, reinsertion will be difficult and sand could lodge within the gun, restricting fuel flow.
Bed pressure taps	Inspect and clean bed pressure taps. This may be required during operation.
Freeboard water sprays	Inspect water sprays. Replace worn nozzles. If hoses are used, inspect and replace worn hoses.
Expansion joints	Inspect all expansion joints on ducting. Remove any ash that has accumulated between the liner and the expansion piece. If expansion joints are metallic, inspect for cracks.
Instrumentation	All instrumentation must be checked and recalibrated where necessary: pressure transmitters, level transmitters, and flow transmitters. Inspect and, where damaged, replace thermocouples and their wells. Replace gauge glasses if scratched or otherwise damaged.
Gaskets	Before closing inspection doors on the incinerator, inspect door gaskets. If damaged, replace.

- Photographs; manufacturer, type, or trade name of castable and brick; and brick part numbers (standard or special shape).
- Changes from the original design, problems during construction or startup, and information regarding the company that performed the work.

When repairs or modifications are made, the following information should be recorded:

- Company (or persons) who performed the work.
- Detailed description of work, including date.
- Material—type, amount used, and cost.
- Labor required—hours and cost.
- Any problems during repairs.
- Source of spare parts—original manufacturer or qualified foundry.

Preventive maintenance is needed once a year when the incinerator is shut down and cleaned. A thorough inspection of all internal parts should be conducted, including the following:

- Burners (tile cracking and slag),
- Side walls (cracking and slag), and
- Ceilings (cracking and slag).

3.0 SUGGESTED READING

ASTM International *Active Standard: D1857-04 Standard Test Method for Fusibility of Coal and Coke Ash.* http://www.astm.org (accessed April 2008).

Appendix A

Combustion Fundamentals

1.0 INTRODUCTION

Combustion is the controlled release of heat caused by the chemical reaction between a fuel and an oxygen source. The incineration of sludge cake is a combustion process. The volatile fraction of the total solids in sludge cake is the fuel that ultimately reacts with oxygen and releases heat. High moisture content distinguishes combustion of sludge cake from that of more common fuels because the amount of heat required to convert water from liquid to vapor is high. A significant portion of the heat released when sludge cake is combusted must be used to evaporate cake moisture. Thus, only the smaller remaining fraction is available to raise the temperature of the products of combustion to proper levels. When the thermal properties of the sludge cake and design of the furnace combine to allow sludge incineration to proceed without the need for an auxiliary fuel, such as natural gas or oil, the combustion process is considered autogenous.

Each wastewater treatment plant (WWTP) incinerator operator strives to achieve autogenous combustion amidst ever-changing biological treatment processes and increasingly stringent environmental regulations.

This appendix is broken down into four major sections. The first section reviews some engineering fundamentals associated with the combustion process. The second section provides an introduction to the basic science of sludge combustion and examines typical compositions of auxiliary fuels and sludges. The third section examines the combustion of the most common auxiliary fuels, natural gas and fuel oil. The fourth section describes the combustion of wastewater treatment plant sludge.

2.0 ENGINEERING FUNDAMENTALS ASSOCIATED WITH COMBUSTION

Because wastewater treatment and combustion each has its own basic terminology and units of measure, it is sometimes difficult to switch between the disciplines. This section will illustrate some important concepts and relationships.

2.1 Describing Physical Quantities in a Mechanical System

Four basic quantities are typically used in a mechanical system:

- Length = feet (ft),
- Time = seconds (sec),
- Force = pounds of force (lbf), and
- Mass = pounds (lb).

Metric conversions for these units are as follows: ft × 0.304 8 = m; lbf × 4.448 = N; and lb × 0.453 6 = kg.

Other quantities, such as velocity, area, and pressure, described in terms of the basic quantities, are known as secondary quantities.

2.1.1 *Mass and Force*

Sir Isaac Newton's famous second law of motion states: "The acceleration produced by a particular force acting on a body is directly proportional to the magnitude of the force and inversely proportional to the mass of the body."

$$\text{Force} \approx \text{Mass} \times \text{Acceleration}$$

Where

force = N (lbf),

mass = kg (lb), and

acceleration = g = local gravity 9.807 m/s^2 (32.174 ft/sec^2).

It is important to note the choice of the word "proportional" over the more commonly used (but incorrect) equal. To express Newton's law as an equality, it is necessary to multiply the right-hand side (RHS) by $1/g_c$, where g_c = universal gravitation constant = 32.174 (ft-lb)/(lbf-sec^2) (1 lbf = 4.448 N).

From the foregoing, the units of g/g_c are lbf/lb (lbf/lb × 9.806 65 = N/kg). Whenever pounds of force must be converted to pounds of mass, it must be multiplied by (g/g_c) (note that ft-lb × 1.356 = N · m).

To illustrate, calculate how much force you would exert on a spring bathroom scale standing on the moon, which has approximately one-sixth of the earth's gravity:

$$\text{lbf} = \text{lb} \times ([32.174/6]\ \text{ft/sec}^2)/[(32.174\ (\text{ft-lb})/(\text{lbf-sec}^2)] = \text{lb}/6$$

Therefore, a mass of 200 lb, would require a force of 200/6 = 33.33 lbf. One of the most common situations that requires use of the ratio (g/g_c) is the determination of pressure, which is typically represented to the force exerted over a specific area by a column of liquid of known height and density:

$$\text{Pressure (lbf/sq ft)} = \text{Height (ft)} \times \text{Density (lb/cu ft)}$$

By multiplying the RHS of the equation by (g/g_c), we get

$$\text{lbf/sq ft} = \text{ft} \times (\text{lb/cu ft}) \times (32.174\ \text{ft/sec}^2)/[(32.174\ \text{ft-lb})/(\text{lbf-sec}^2)]$$

Or, upon simplification, lbf/sq ft = lbf/sq ft (lbf/sq ft × 47.88 = Pa and lb/cu ft × 16.02 = kg/m^3).

With this understanding of force and mass, the four basic quantities described above can be used to define all other physical quantities within a mechanical system, as illustrated in Table A.1 by way of common examples of secondary quantities.

2.1.2 *Absolute Pressure and Temperature*

The gases in combustion systems are typically at pressures and temperatures different from ambient conditions. Therefore, it is necessary to use absolute pressure and temperature scales, where the lowest value is zero and not a negative number, to adjust properties to actual process conditions.

2.1.2.1 Pressure

Pressure measurements that are not in absolute are typically referred to as gage or gauge pressures. Standard atmospheric pressure is 14.696 lbf/sq in absolute (psia). To convert any gage pressure (psig) to absolute (psia), simply add 14.696 (note that psig × 6.895 = kPa). To convert this pressure to lbf/sq ft (psfa), simply multiply by 144 (note that 1 lbf/sq ft = 0.0005 Pa). The following example converts a gage pressure of 10 psig to an absolute pressure (psfa):

$$(10 \text{ psig} + 14.696) \times (144 \text{ sq in/sq ft}) = 3556.2 \text{ psia}$$

(Note that sq in × 0.000 645 2 = m^2)

Other pressure units commonly encountered are
Inches of water column = in H_2O (kPa),
Inches of mercury = in Hg (kPa),
Ounces per square inch = osi, and
Head (pumps), measured in feet of water column = ft H_2O (meters [m] of head).

Densities of water and mercury used for converting pressure are as follows: density of water at 68°F equals 62.32 lb/cu ft and density of mercury at 68°F equals 848.71 lb/cu ft.

2.1.2.2 Temperature

Even though in this appendix U.S. customary units are used, the metric temperature scale is quite commonly used in some venues. U.S. customary temperature units are expressed in degrees Fahrenheit (°F); metric temperatures are expressed in degrees Celsius (°C). Converting between Fahrenheit and Celsius temperatures is done as follows: °F = 1.800 (°C) + 32 and °C = 0.5556 (°F − 32).

Absolute temperature scales are degrees Rankine (°R) for Fahrenheit and degrees Kelvin (K) for Celsius. Determining the absolute temperature is done by adding a

TABLE A.1 Common secondary quantities in mechanical systems.

Item	Definition	Dimensions*
Area	$(\text{Length})^2$	m^2 (sq ft)
Volume	$(\text{Length})^3$	m^3 (cu ft)
Velocity	Length/time	m/s (ft/sec)
Acceleration	Length/$(\text{time})^2$	m/s^2 (ft/sec^2)
Pressure	Force/area = Force/$(\text{length})^2$	$N \cdot m^2$ (lbf/sq ft)
Density	Mass/volume = Mass/$(\text{length})^3$	kg/m^3 (lb/cu ft)

* sq ft × 0.092 90 = m^2; cu ft × 0.028 32 = m^3; ft/sec × 0.304 8 = m/s.

specific number of degrees to correctly reference temperatures to absolute zero as follows: °R = °F + 459.67 and K = °C + 273.15.

2.2 Fundamental Combustion Concepts

This section provides a discussion of various fundamental concepts that are essential to a basic understanding of the combustion process. Included are an introduction to combustion elements and atomic weights; molecules and molecular weights; the ideal gas law; and the basic compositions of air and typical commercial fuels.

2.2.1 Combustion Elements and Atomic Weights

To date, scientists have identified 109 elements. Each element has a specific atomic weight. Fortunately, most of these elements are not involved in sludge combustion. The five most common elements encountered in sludge combustion are carbon, hydrogen, oxygen, nitrogen, and sulfur. Table A.2 summarizes abbreviations and atomic weights for these elements.

2.2.2 Molecular Weights

Carbon and sulfur are solids at room temperature, whereas hydrogen, oxygen, and nitrogen are gases. A molecule of either carbon or sulfur contains a single atom. A molecule of hydrogen, oxygen, or nitrogen contains two atoms. These are sometimes called diatomic (di = 2) gases. When writing the abbreviation for these gases, the subscript 2 is used to indicate a gas molecule. While the correct atomic weight should always be used in calculations, rounded off values are more commonly used in narrative discussions and will be used in the example calculations presented in this appendix.

The molecular weights of solids, gases, or liquids that are a combination of the five elements shown in Table A.2 can be easily calculated. Several illustrative examples of familiar compounds are shown in Tables A.3 and A.4.

TABLE A.2 Abbreviations and atomic weights of common elements in combustion.

Element	Abbreviation	Atomic weight
Carbon	C	12.01 (~12)
Hydrogen	H	1.008 (~1)
Oxygen	O	16.00 (~16)
Nitrogen	N	14.005 (~14)
Sulfur	S	32.06 (~32)

2.2.3 *Moles*

In combustion and other subjects generally related to chemistry, a basic unit of mass called the mole is commonly used. A mole of a substance is a mass equal to the molecular weight. When working in U.S. customary units, the basic unit is a pound-mole (lb-mole); in metric units, the basic unit is a gram-mole (g-mol). A pound-mole of oxygen would weigh 32 lb, and 1 lb-mole of water would weigh 18 lb. A gram-mole of carbon dioxide would weigh 44 g. This appendix uses U.S. customary units for the examples; therefore, only pound-moles will be used.

The product of the number of moles (n) and the molecular weight (MW) is equal to the mass (m):

$$m = n \times \text{MW}$$

$$n = m/\text{MW}$$

Because molecular weight is the same as the number of pounds in a pound-mole:

$$\text{lb} = \text{lb-mole} \times (\text{lb}/\text{lb-mole})$$

$$\text{lb-mole} = \text{lb}/(\text{lb}/\text{lb-mole}) \text{ (note that 1 lb/lb mole} = 1 \text{ g/g-mol)}$$

TABLE A.3 Molecular weights of common diatomic gases.

Gas	Abbreviation	Molecular weight
Hydrogen	H_2	2.016 (~2)
Oxygen	O_2	32.00 (~32)
Nitrogen	N_2	28.01 (~28)

TABLE A.4 Nominal molecular weights of common compounds in combustion.

Compound	Formula	Calculation	Molecular weight
Water	H_2O	2 + 16	18
Hydrogen peroxide	H_2O_2	2 + 32	34
Carbon dioxide	CO_2	12 + 32	44
Carbon monoxide	CO	12 + 16	28
Hydrogen sulfide	H_2S	2 + 32	36
Sulfur dioxide	SO_2	32 + 32	64

For example, knowing that the molecular weight of CO_2 is 44 lb / lb-mole and the molecular weight of H_2O is 18 lb / lb-mole:

$$2 \text{ lb-mole of } CO_2 = 2 \text{ lb-mole} \times (44 \text{ lb/lb-mole}) = 88 \text{ lb}$$

$$72 \text{ lb of } H_2O = 72 \text{ lb}/(18 \text{ lb/lb-mole}) = 4 \text{ lb-mole}$$

2.2.4 *Ideal Gas Law*

In the sludge incineration process, the products of combustion and some auxiliary fuels are gases. Therefore, it is important to understand the behavior of gases. Avogadro's law states that equal volumes of different ideal gases at the same pressure and temperature contain the same number of molecules. Because the molecular weight gives the relative mass of individual molecules, it follows that equal volumes of different ideal gases at the same pressure and temperature contain the same number of moles. An ideal gas may be defined as a substance that behaves according to the following equation:

$$P \times V = n \times R \times T$$

Where

P = absolute pressure, lbf / sq ft;

V = volume, cu ft;

n = number of lb-mole;

R = universal gas constant = 1545.32 (ft-lbf)/[(lb − mole) °R] (note that 1 ft-lbf = 1.356 J); and

T = absolute temperature, °R.

To illustrate the concept, the volume of 1 lb-mole of nitrogen at atmospheric pressure (14.696 psia) and a temperature of 60°F is calculated as follows:

Absolute pressure = 14.696 psia × (144 sq in/sq ft) = 2116.2 lbf/sq ft
Absolute temperature = 60°F = 60 + 459.67 = 519.67°R

Rearranging the ideal gas equation:

$V = n \times R \times T/P$
V = (1 lb-mole) × 1545.32 (ft-lbf)/[(lb-mole)°R] × 519.67°R/(2116.2 lbf/sq ft)
V = 379.5 cu ft

As a second example, the volume of 64 lb of oxygen at atmospheric pressure and a temperature of 60°F is calculated as follows:

Molecular weight of oxygen = 32
64 lb/(32 lb/lb-mole) = 2 (lb-mole)
V = (2 lb-mole) × 1545.32 (ft-lbf)/[(lb-mole)°R] × 519.67°R/(2116.2 lbf/sq ft)
V = 759.0 cu ft

There are two additional ideal gas laws that are important. These are commonly referred to as Boyle's law and Charles' law. Boyle's law states that the volume of a perfect, or ideal gas, varies inversely with the absolute pressure. Written in mathematical terms,

$$V_1/P_1 = V_2/P_2$$

Charles' law states that the volume of a perfect gas varies directly with the absolute temperature. Written in mathematical terms,

$$V_1 \times T_1 = V_2 \times T_2$$

When working with an ideal gas undergoing a change in a process, we use the subscript 1 (V_1 or T_1) to denote the beginning and subscript 2 (V_2 or T_2) to denote the end of the process. If neither the mass nor gas changes during the process (n and R remain unchanged), the above equations can be rewritten as follows:

$$P_1 \times V_1/T_1 = n_1 \times R_1$$
$$P_2 \times V_2/T_2 = n_2 \times R_2$$

Because $n_1 \times R_1 = n_2 \times R_2$, $P_1 \times V_1/T_1 = P_2 \times V_2/T_2$

Thus, another definition of an ideal gas is one that follows Boyle's and Charles' laws.

2.2.5 *Composition and Properties of Air*

The air in the atmosphere that surrounds us is composed of several gases, including some not previously discussed. Table A.5 gives the normal composition of bone dry air (water vapor not included), both in volume fraction and weight fraction.

For combustion problems, it is generally acceptable to consider the small fractions of argon and carbon dioxide as nitrogen, thus giving a bone dry air composition of 79.05% nitrogen and 20.95% oxygen by volume (76.86% nitrogen and 23.15% oxygen by weight). The molecular weight of bone dry air is 28.996. From this, the density of air at any pressure and temperature can be calculated using the ideal gas law.

The term relative humidity often is used to describe atmospheric moisture conditions. Relative humidity is defined as percentage of moisture in the air compared with the maximum amount the air could hold at saturation at the same pressure and temperature. The term psychrometrics is used to describe the science of air-water vapor mixtures. In combustion problems, absolute humidity is typically used. Absolute humidity has the units of lb H_2O/lb dry air. Table A.6 gives an abbreviated range of typical values of absolute humidity found in combustion problems, tabulated in terms of relative humidity and temperature. In Table A.6, all tabulated values are expressed as lb H_2O/lb dry air (note that lb/lb × 1000 = g/kg).

As an illustrative example based on Table A.6, the absolute humidity of air at atmospheric pressure, 90°F and 75% relative humidity is calculated as follows:

$$\text{Absolute humidity} = 0.75 \times 0.031\,18 \text{ (lb } H_2O/\text{lb dry air)} = 0.023\,39 \text{ lb } H_2O/\text{lb dry air}$$

Table A.5 Normal composition of dry outdoor air*.

Component	Fraction by volume	Fraction by weight
Nitrogen	0.7809	0.7553
Oxygen	0.2095	0.2315
Argon	0.0093	0.0128
Carbon dioxide	0.0003	0.0004
Totals	1.0000	1.0000

*Air also contains slight traces of neon, helium, krypton, ozone, and other components.

TABLE A.6 Properties of moist air at standard atmospheric pressure.

Temperature	100% Relative humidity	80% Relative humidity	70% Relative humidity	60% Relative humidity
40°F	0.005 213	0.004 170	0.003 649	0.003 128
50°F	0.007 658	0.006 126	0.005 361	0.004 595
60°F	0.011 08	0.008 86	0.007 76	0.006 65
70°F	0.015 82	0.012 66	0.011 07	0.009 49
80°F	0.022 33	0.017 86	0.015 63	0.013 40
90°F	0.031 18	0.024 94	0.021 83	0.018 71
100°F	0.043 19	0.034 55	0.030 23	0.025 91

Both the molecular weight and pressure of the dry gas have an effect on the maximum quantity of water vapor present at saturation for a given temperature. However, these changes are complex and outside the scope of this text. For many sludge incinerator evaluations, the effects of changes in pressure and molecular weight on flue gas psychrometrics can be ignored and data for air can be used without serious error.

2.2.6 Composition of Typical Commercial Fuels

Carbon and hydrogen are seldom burned in their pure forms. Most practical, commercial fuels are mixtures of chemical compounds called hydrocarbons (combinations of

TABLE A.7 Composition of typical commercial fuels and combustible compounds.

Compound	Chemical formula
Natural gas (methane)	CH_4
Acetylene	C_2H_2
Propane	C_3H_8
Butane	C_4H_{10}
Gasoline (octane)	C_8H_{18}
Methanol	CH_3OH
Ethanol	C_2H_5OH
Cellulose	$C_6H_{10}O_5$

TABLE A.8 Composition of typical commercial fuel oils.

	Weight percentage					
Oil no.*	C	H	O	N	S	Ash
1	86.04	13.66	0.00	0.05	0.25	0.00
2	86.65	12.78	0.00	0.05	0.52	0.00
4	87.15	11.71	0.00	0.00	1.09	0.05
5	87.07	11.15	0.00	0.00	1.73	0.05
6	87.08	10.60	0.00	0.00	2.07	0.25

*There is no no. 3 oil.

carbon and hydrogen). Gasoline is made up of several hydrocarbons but most closely resembles octane. Some other gasoline additives such as the alcohols methanol and ethanol also contain some oxygen. Table A.7 lists some of the more common fuels. Pure cellulose is added for reference. For fuels composed of one primary constituent, this constituent is shown in parentheses with its formula.

Fuel oils are graded by number, with no. 1 being the cleanest and least viscous and no. 6 (bunker C) being the dirtiest and most viscous. Table A.8 lists typical average values.

3.0 BASIC SCIENCE OF COMBUSTION

This section builds on the fundamental concepts presented in the previous section to provide a more complete understanding of the combustion process. Included are an introduction to combustion reactions and equations, stoichiometry, higher and lower heating value, and the composition and properties of common auxiliary fuels used in sludge combustion.

3.1 Combustion Reactions and Equations

The terms combustion reactions and combustion equations are used interchangeably.

REACTANTS EQUALS PRODUCTS. An equal (=) sign means that the quantity (lb) on the left-hand side (LHS) is equal to the quantity (lb) on the RHS. In addition to the total pounds being the same on both sides, the quantities (lb) of the individual elements (C, H, O, N, and S) must also be equal. This is generally called "balancing the equations." The illustrative examples below have been selected so that they can be verified by inspection.

Some conventions are used when writing these equations. Whenever the reactions involve molecules, the molecular designations such as O_2 and H_2O are used. In writing the example equations, only whole numbers (no decimals or fractions) are used. For convenience, the LHS of the equation and the RHS of the equation are abbreviated as such. Tables A.9 through A.12 provide examples of the mass balance between the LHS and RHS of several common combustion reactions.

3.2 Stoichiometry

Stoichiometry refers to the science of determining the precise combining proportions of elements and compounds involved in reactions and balancing the reactants with the products on an elemental basis. When the reaction equation is balanced, the quantities of each reactant and product are considered to be in stoichiometric proportion with one another. For example, in the combustion of carbon (see Table A.9), 1 lb-mole of O_2 is stoichiometrically required for the combustion of 1 lb-mole of carbon. Expressed in terms of mass, 32 lb of oxygen are stoichiometrically required for the combustion of 12 lb of carbon (32/12 = 2.67 lb O_2/lb carbon). This is referred to as the stoichiometric ratio, which is the chemically correct ratio of oxygen (or air) to fuel that would produce a mixture capable of perfect combustion with no unused

TABLE A.9 Combustion of carbon ($C + O_2 = CO_2$).

Element	lb (LHS)	lb (RHS)
Carbon	12	12
Oxygen	32	32
Totals	44	44

TABLE A.10 Combustion of hydrogen ($2H_2 + O_2 = 2H_2O$).

Element	lb (LHS)	lb (RHS)
Hydrogen	4	4
Oxygen	32	32
Totals	36	36

TABLE A.11 Combustion of methane ($CH_4 + 2O_2 = CO_2 + 2H_2O$).

Element	lb (LHS)	lb (RHS)
Carbon	12	12
Hydrogen	4	4
Oxygen	64	64
Totals	80	80

TABLE A.12 Combustion of cellulose ($C_6H_{10}O_5 + 6O_2 = 6CO_2 + 5H_2O$).

Element	lb (LHS)	lb (RHS)
Carbon	72	72
Hydrogen	10	10
Oxygen	272	272
Totals	354	354

fuel or oxygen (or air). In practice, to ensure completeness of combustion, more oxygen is provided than the theoretical stoichiometric amount. This extra oxygen is referred to as "excess oxygen" or "excess air" (XSA or XS air) when air is used as the source of oxygen for combustion.

Excess air is the fractional amount supplied in addition to the air required for stoichiometric combustion. Excess air cannot be determined without a measurement of the unreacted oxygen in the flue gas unless flow rates and heating values are known for all fuels entering a furnace, with the total airflow rate. Excess air is calculated by the formula:

$$\text{XS air} = \%O_{2,\,dry} / (20.9\% - \%O_{2,\,dry})$$

where $\%O_{2,dty}$ = the flue gas oxygen concentration (volume, dry basis).

This simplified formula is accurate for high carbon fuels but introduces a slight error for high hydrogen fuels (e.g., natural gas). However, the overall accuracy is sufficient for it to be accepted by the U.S. Environmental Protection Agency for emission calculations.

3.3 Higher and Lower Heating Values

In U.S. customary units, the heating value of a fuel—sometimes referred to as the heat of combustion—is measured in British thermal units (Btu × 1.055 = kJ). A Btu is defined as the amount of heat needed to heat 1 lb of water 1°F. The terms higher heating value (HHV) and lower heating value (LHV) represent different accounting means by which the heat of vaporization of the water vapor formed by the combustion of hydrogen in the fuel is accounted. Some texts refer to HHV as gross heat and LHV as net heat. The HHV, which includes the heat of vaporization, is the most common unit used in the United States. For a 60°F base temperature, the difference between HHV and LHV for a bone-dry fuel is ~1060 Btu/lb of water vapor formed by the combustion of hydrogen in the fuel (Btu/lb × 2.326 = kJ/kg).

3.4 Common Auxiliary Fuels

There are many auxiliary fuels that can assist in the combustion of sludge. However, natural gas and no. 2 fuel oil are the most common. Natural gas is just what its name implies, a natural gas that varies slightly in composition throughout the United States. The main constituent of natural gas is methane (CH_4); therefore, calculations will assume that natural gas is pure methane with an HHV of 23 875 Btu/lb. This measure is equivalent to an HHV of 1009 Btu/cu ft (60°F and 14.696 psia) (Btu/cu ft × 37.26 = kJ/m^3). For computations, natural gas is typically assumed to have an HHV of 1000 Btu/cu ft. Similarly, no. 2 fuel oil exhibits some variation throughout the United States. In some locations, state and local air environmental regulations govern the maximum quantity of sulfur. Calculations in this appendix are based on a fuel oil that is free of nitrogen and sulfur, with a composition of 87% carbon and 13% hydrogen and an HHV of 19 500 Btu/lb. The average density of no. 2 fuel oil is 7.21 lb/gal (lb/gal × 0.119 8 = kg/L).

4.0 COMBUSTION OF AUXILIARY FUELS

As with any combustible compound, auxiliary fuels require oxygen to complete their combustion. In most cases, this oxygen comes from air, which also contains nitrogen and water vapor. For simplicity these initial illustrative stoichiometric calculations assume that the combustion air has zero moisture.

The stoichiometric combustion of methane proceeds as follows:

$$CH_4 + 2O_2 = CO_2 + 2H_2O$$

Starting with 1 lb-mole of CH_4:

$$16 \text{ lb } CH_4 + (2 \times 32 \text{ lb } O_2) = 44 \text{ lb } CO_2 + (2 \times 18 \text{ lb } H_2O)$$

Because the oxygen comes from air, which is a mixture of oxygen and nitrogen, one must calculate the total mass of nitrogen contained in the combustion air. From the composition of air (weight fractions), the mass of nitrogen can be calculated and, hence, the total mass of air, as follows:

$$64 \text{ lb } O_2 \times 0.7685 \text{ lb } N_2/0.2315 \text{ lb } O_2 = 212.46 \text{ lb } N_2$$
$$64 \text{ lb } O_2 + 212.46 \text{ lb } N_2 = 276.46 \text{ lb air}$$

Therefore, the equation for the combustion of methane in air is as follows:

$$16 \text{ lb } CH_4 + 276.46 \text{ lb air} = 44 \text{ lb } CO_2 + 36 \text{ lb } H_2O + 212.46 \text{ lb } N_2$$

The stoichiometric combustion of 1 lb no. 2 fuel oil proceeds as follows:
Reaction inputs (LHS):

$$1.00 \text{ lb oil} = 0.87 \text{ lb C} + 0.13 \text{ lb } H_2$$

Stoichiometric oxygen required =

$$(0.87 \text{ lb C}/12 \text{ (lb/lb-mole C)} \times 1.0 \text{ (lb-mole } O_2\text{/lb-mole C)} \times$$
$$32 \text{ (lb/lb-mole } O_2) = 2.32 \text{ lb } O_2\text{/lb oil for carbon combustion} +$$
$$(0.13 \text{ lb } H_2/2 \text{ (lb/lb-mole } H_2) \times 0.5 \text{ (lb-mole } O_2\text{/lb-mole } H_2) \times$$
$$32 \text{ (lb/lb-mole } O_2) = 1.04 \text{ lb } O_2\text{/lb oil for hydrogen combustion}$$
$$= 3.36 \text{ lb } O_2\text{/lb oil}$$

Reaction products (RHS) =

$$(0.87 \text{ lb C}/12 \text{ (lb/lb-mole C)} \times 1.0 \text{ (lb-mole CO}_2\text{/lb-mole C)} \times$$
$$44 \text{ (lb/lb-mole O}_2\text{)} = 3.19 \text{ lb CO}_2\text{/lb oil from carbon combustion} +$$
$$(0.13 \text{ lb H}_2/2 \text{ (lb/lb-mole H}_2\text{)} \times 1.0 \text{ (lb-mole H}_2\text{O/lb-mole H}_2\text{)} \times$$
$$18 \text{ (lb/lb-mole H}_2\text{O)} = 1.17 \text{ lb O}_2\text{/lb oil from hydrogen combustion}$$
$$= 4.36 \text{ lb (CO}_2 + \text{H}_2\text{O)/lb oil total}$$

Overall equation based on oxygen =

$$1 \text{ lb oil} + 3.36 \text{ lb O}_2 = 3.19 \text{ lb CO}_2 + 1.17 \text{ lb H}_2\text{O}$$

Calculation of nitrogen and stoichiometric air:

$$N_2 = 3.36 \text{ lb O}_2 \times 0.7685 \text{ lb N}_2/0.2315 \text{ lb O}_2 = 11.15 \text{ lb N}_2\text{/lb oil}$$
$$\text{Air} = 3.36 \text{ lb O}_2 + 11.15 \text{ lb N}_2 = 14.51 \text{ lb air/lb oil}$$

Overall equation based on air =

$$1 \text{ lb oil} + 14.51 \text{ lb air} = 3.19 \text{ lb CO}_2 + 1.17 \text{ lb H}_2\text{O} + 11.15 \text{ lb N}_2$$

Adding 20% excess air:

$$\text{XS air} = 0.2 \times 14.51 \text{ lb air/lb oil} = 2.90 \text{ lb XS air/lb oil}$$
$$= 0.2 \times 3.36 \text{ lb O}_2\text{/lb oil} = 0.67 \text{ lb O}_2\text{/lb oil}$$
$$N_2 \text{ in XS air} = 0.2 \times 11.15 \text{ lb N}_2\text{/lb oil} = 2.23 \text{ lb N}_2\text{/lb oil}$$

Overall equation based on 20% XS air:

$$1 \text{ lb oil} + 17.41 \text{ lb air} = 3.19 \text{ lb CO}_2 + 1.17 \text{ lb H}_2\text{O} + 0.67 \text{ lb O}_2 + 13.38 \text{ lb N}_2$$
$$18.41 \text{ lb reactants} = 18.41 \text{ lb products (17.24 lb dry products)}$$

Calculating the number of moles of each component of the flue gas is done by dividing the calculated mass by the respective molecular weight of each component, as follows:

$$3.19 \text{ lb CO}_2/44 \text{ (lb/lb-mole CO}_2\text{)} = 0.0725 \text{ lb-mol CO}_2 = 11.4\%$$
$$1.17 \text{ lb H}_2\text{O}/18 \text{ (lb/lb-mole H}_2\text{O)} = 0.0650 \text{ lb-mol H}_2\text{O} = 10.2\%$$
$$0.67 \text{ lb O}_2/32 \text{ (lb/lb-mole O}_2\text{)} = 0.0209 \text{ lb-mol O}_2 = 3.3\%$$
$$13.38 \text{ lb N}_2/28 \text{ (lb/lb-mole N}_2\text{)} = 0.4779 \text{ lb-mol N}_2 = 75.1\%$$
$$\text{Total} = 0.6363 \text{ lb-mol} = 100.0\%$$

Calculating the volume at standard conditions for each component of the flue gas:

$$0.0725 \text{ lb-mole } CO_2 \times 379.5 \text{ cu ft/(lb-mole)} = 27.51 \text{ cu ft } CO_2 = 11.4\%$$
$$0.0650 \text{ lb-mole } H_2O \times 379.5 \text{ cu ft/(lb-mole)} = 24.67 \text{ cu ft } H_2O = 10.2\%$$
$$0.0209 \text{ lb-mole } O_2 \times 379.5 \text{ cu ft/(lb-mole)} = 7.95 \text{ cu ft } O_2 = 3.3\%$$
$$0.4779 \text{ lb-mole } N_2 \times 379.5 \text{ cu ft/(lb-mole)} = 181.36 \text{ cu ft } N_2 = 75.1\%$$
$$\text{Total} = 241.48 \text{ cu ft} = 100.0\%$$

These calculations demonstrate that volume percent and mole percent are equal. The volume and mole percents tabulated above are calculated on a wet basis. Often, it is necessary to calculate volume percents on a dry basis for emissions reporting and other purposes. The shortcut formula for excess air requires dry volume percent oxygen. This is calculated by subtracting the volume of H_2O from the total and recomputing the relative percentage of the remaining components as follows:

$$\text{Wet volume } \% \, O_2 = 7.95 \text{ cu ft } O_22/241.48 \text{ cu ft} \times 100 = 3.29\% \, O_{2,\,wet}$$
$$\text{Dry volume } \% \, O_2 = 7.95 \text{ cu ft } O_2/(241.48 \text{ cu ft} - 24.67 \text{ cu ft}) \times 100 = 3.67\% \, O_{2,dry}$$

From the previous formula for XS air:

$$\text{XS air} = 3.67\% \, O_{2,\,dry}/(20.9\% - 3.67\% \, O_{2,\,dry}) = 21.3\%$$

The actual basis for the combustion calculation was 20% XS air. As can be seen, the above shortcut method introduced a slight error. However, as stated previously, it is generally adequate for emission rate computations.

4.1 Adiabatic Flame Temperature

The adiabatic flame temperature is the temperature that would be reached with perfect stoichiometric combustion (0% excess air), zero heat loss, and no disassociation of flue gas molecules. The adiabatic flame temperatures of natural gas and no. 2 oil are between 1927 and 2065°C (3500 and 3750°F). These extremely high temperatures for dry fuels are of academic interest but do not play a role in sludge incineration calculations.

4.2 Theoretical Temperature of the Products of Combustion

The theoretical temperature of the products of combustion (TTPC) is the temperature (calculated) that would be reached if a wet or dry fuel was burned with zero heat loss. In some cases, especially for a high moisture sludge cake with no auxiliary fuel, the TTPC may be well below the temperatures required to produce actual burning and a flame. For any given fuel or fuel combination, the flame temperature at zero heat loss is a unique function of the combined effects of air preheat temperature, fuel temperature, XS air (oxygen in flue gas), and the moisture content of the fuel.

4.3 Availability of Heat

In the previous sections, the gross HHV of a few different fuels were identified. The next section will deal with the subject of available heat. Available heat can be defined as the HHV of the fuel or waste being combusted minus heat losses associated with the flue gas, including the heat of vaporization associated with the flue gas moisture. The percent available heat is the available heat divided by the heat input. Available heat must always be defined at specific inlet and outlet temperatures and excess air level. When associated with auxiliary fuels, it generally represents the heat available for transfer to another body.

The concept of available heat is also useful to characterize wet sludge cake. If the available heat (temperatures and XS air level specified) from the flue gas resulting from the combustion of sludge cake is negative, then addition of heat to the process is required. If the available heat is zero, then the sludge cake is autogenous at specified conditions. If the available heat is a positive number, then either heat must be extracted to meet the specified conditions or the conditions must be changed so as to result in zero availability. This may involve reducing the inlet temperature, increasing the outlet temperature, increasing the XS air level, or some combination of these three methods.

The calculation of the available heat from common auxiliary fuels will serve as an introduction to the broader concept of heat and material balances for sludge incineration systems. For this, it is necessary to introduce the concept of enthalpy. The enthalpy of a flue gas represents the total heat content, expressed as Btu/lb, of a gas above a baseline or reference temperature which, for this text, is 15°C (60°F). Enthalpy is the mathematical product of the difference between the actual gas temperature and the base temperature times the average specific heat between those two temperatures:

$$h = C_p \times (T_a - T_b)$$

Where

h = enthalpy, Btu/lb;
C_p = average specific heat at constant pressure (between T_a and T_b), Btu/(lb-°F);
T_a = actual temperature, °F; and
T_b = base temperature, °F.

This heat is often referred to as sensible heat. The specific heat of common flue gases over temperatures of general interest in sludge incineration (60 to 2200°F) is not constant and there are numerous equations that have been developed to quantify this variation. To determine the average specific heat these equations for specific heat must be integrated over the temperature range of interest.

For this appendix, this often complex mathematical integration step will be eliminated and the following fourth power polynomial equations for enthalpy (60°F base) will be used. The enthalpy (h) at any temperature (T) between 60 and 2200°F may be calculated as follows:

$$h = X_0 + X_1 \times T + X_2 \times T_2 + X_3 \times T_3 + X_4 \times T_4$$

where X_0, X_1, X_2, X_3, and X_4 are constants used in the equation.

Polynomial equations such as the one above are easily programmed into a spreadsheet. The five equation constants are unique for each gaseous compound and are tabulated in Table A.13 for the more common gaseous compounds found in flue gases.

Application of these constants in the basic equation gives the enthalpy of each compound at the selected temperature. Table A.14 summarizes the computed values of enthalpy for the more common gaseous compounds found in flue gases at temperatures from 60 to 2200°F.

It is important not to use these simplified polynomial equations outside their range of applicability (60 to 2200°F), as significant errors may be introduced. Figure A.1 depicts the variations in enthalpy calculated from these equations in graphical format over their range of applicability.

The heat of vaporization of water at 60°F is taken as 1059.7 Btu / lb. When ash is a constituent in the flue gas that must be considered, it does have a fairly constant spe-

TABLE A.13 Enthalpy constants for common flue gases (60°F base).

Constant	CO_2	O_2	N_2	H_2O	SO_2
X_0	−11.781 55	−12.985 06	−14.846 22	−27.141 43	−8.687 84
X_1	0.192 338	0.215 133	0.247 333	0.451 462	0.142 168
X_2	6.815×10^{-5}	2.151×10^{-5}	1.230×10^{-6}	1.399×10^{-5}	4.466×10^{-5}
X_3	-1.890×10^{-8}	-1.731×10^{-9}	8.612×10^{-9}	1.565×10^{-8}	-1.425×10^{-8}
X_4	2.357×10^{-12}	-2.947×10^{-13}	-1.928×10^{-12}	-2.760×10^{-12}	1.947×10^{-12}

TABLE A.14 Enthalpy of common flue gases (60°F base).

Temperature °F	CO_2	O_2	N_2	H_2O	SO_2
60	0.00	0.00	0.00	0.00	0.00
100	8.12	8.74	9.91	18.16	5.96
500	99.21	99.72	110.08	203.87	71.90
1000	232.16	221.63	240.40	451.20	165.84
1500	378.19	350.78	378.22	720.38	266.83
2000	531.97	484.76	522.78	1012.80	371.49
2200	595.13	539.08	581.77	1135.79	414.18

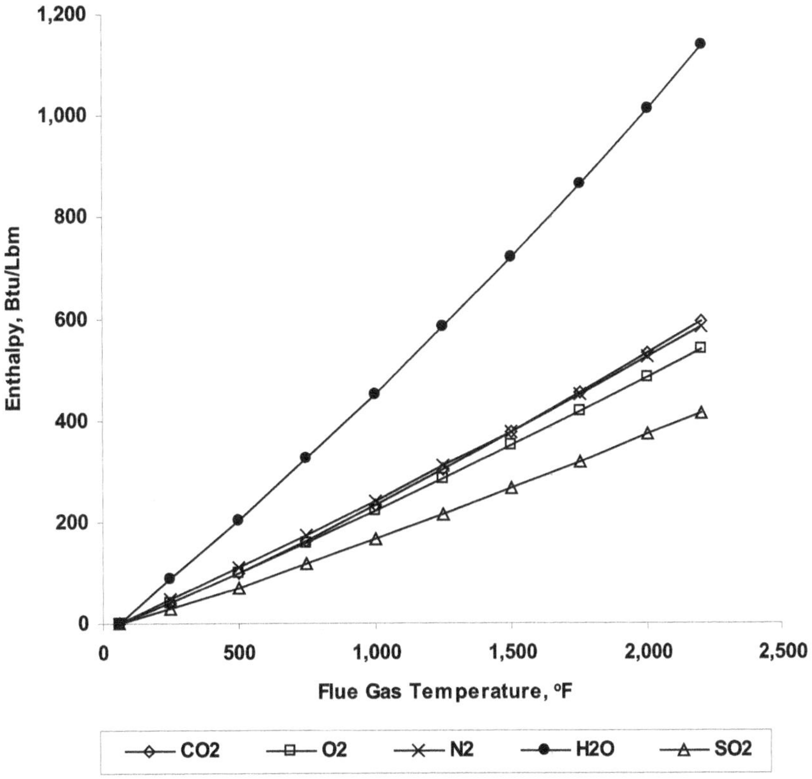

FIGURE A.1 Enthalpy of common flue gases (60°F base).

cific heat (~0.22 Btu/lb°F) over a wide range. The enthalpy of ash, referenced to 60°F, may be computed as follows:

$$h_{(ash)} = 0.22 \times (T_a - T_b) = 0.22 \times (T_a - 60) = (0.22 \times T_a - 13.2)\ \text{Btu/lb}$$

The following heat and material balance calculates the availability of methane at 1400°F and 25% XS air. In this example, the air is bone dry at a temperature of 60°F, hence provides neither sensible heat nor heat of vaporization into the system. The calculations are shown in the general format of an accounting balance sheet to make the calculation steps easy to follow and understand by inspection. Tables A.15 and A.16 provide material and heat balances, respectively, for the combustion of natural gas (CH_4) at an exhaust temperature of 1400°F and 25% XS air.

The next set of calculations represents fuel oil at 40% XS air and 1600°F. In this example, the air temperature is 125°F and has an absolute humidity of 0.01 lb H_2O/lb bone dry air. These are presented in Tables A.17 and A.18.

TABLE A.15 Material balance for combustion of 100 lb/hr of natural gas (CH_4) at 60°F inlet and 1400°F exhaust temperature, 25% XS air .

	Carbon	Hydrogen	Oxygen	Nitrogen	Sulfur	Total
Inputs, lb/hr[a]						
Fuel	75.00	25.00	0.00	0.00	0.00	100.00
Stoichiometric air			400.00	1327.86		1727.86
XS air			100.00	331.97		431.97
Air moisture		0.00	0.00			0.00
Totals	75.00	25.00	500.00	1659.83	0.00	2259.83
Outputs, lb/hr						
CO_2	75.00		200.00			275.00
O_2			100.00			100.00
N_2				1659.83		1659.83
H_2O		25.00	200.00			225.00
SO_2			0.00		0.00	0.00
Totals	75.00	25.00	500.00	1659.83	0.00	2259.83
Outputs, lb-mole/hrb	CO_2	O_2	N_2	H_2O	SO_2	Total
lb-mole/hr	6.25	3.13	59.28	12.50	0.00	81.16
Volume % (wet)	7.70%	3.86%	73.04%	15.40%	0.00%	100.00%
Volume % (dry)	9.10%	4.56%	86.34%		0.00%	100.00%

[a] lb/hr × 0.453 6 = kg/h.
[b] lb-mole/hr × 0.4536 = kg-mol/h.

TABLE A.16 Heat balance for combustion of 100 lb/hr of natural gas (CH_4) at 60°F inlet and 1400°F exhaust temperature, 25% XS air.

	Heat of combustion	Heat of vaporization	Sensible heat	Total
Heat inputs, Btu/hr*				
Fuel	2 387 500			2 387 500
Air			0	0
Air moisture		0		0
Totals	2 387 500	0	0	2 387 500
Heat outputs, Btu/hr				
CO_2			95 767	95 767
O_2			32 448	32 448
N_2			581 030	581 030
H_2O		238 433	149 551	387 983
SO_2			0	0
Totals	0	238 433	858 795	1 097 228
Difference (Input − Output), Btu/hr				1 290 272
Percent availability (Difference/Input)				54.04%

*Btu/hr × 0.2931 = W.

TABLE A.17 Material balance for combustion of 100 lb/hr of no. 2 oil at 125°F inlet and 1600°F exhaust temperature, 40% XS air.

	Carbon	Hydrogen	Oxygen	Nitrogen	Sulfur	Total
Inputs, lb/hr						
Fuel	87.00	13.00	0.00	0.00	0.00	100.00
Stoichiometric air			336.00	1115.40		1451.40
XS air			134.40	446.16		580.56
Air moisture		2.26	18.06			20.32
Totals	87.00	15.26	488.46	1561.56	0.00	2152.28
Outputs, lb/hr						
CO_2	87.00		232.00			319.00
O_2			134.40			134.40
N_2				1561.56		1561.56
H_2O		15.26	122.06			137.32
SO_2			0.00		0.00	0.00
Totals	87.00	15.26	488.46	1561.56	0.00	2152.28
Outputs, lb-mole/hr	CO_2	O_2	N_2	H_2O	SO_2	Total
lb-mole/hr	7.25	4.20	55.77	7.63	0.00	74.85
Volume % (wet)	9.69%	5.61%	74.51%	10.19%	0.00%	100.00%
Volume % (dry)	10.79%	6.25%	82.96%		0.00%	100.00%

TABLE A.18 Heat balance for combustion of 100 lb/hr of no. 2 oil at 125°F inlet and 1600°F exhaust temperature, 40% XS air.

	Heat of combustion	Heat of vaporization	Sensible heat	Total
Heat inputs, Btu/hr				
Fuel	1 950 000			1 950 000
Air			1 581	1 581
Air moisture		21 533		21 533
Totals	1 950 000	21 533	1 581	1 973 114
Heat outputs, Btu/hr				
CO_2			130 291	130 291
O_2			50 705	50 705
N_2			635 046	635 046
H_2O		145 518	106 702	252 220
SO_2			0	0
Totals	0	145 518	922 744	1 068 262
Difference (Input − Output), Btu/hr				904 852
Percent availability (Difference/Input)				45.86%

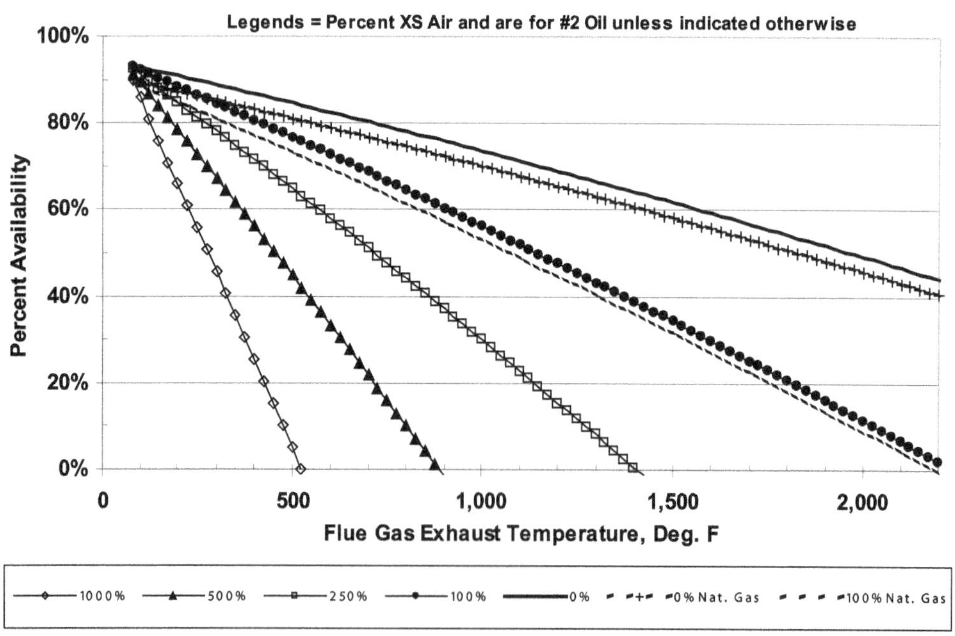

FIGURE A.2 Availability of heat for no. 2 fuel oil and natural gas (CH_4).

Figure A.2 provides a graphical representation of the availability of heat for no. 2 oil and natural gas (CH_4). Curves are shown for various XS air levels and show the percent availability as a function of flue gas exhaust temperature. All of the curves were developed using an inlet air temperature of 60°F and bone dry air. The adiabatic flame temperature, or TTPC, is the temperature at zero percent (0%) availability.

When the concept of available heat is expanded to sludge cake and the results are plotted graphically, it provides the means to quickly visualize the effects of different parameters. Figure A.3 demonstrates the effects of varying sludge solids and combustion air preheat for one particular type of sludge. Unique calculations are required to match the properties of sludge at each application, so Figure A.3 should be used only as an example of the general trends and shape of the availability curves for

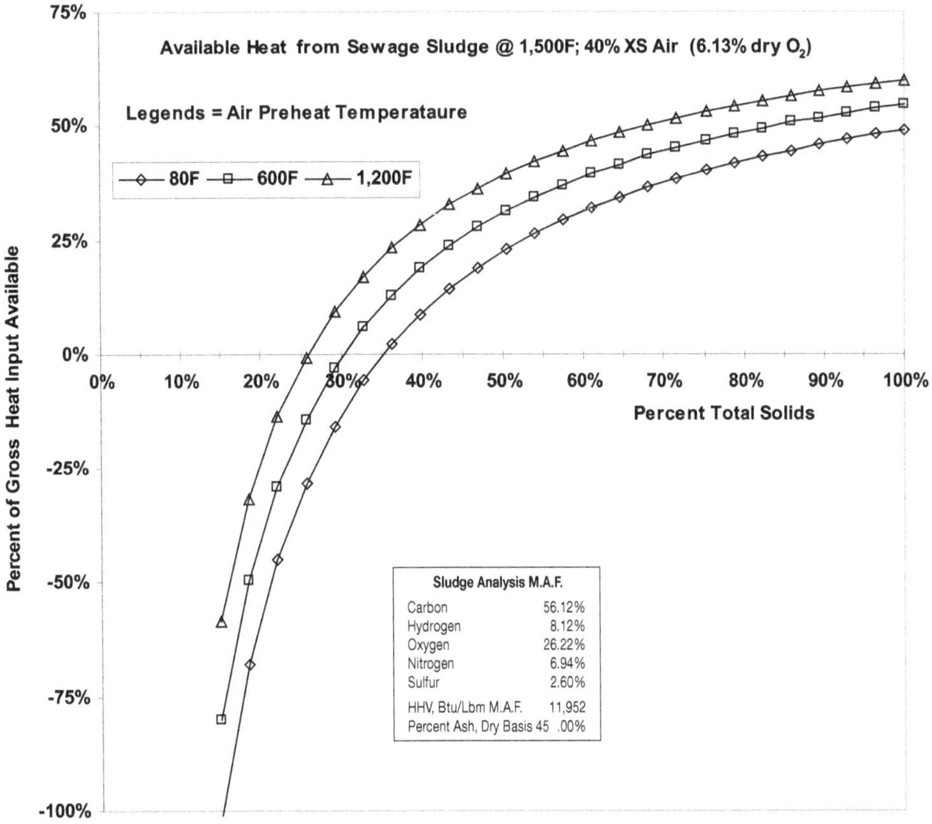

FIGURE A.3 Example of availability of heat for sludge combustion.

sludge. Gross heat input includes not only the heat of combustion of the sludge volatiles but also the sensible heat in the preheated combustion air. Autogenous combustion is achieved at zero percent (0%) availability. When availability is less than 0%, additional heat input is required to bring the products of combustion to temperature. In some cases this can be done by increasing inlet air temperature, reducing exhaust temperature, or reducing the XS air level, all within certain limits. When this is not possible because of process or mechanical considerations, it is necessary to provide auxiliary fuel to the combustion system. When availability is greater than 0%, either the exhaust temperature or XS air must be increased or the inlet air temperature must be reduced to achieve a thermal balance. Changes in sludge solids below the autogenous point have a pronounced effect on availability. However, increasing the percent sludge solids above the autogenous value can be characterized by diminishing returns.

Appendix B

Incineration Subsystems

1.0 INTRODUCTION

This section provides background information on sludge feed systems, blowers and fans, and burners and supplemental fuel systems.

2.0 FEED SYSTEMS

There are three common types of feed systems: cake pumps, belt conveyors, and screw conveyors. Fluid bed incinerators are typically fed in or just above the sand bed, although some systems have feed ports in the freeboard. The fluid bed feed system needs to be designed to discharge a higher pressure than the fluidizing blower. Multiple-hearth furnaces (MHFs) are typically fed at the top.

2.1 Cake Pumps

There are two types of cake pumps used. The most common is the high-pressure piston pump because it is able to develop nearly 6895 kPa (1000 psi) of material pressure and can convey material up to 152.4 m (500 ft), depending on the solids characteristics. In general, the drier the cake, the more difficult it is to pump. For shorter distances, a progressing cavity pump may be considered.

2.1.1 *High-Pressure Piston Pump*

The high-pressure piston pump consists of a hydraulic unit, auger feeder, and pump. The hydraulic unit uses a pump to develop hydraulic pressure which is used to operate the auger feeder and pump. The auger feeder has two screws that convey the solids from a collection bin to the pistons. Because the feed and fill connections are near the discharge, poppet valves are used to control the fill and discharge cycle of the pistons. High-pressure piston pumps are available with one or two discharge connections. Because most incinerators have two feed points, the cake pumps typically are equipped with two discharge connections with independent piping systems to the incinerator. The advantage of a two-pipe system is that the solids and heat are more evenly distributed, resulting in smaller temperature gradients within the incinerator. A photograph of a high-pressure piston pump is presented in Figure B.1.

The sizing of the cake pumps needs to be based on the maximum and minimum feed rates anticipated. The operator typically has control over the timing of the discharge but not the volume. If the pump is too large, the volume may overwhelm the incinerator and cause temperature gradients.

2.1.2 *Flow Measurement*

Cake pumps can have two types of flow measurement: a pulsating flow meter or a proprietary system that measures the volume displaced. Both systems have been calibrated to within 5% accuracy.

2.1.3 *Pipeline Lubrication System*

For high-pressure applications, a lubrication pump, injection ring, and pressure sensor are added to minimize the pressure. The system is designed to start on high pressure and inject a thin film of water along the interior of the pipe to reduce friction losses. The water can reduce by 50% the operating pressure of the piping system.

FIGURE B.1 A high-pressure piston pump with single discharge (courtesy of Metropolitan Sewer District of Greater Cincinnati).

Other fluids such as polymer or mineral oil could also be used as a lubricant. Interlocks are required to prevent the lubrication pump from operating when the piston pump is stopped or in the idle mode (the hydraulic pump is in recirculation mode and no solids are being pumped).

2.1.4 *Piping System*

The piping system needs to be designed for high-pressure application. Carbon steel piping with welded joints or victaulic fittings are common. Because the piping system is operating at high pressures, a piping anchor near the pump is recommended to limit the forces that are transferred to the piping system.

2.1.5 *Progressing Cavity Pumps*

The progressing cavity cake pumps consist of a bridge-breaker and high-pressure progressing cavity pump. The pressure developed depends on the number of pumping stages. Multistage pumps can deliver discharge pressures greater than 2068 kPa (300 psi). The progressing cavity pump has only one discharge so two pumps may be required in large installations. A photograph of a progressing cavity solids pump is presented in Figure B.2.

FIGURE B.2 A progressing cavity pump with bridge-breaker (courtesy of Burl, Washington).

2.2 Screw Conveyors

Screw conveyors are compact and may be used to convey solids horizontally and vertically. They rotate in a stationary trough. There are two types of screw conveyors: shaftless (see Figure B.3) and shafted. Both types are commonly used in bulk material conveying applications. Unlike belt conveyors, screw conveyors may be enclosed, reducing the potential for odors. Consistency of the material needs to be considered, however; systems with a sticky consistency have had problems, particularly in vertical applications.

2.3 Belt Conveyors

Belt conveyors have been used in many applications to convey solids from dewatering to incineration, especially for MHFs because of low pressure of the feed location. Belt conveyors come in varying types; flat belts are the most common but ribbed belts are sometimes used, especially for inclines. Weight scales can be installed to measure the quantity of solids conveyed. Because belt conveyors are open, they can result in nuisance odors and require additional operator awareness and safety precautions.

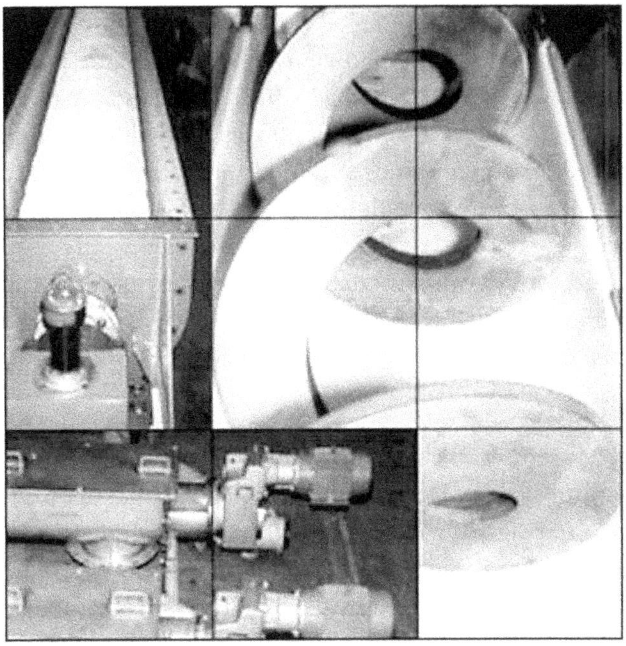

FIGURE B.3 A shaftless screw conveyer (courtesy of Schwing Bioset, Inc.).

3.0 BLOWERS AND FANS

Blowers and fans are used to supply combustion air (positive pressure in burner or incinerator) to remove exhaust gases (negative pressure in incinerator). Some MHFs have been retrofitted to include an exhaust gas recirculation system that conveys hot exhaust gas from exhaust breeching to lower hearths for NOx control.

3.1 Blowers

Blowers are used in fluidized bed incinerators to provide fluidizing air and purge air.

3.1.1 Fluidizing Air Blower

Fluidizing air systems consist of multistage centrifugal blowers, intake filter silencers, and a control system. Fluidizing air blowers are capable of producing pressures of 34 to 55 kPa (5 to 8 psig). Intake filters clean the air and remove debris that could damage or clog the blower. The silencer reduces noise transmission through the filter. The multistage blower consists of a series of radial and backward inclined blades that provide the pressure and shape of the curve. Blowers with more backwardly inclined blades have a flatter blower discharge pressure and flow curve. The turndown on a blower is controlled by inlet valves or vanes and is typically approximately 60% of the rated blower capacity but may be limited by the fluidizing requirement of the fluid bed. The incinerator supplier may recommend a turndown of no greater than 90% to ensure adequate fluidization.

Some fluid bed systems use the fluidizing blower as the only motive force in the system and are sized for the pressure drop across the incinerator, heat recovery, air pollution control, and stack. If a waste heat boiler is used, an induced draft fan is typically added to keep a negative pressure in the waste heat boiler.

3.1.2 Purge Air Blower

Purge air blower systems consist of positive-displacement blowers, intake filter silencers, and a control system. Purge air blowers are capable of producing pressures greater than 70 kPa (10 psig). The advantage of using a positive-displacement blower for purge air is that they are able to produce constant airflow rates at varying operating pressures. Flowrates are changed by adjusting the speed of the blower, typically through an adjustable-frequency drive.

Purge air blowers provide protection of equipment and are independent of the fluidizing air system to ensure airflow in the event the blower is stopped. The air is fed to the annular space in the fuel guns, high-pressure spray nozzles, sight glasses, and at instrument locations to prevent thermal damage.

3.2 Fans

Fans serve the same function as blowers but develop less differential pressure (fans are typically rated in terms of inches per water column and blowers in pounds per square inch).

3.2.1 Combustion Air Fan

Combustion air fans are used to supply air to burners or to combustion air ports on MHFs. Centershaft air fans supply air to cool the centershaft, which is subject to hot temperatures and is often used as a source for combustion air in MHFs. Outside or building air is often used as the air supply.

3.2.2 Induced Draft Fans

Induced draft fans are used to keep a negative pressure in the incinerator. The fans are installed after the air pollution control system and before the stack. As a result, the exhaust gas is at a temperature between 30 to 80°C (100 and 180°F), saturated with water, and abrasive because of the small amount of particulate matter that may pass through. The water that condenses can form acids that attack the fan blades. As a result, the casing is typically constructed of 316L stainless steel, Hastelloy, Inconel, or other corrosion- and abrasion-resistant material. Designers should consider providing trapped drains to remove condensate and an inspection port to view the condition of the blades and housing.

3.2.3 *Recirculation Fans*

Recirculation fans have been used to recirculate hot gases from the exhaust breeching of MHFs to the lower hearths. This recirculation has the advantage of providing a more even heat distribution throughout the incinerator and reduces the oxygen content within the incinerator. As a result, these fans operate at temperatures of 427 to 649°C (800 to 1200°F) and are subject to severe soot and abrasive loadings. The high temperature needs to be considered both from a material expansion and personnel safety perspective.

4.0 AUXILIARY FUEL SYSTEMS

Auxiliary fuel systems provide fuel to burners or to fluid bed injection guns. For MHFs, the auxiliary fuel is fed to burners and, for fluid bed incinerators, auxiliary fuels may be fed to burners or to fuel guns that inject the fuel directly to the bed. Fuel oil and natural gas are the most common types of auxiliary fuel, but digester gas and coal have also been used successfully.

4.1 Fuel Oil

Fuel oil systems consist of a storage tank day tank, positive-displacement pump(s), piping and flow measurement, and control and relief valves.

Burner pumps are typically constant speed with an oil return system. The oil injection pumps supplying the oil guns are typically variable speed to control the bed temperature. Regulations require the bed temperature to be higher than 620°C (1150°F) before direct injection of oil is permitted.

For fluid bed incinerators, oil guns are typically installed in the middle of the sand bed to ensure complete mixing. The oil gun is an injection nozzle that extends into the bed to evenly distribute the heat. The number of guns is dependent on the size and can be as high as 12 guns for large units. Each injection nozzle typically has an isolation valve, quick coupling, flow meter, and check valve. A block and bleed is provided to prevent oil feed into an offline incinerator if the operator does not remove the feed gun. The nozzle fits into the annular space of a larger pipe and purge air is provided to the annular space for cooling.

4.2 Gas Systems

Most systems use natural gas, but some use digester gas or propane for a pilot on the burner. Gas systems include a valve train assembly that includes pressure-reducing valves, block and bleed assembly, flow and pressure measurement, and pressure-regulating valves. Gas is provided to burners or for fluid bed incinerators to gas guns that inject gas directly to the bed. Regulations require the bed temperature to be higher than 732°C (1350°F) before direct injection of gas is permitted.

For fluid bed incinerators, gas guns are typically installed at the bottom of the sand bed, which allows combustion to complete in the bed and not the freeboard. The gas gun is an injection nozzle that extends into the bed to evenly distribute the heat. The number of gas guns is dependent on size and can be as high as 16 guns for large units.

For fluid bed incinerators, the gas supply pressure must be higher than the fluidizing air blower maximum discharge pressure to ensure flow. Facilities that replace a multiple hearth with a fluid bed incinerator typically require a pressure increase.

4.3 Other Fuels

Coal has also been used successfully as an auxiliary fuel for a fluid bed incinerator. Coal is an inexpensive supplemental fuel that has been used in facilities that are generating electricity and want to keep a constant heat rate. Coal handling and regulatory issues must be considered.

Appendix C

Determination of Wastewater Solids Incineration Related Costs

The management of wastewater solids has always been an issue of concern for wastewater treatment plants (WWTPs), especially because these costs substantially contribute to the overall costs of operating and maintaining a WWTP. Moreover, many utilities find it difficult to track all of the costs associated with wastewater solids management. Representative utility-wide costs from which to benchmark operations are difficult to find.

Some of the questions frequently asked by WWTPs that practice incineration are

- How can I determine the actual cost to incinerate wastewater solids at my plant?
- Which costs should be included in this analysis?
- How do my incineration costs compare to incineration costs at similar WWTPs?

Appendix C provides a framework for WWTPs that practice incineration to accurately determine their costs to incinerate.

1.0 COMPOSITION OF WASTEWATER SOLIDS INCINERATION COSTS

Incineration costs consist of both amortized capital costs and operation and maintenance (O&M) costs. The components of these two costs are as follows:

- Amortized capital cost
    - Initial construction costs and
    - Subsequent contractor-related replacement and repair costs.
- Operation and maintenance costs
    - Direct labor and benefits for WWTP incineration personnel;
    - Incineration-related electrical, natural gas, fuel oil, and scrubber water costs;
    - Maintenance work performed by WWTP personnel; and
    - Ash disposal costs.

2.0 STARTING POINT

Collecting accurate incineration-related wastewater solids management costs is a daunting task that presents numerous challenges. The first step in this process is to select a starting point in the solids processing train.

For the majority of WWTPs, the starting point can be assumed to be immediately after dewatering because wastewater solids can either be incinerated, landfilled, land applied (depending on quality), or placed in a surface disposal site.

Because incinerator ash is a byproduct of wastewater solids incineration, its reuse and disposal cost is included in the overall incineration cost determination. The starting point to determine the ash-related disposal cost is typically at the point the ash leaves the incinerator.

3.0 STANDARD MULTIPLE-HEARTH FURNACE SYSTEM EXAMPLE

The following is a summary of the costs incurred by the Northeast Ohio Regional Sewer District (NEORSD) in 2004 to incinerate wastewater solids at its Westerly WWTP (Dominak, 2005). Westerly has two multiple-hearth furnaces (MHFs) that were originally constructed in 1978 and underwent a significant renovation in 1994 to 1995. Wastewater solids are injected to hearth number 4, and the top hearths are used as an afterburning zone to minimize emissions of potentially toxic organic compounds. Table C.1 contains a summary of some of the differences between wastewater solids and wastewater solids incineration-related costs at the NEORSD's Southerly and Westerly WWTPs.

TABLE C.1 Differences between wastewater solids and wastewater solids incineration-related costs at the NEORSD's Southerly and Westerly WWTPs.

	Southerly WWTP	Westerly WWTP
Biosolids incinerated	75 503 wet tons[a]	16 444 wet tons
Percent solids	47%	33%
Dewatering device	High solids centrifuges	High solids centrifuges
Biosolids conditioning	Thermal (Zimpro)	Chemical
Heat exchangers	No	Yes
Waste heat boilers	Yes	No
Steam production natural gas savings	120 702 MCF[b]	0
Steam production natural gas savings	$627,110	0
Ash disposal method	Beneficially reused	Landfill
Ash disposal costs	$115,000	$98,717
Total incineration O&M costs	$1,559,747	$902,961
Total incineration O&M unit cost	$21.19/wet ton incinerated	$54.92/wet ton incinerated
Net incineration O&M costs[c]	$972,637	NA[d]
Net incineration O&M unit cost	$12.88/wet ton incinerated	NA

[a] wet ton × 0.907 = metric tonne.

[b] MCF = 1000 cu ft = 28 m^3.

[c] Net incineration O&M costs = total incineration O&M costs—natural gas related cost savings attributable to the steam produced by the waste heat boilers.

[d] NA = not applicable.

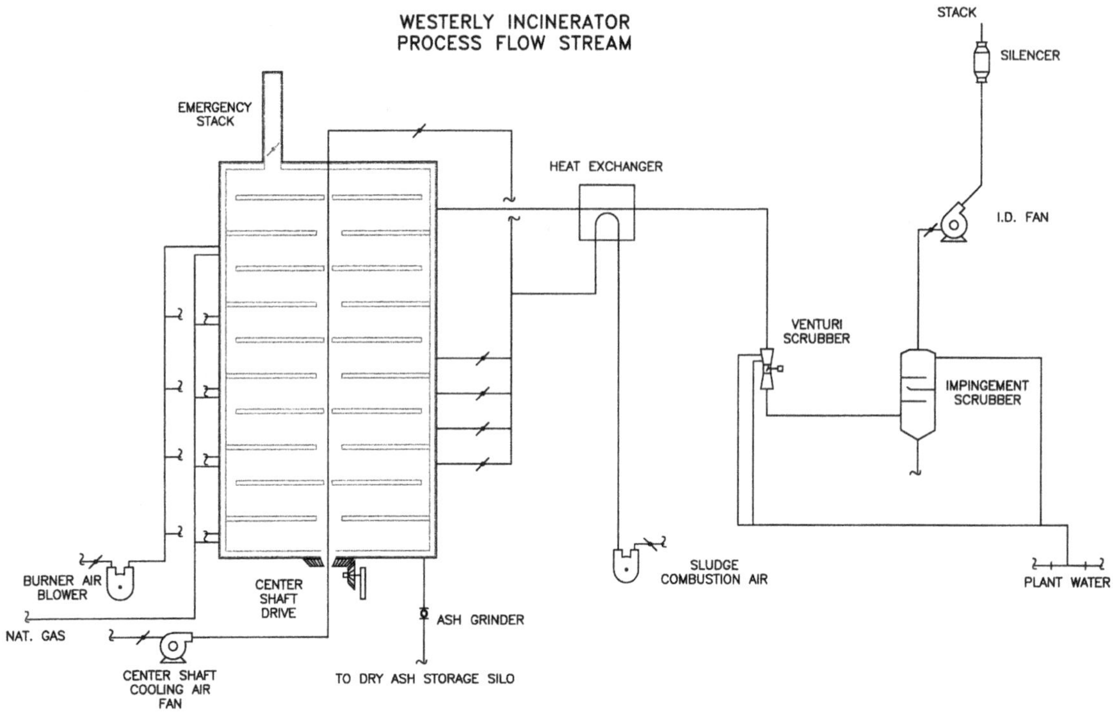

Figure C.1 Westerly WWTP incinerator process flow stream (I.D. = induced draft).

Figure C.1 shows the components of Westerly's MHF system and Table C.2 provides all pertinent details for NEORSD.

3.1 Natural Gas

Natural gas used by the incineration process is metered.

$$64\,792 \text{ MCF} \times \$6.18/\text{MCF} = \$400{,}415$$
$$\text{MCF} = 1000 \text{ cu ft } (28 \text{ m}^3)$$

3.2 Electricity

Electricity used by the incinerator process is not separately metered. However, the total horsepower of the incineration equipment is known as are the number of hours per year in operation (see Table C.3).

$$275 \text{ total hp} \times 0.746 \text{ kW/hp} \times 6694 \text{ hours of operation} \times \$0.060/\text{kWh} = \$82{,}396$$

TABLE C.2 Pertinent information for Northeast Ohio Regional Sewer District case study.

Year of cost determination	2004
Quantity of wastewater solids incinerated	16 441 wet tons (5426 dry tons)[a]
Dewatering devices	High solids centrifuges
Percent solids in dewatered wastewater solids	~ 33%
Year MHFs constructed	1978
Year MHIs placed into service	1983 (startup was delayed for almost five years because of construction-related problems associated with the rest of the plant. This delayed the production of the specified wastewater solids.)
Year of significant MHF system renovations	1994–1995
Service life of all major equipment	20 years
Total number of incinerator operators	4 at 2080 hr/yr
Normal operator overtime	15% (312 hr/yr)
Shift manager	1 at 33% of time (686 hr/yr)
Employee benefits	47.66% of hourly rate
Natural gas cost	$6.18/MCF[b]
Electricity cost	$0.060/kWh
Quantity of ash	2005 dry tons
Ash disposal method	Municipal solids waste landfill
Ash hauling/tipping/solid waste fee	$49.24 per wet ton
Source of scrubber water	Plant process water (nonpotable)

[a] wet ton × 0.907 = metric tonne.
[b] MCF = 1000 cu ft = 28 m^3.

3.3 Labor

A total of four trained operators are assigned to the incineration process. Only one operator, however, is scheduled per shift. In addition, there is a shift manager who spends approximately 33% of his/her time per year on incineration-related issues.

Determining their annual compensation is relatively straightforward (i.e., hours worked per year times hourly rate). However, another factor that has to be taken into account is the cost of benefits, including vacation, sick leave, retirement, medical, and health insurance. For this case, the benefits rate has been determined using U.S. Environmental Protection Agency guidelines to be 47.66% of the hourly rate.

TABLE C.3 Total equipment horsepower requirements for two incinerators at Westerly WWTP[a].

	% Operated Estimated power, hp[b]		
Item	Rated	Estimated	Used
1. Quincy air compressor	25	75	19
2. Center drive shaft	20	75	15
3. Induced draft fan	250	75	188
4. Ash bucket elevator	2	75	2
5. Lower ash screw	2	75	2
6. Upper ash screw	2	75	2
7. Upper ash screw	1	75	1
8. Ash roll crusher	10	75	8
9. Burner air fan	75	75	56
10. Combustion air fan	50	75	38
11. Center shaft cooling fan	25	75	19
12. Stationary and rotary blowers	0.125	75	0
13. Long ash screw conveyor	3	75	2
14. Ash bin truck conveyor	3	75	2
15. Soot blower compressor (Operated 10 min/d)	40	75	0
16. Oil cooling fan (center shaft)	0.5	75	0
17. Building drainage sump	150	50	75
18. Process water pump	250	50	125
Totals	909		551

[a] In accordance with a permit condition, the district is only allowed to burn wastewater solids in one of Westerly's incinerators at a time. As a result, a decision was made to set the average incineration related horsepower at 275. Total hp = 275.

[b] hp × 745.7 = W.

3.3.1 *Operators*

The cost of the four operators is calculated below:

Four operators × \$18.20/hr × 1.4766 (benefits) × 2080 hr/yr = \$223,593

3.3.2 *Overtime*

Instead of adding additional operators to the payroll, each of Westerly's WWTP incinerator operators typically works an additional 312 hours (15%) overtime per year. For overtime pay, the fringe benefit multiplier is reduced from 1.4766 to 1.15, which only includes retirement and Medicare. The cost for this overtime is $26,121.

3.3.3 *Shift Manager*

One shift manager × $31.72/hr × 1.4766 benefits × 686 hr/yr = $32,131

3.4 Scrubber Water

Plant process water (i.e., nonpotable water) is supplied to the scrubbers. The only O&M cost associated with the scrubber water is the electrical cost to pump it to the scrubbers. This cost is included in the electricity cost determination contained above.

3.5 Ash Disposal

The Westerly WWTP contains a dry ash handling system. As a result, the ash does not contain any moisture. The dry ash is stored in a silo and hauled to a municipal solids waste landfill two to three times per week. The district has a contract with a local contractor for disposal of the ash at a municipal solid waste landfill located approximately 120 km (75 miles) southeast of the plant. The ash disposal costs, which include hauling costs, landfill tipping fees, and state and local solid waste taxes, averaged $49.24/ton during 2004 (2005 dry tons × $49.24/ton = $98,717).

3.6 Maintenance by Plant Personnel

Plant maintenance is probably the most difficult incineration cost to determine because of variability in accounting methods used by WWTPs. The Westerly WWTP uses a SPL (formally Synergen) computerized maintenance management system program to keep track of the equipment costs and plant personnel related labor costs associated with all of the maintenance work conducted in its incineration complex. Other facilities may use an alternate means of allocating maintenance expenses to various plant processes. Westerly WWTP's maintenance costs $39,589.

3.7 Amortized Capital Costs

Incineration capital improvements need to be amortized over the life of the equipment to determine the annual costs (see Table C.4). A carrying charge factor may be used in cases where incineration facilities have been financed by a combination of cash reserves and long-term debt. Such a carrying charge, expressed as a percentage, typically includes interest on debt and a return on equity.

Table C.4 Amortized capitalized expenses*.

Year	Item 1	Item 2	Item 3	Item 4	Item 5	Item 6	Item 7	Item 8	Totals
1994					$211,790	$8,091	$15,400		$235,281
1995					$211,790	$8,091	$15,400		$235,281
1996					$211,790	$8,091	$15,400		$235,281
1997	$12,000				$211,790	$8,091	$15,400		$247,281
1998	$12,000	$1,530	$4,295		$211,790	$8,091	$15,400		$253,106
1999	$12,000	$1,530	$4,295		$211,790	$8,091	$15,400		$253,106
2000		$1,530	$4,295	$8,333	$211,790	$8,091	$15,400	$9,000	$249,439
2001		$1,530	$4,295	$8,333	$211,790	$8,091	$15,400	$9,000	$249,439
2002		$1,530	$4,295	$8,333	$211,790	$8,091	$15,400	$9,000	$258,439
2003		$1,530	$4,295		$211,790	$9,682	$15,400	$9,000	$251,697
2004		$1,530	$4,295		$211,790	$9,682	$15,400	$9,000	$251,697
2005		$1,530	$4,295		$211,790	$9,682	$15,400	$9,000	$251,697
2006		$1,530	$4,295		$211,790	$9,682	$15,400	$9,000	$251,697
2007		$1,530	$4,295		$211,790	$9,682	$15,400	$9,000	$251,697
2008		$1,530	$4,295		$211,790	$9,682	$15,400	$9,000	$251,697
2009		$1,530	$4,295		$211,790	$9,682	$15,400	$9,000	$251,697
2010		$1,530	$4,295		$211,790	$9,682	$15,400	$9,000	$251,697
2011		$1,530	$4,295		$211,790	$9,682	$15,400	$9,000	$251,697
2012		$1,530	$4,295		$211,790	$9,682	$15,400	$9,000	$251,697
2013		$1,530	$4,295		$211,790	$9,682	$15,400	$9,000	$251,697
2014		$1,530	$4,295		$211,790	$9,682	$15,400	$9,000	$251,697
2015		$1,530	$4,295		$211,790	$9,682		$9,000	$236,297
2016		$1,530	$4,295		$211,790	$9,682		$9,000	$236,297
2017		$1,530	$4,295		$211,790	$9,682		$9,000	$236,297
2018					$211,790	$9,682		$9,000	$230,472

* Item 1 = incinerator cleaning, Item 2 = incinerator equipment rehabilitation, Item 3 = incinerator equipment rehabilitation, Item 4 = incinerator cleaning, Item 5 = renovation of incineration system, Item 6 = purchase of ash trailers, Item 7 = purchase and installation of total hydrocarbon monitors (initial installation), and Item 8 = purchase and installation of total hydrocarbon monitors (replacement installation).

Because Westerly's MHFs were originally constructed in 1978 and the significant equipment had a service life of 20 years, the original construction costs are not included in the 2004 cost determination. However, all costs associated with the 1994 to 1995 improvement program, with refractory replacement work—such as installation of two different sets of total hydrocarbon monitors and the purchase of ash trucks—are included in the amortized capital cost calculation (see Table C.3 for details). Amortized capital costs amounted to $251,697.

3.8 Incineration Cost Summary

Total incineration costs are listed below:

Total incineration costs (includes O&M and amortized capital costs) = $1,154,658
Incineration O&M (does not include amortized capital costs) = $902,961

3.9 Wastewater Solids Management Costs

Wastewater solids unit management costs are typically determined on an O&M wet weight basis. However, these costs can also be determined on a total wet, total dry, and O&M dry weight basis.

Wastewater solids incinerated in 2004: 16 441 wet tons (5426 dry tons)
Total unit disposal cost: $1,154,658 / 16 441 wet tons = $70.23 per wet ton
$1,154,658 / 5426 dry tons = $212.80 per dry ton
O&M unit disposal cost: $902,961 / 16 441 wet tons = $54.92 per wet ton
$902,961 / 5426 dry tons = $166.41 per wet ton
Note that 1 wet ton = 0.907 metric tonne

4.0 BENCHMARKING—COMPARISON WITH INCINERATION COSTS AT OTHER WASTEWATER TREATMENT PLANTS

In a perfect world, comparing a WWTP's total incineration related O&M costs against a similar WWTP's incineration-related O&M costs should reveal which is more cost effectively incinerating its wastewater solids. However, because there are several site-specific factors that can affect O&M costs, a direct WWTP-to-WWTP comparison simply is not realistic. For example,

- Electricity; fuel oil; natural gas; and potable water unit costs; and incinerator ash hauling, tipping, and solid waste fees can vary greatly, depending on the area in the country in which the WWTP is located.
- The amount of fuel required to combust the wastewater solids depends on the kilojoule (British thermal unit) content and the moisture content of the wastewater solids, which are determined by the WWTP's influent, the solids conditioning system (e.g., digestion, chemical, thermal), and the dewatering devices used (e.g., belt filter press, plate and frame press, low solids centrifuge, high solids centrifuge).
- The type of incinerator being used (e.g., fluid bed, multiple hearth, electric arc).
- Whether a waste heat recovery or a waste heat steam generation system is used.
- Wage rates, benefits, and other miscellaneous costs.
- The quantity of wastewater solids being incinerated.

For example, in 2004, the NEORSD's Southerly WWTP incurred an incineration-related O&M unit disposal cost of $21.19 per wet ton, whereas its Westerly WWTP incurred costs of $54.92 per wet ton (see Table C.1 for a comparison of cost differences). The greatest benefit of knowing incineration costs is the ability to track them over time and to use this information to reduce costs.

Figure C.2 is a graph of Westerly WWTP's total incineration-related O&M costs from 1996 to 2004. Figure C.3 is a graph of Westerly WWTP's wastewater solids incineration costs on a unit weight basis.

Given the dearth of information concerning actual costs associated with the operation of wastewater solids incinerators, it is recommended that WWTPs enact a process by which their incineration-related total and unit O&M costs are calculated on a regular basis.

The approach to developing amortized capital costs described above is similar to how accountants might depreciate an incineration "asset" over its useful life. That is, an asset (e.g., pump, reactor, electrical control panel) is typically depreciated, or amortized, by dividing its original installed cost by the number of years assigned as its useful life. This approach ignores the method used to finance the asset and any related interest costs.

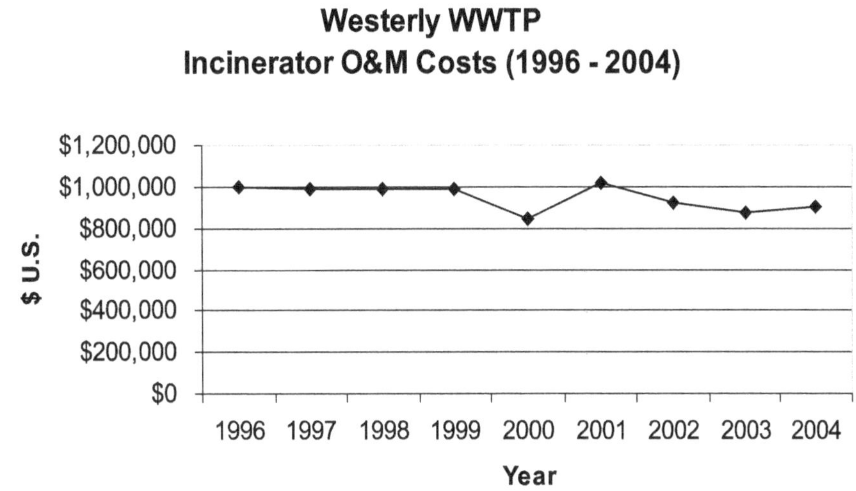

FIGURE C.2 Westerly WWTP incinerator O&M costs (1996–2004).

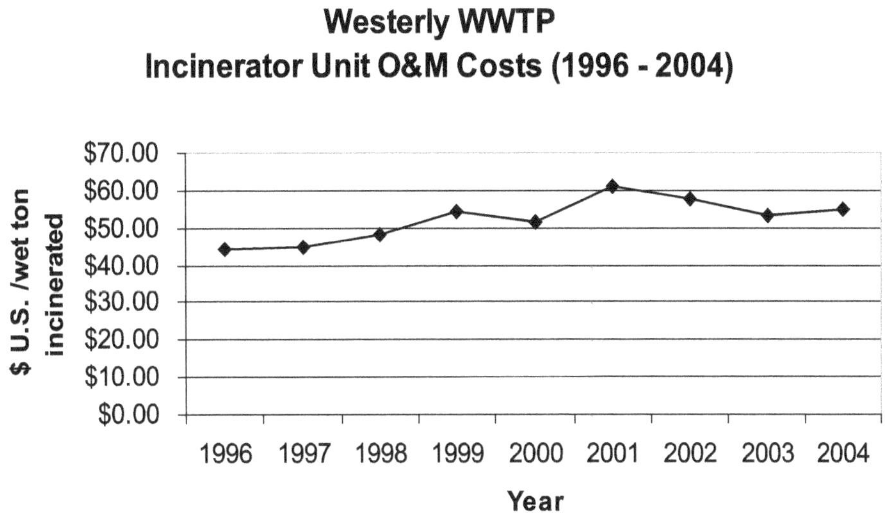

FIGURE C.3 Westerly WWTP incinerator unit O&M costs.

TABLE C.5 Factors that might be used to amortize capital costs.

Useful life	3.0%	3.5%	4.0%	4.5%	5.0%	5.5%	6.0%
5	0.218	0.221	0.225	0.228	0.231	0.234	0.237
10	0.117	0.120	0.123	0.126	0.130	0.133	0.136
15	0.084	0.087	0.090	0.093	0.096	0.100	0.103
20	0.067	0.070	0.074	0.077	0.080	0.084	0.087
25	0.057	0.061	0.064	0.067	0.071	0.075	0.078
30	0.051	0.054	0.058	0.061	0.065	0.069	0.073
40	0.043	0.047	0.051	0.054	0.058	0.062	0.066
50	0.039	0.043	0.047	0.051	0.055	0.059	0.063

Example: Installed cost = $1,000,000 useful life = 10 years interest rate = 4.5%
$1,000,000 × 0.126 = $126,000 annual amortized capital cost

Depending on the intended use of the incineration cost information, it may be appropriate to include interest when amortizing capital costs. If a facility is using cost information to track wastewater solids management costs or compare the cost of various technologies, it is acceptable to exclude interest costs and calculate amortized capital costs as described above. However, if a facility is using cost analysis as part of an annual budget presentation, it may be preferable to acknowledge that bond interest is indeed part of the total financial cost of wastewater solids incinerators.

Table C.5 presents some factors that might be used to amortize capital costs in a manner that includes interest expense.

5.0 REFERENCE

Dominak, R. P.; et al. (2005) *Long-Term Residuals Management Plan for the Northeast Ohio Regional Sewer District;* Northeast Ohio Regional Sewer District: Cleveland, Ohio.

6.0 SUGGESTED READINGS

Harder, S.; Dominak, R. P. (2006) Biosolids Economics: A Case For Careful Analysis. *Proceedings of Water Environment Federation Residuals & Biosolids Management Conference* [CD-ROM]; Cincinnati, Ohio, March; Water Environment Federation: Alexandria, Virginia.

Dominak, R. P.; Stone, L. A. (2003) Long-Term Residuals Management Options for the Northeast Ohio Regional Sewer District. *Proceedings of the 76th Annual Water Environment Federation Technical Exposition and Conference* [CD-ROM]; Los Angeles, California, Oct 11–15; Water Environment Federation: Alexandria, Virginia.

Dominak, R. P.; Stone, L. A. (2002) Residuals Disposal Costs—A Detailed Analysis. *Proceedings of the 75th Annual Water Environment Federation Technical Exposition and Conference* [CD-ROM]; Chicago, Illinois, Sept 29–Oct 2; Water Environment Federation: Alexandria, Virginia.

Appendix D

Case Studies

1.0 HARTFORD WATER POLLUTION CONTROL FACILITY

1.1 Incinerator Upgrades and Improvements

When the Connecticut General Assembly created the Metropolitan District (MDC) in 1929, one of its most important missions was to provide a quality wastewater collection, treatment, and disposal system for the region. In the early 1970s, in conjunction with the growth of the environmental movement in the United States and the advent of the federal Clean Water Act, the Hartford Water Pollution Control Facility (HWPCF) was upgraded to provide secondary treatment. Included in the upgrade was the construction of three multiple-hearth furnaces (MHFs) to burn the approximately 45 tonnes (50 tons) of dry sludge produced by the plant each day.

When the incinerators went online in 1972, they performed extremely well, meeting and often exceeding all federal and state air emissions standards in place. Because of an ongoing preventative maintenance program, improvements to sludge dewatering technology used at the facility and other enhancements, the facility continued to perform effectively through the 1990s.

1.2 Metropolitan District/Department of Environmental Protection Initiative to Upgrade Incinerators

Facing the need to renew incinerator permits issued by the Department of Environmental Protection (DEP), the MDC began to reexamine its entire sludge processing operation to identify the most effective and efficient way to comply with increasingly stringent stack emission regulations and reduce facility odors. As part of that reexamination, a preliminary determination was made that incineration should continue as the primary means of sludge disposal. This decision required significant upgrades to the MHFs to meet air emissions standards and reduce odors in addition to maintaining the integrity of the units and increasing throughput to achieve the most efficient operation possible.

Working in concert with the DEP, the MDC hired Montgomery Watson, an engineering consulting firm nationally known for its expertise in air pollution control and work with MHFs. The engineering consultants examined all of the proven technology used for incineration throughout the country to identify what is referred to in the industry as the best available control technology (BACT). Incinerator number three received immediate corrective repairs to make it more reliable so it could operate while two other incinerators underwent major upgrades. Overall, two incinerators were upgraded, which improved overall operations, increased capacity, reduced natural gas consumption, and increased incinerator reliability. Most importantly, these upgrades have yielded significant and unprecedented reductions in stack emissions.

Specific improvements designed, constructed, and placed in operation to date are categorized in Figure D.1.

1.2.1 *Incinerator Feed System*

To improve delivery of the sludge material to be burned, engineers designed a totally enclosed piped system that delivers sludge to two strategically efficient locations

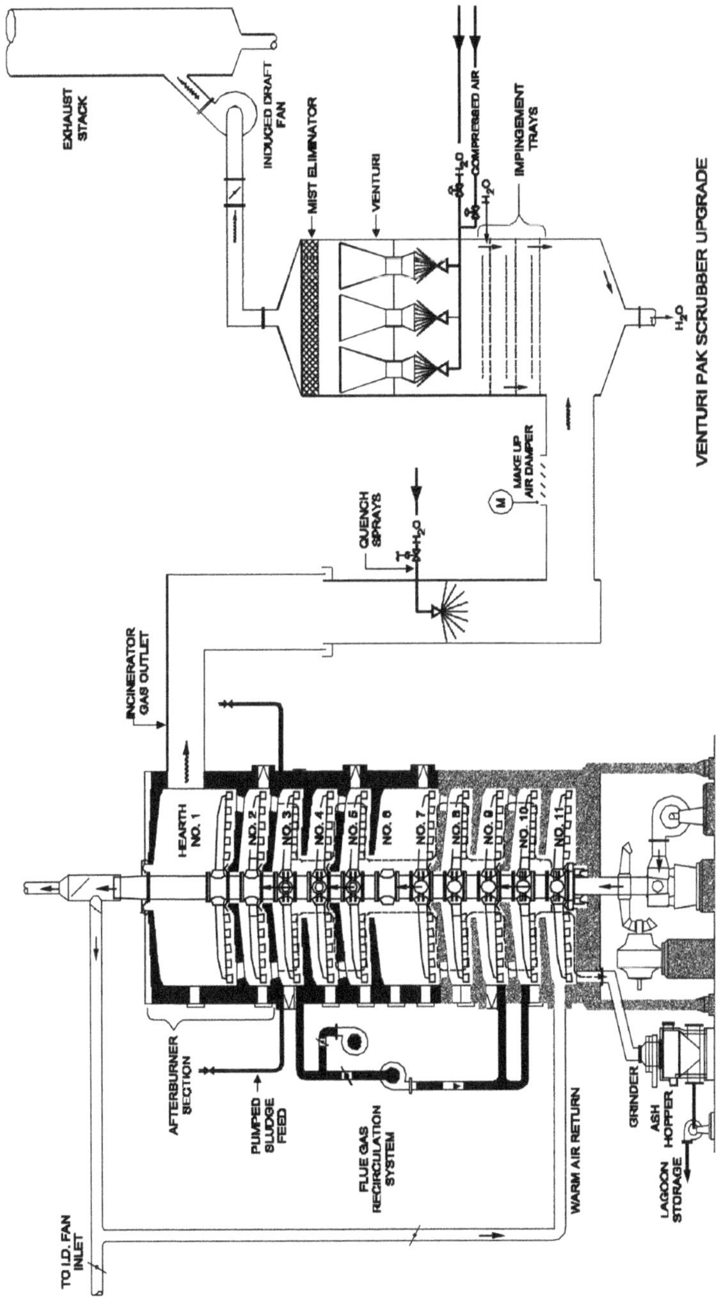

FIGURE D.1 Incinerator number one after reconfiguration (I.D. = induced draft).

within the incinerator to ensure optimum combustion—a key factor in reducing emissions and odors.

1.2.2 *Incinerator Interior Modifications*

The five-story tall incinerators needed new, reconfigured brickwork, a state-of-the-art gasflow recycling system, new burners, upgraded mechanical equipment, and a computer control system. These interior improvements have enhanced the effectiveness and consistency of the combustion process and have reduced fuel requirements.

1.2.3 *Air Pollution Control System*

The existing pollutant scrubbing systems were replaced. A new gas cooling system, state-of-the-art scrubbers, exhaust fans, air compressors, and controls were installed. Substantially upgraded and returned to operation, MDC, its consulting engineers, and the DEP performed stack emissions testing. The results of these tests, as they relate to the DEP's five criteria pollutants, indicated that emissions are now significantly below permit requirements. There are several criteria pollutants: total suspended particulate (TSP); nitrogen oxides (NOx); sulfur oxides (SOx); total hydrocarbons (THC); and carbon monoxide (CO).

In addition to meeting emissions requirements for the five criteria pollutants, the incinerator operation must meet other numerous DEP and U.S. Environmental Protection Agency (U.S. EPA) air regulations. Test results to date indicate that all applicable DEP and U.S. EPA regulations are being met or exceeded. Stack emissions are barely visible, and the operation is essentially odor-free. Since completing the majority of the planned upgrades, passers-by have told MDC staff that it is difficult to tell whether the two upgraded incinerators are even online.

Incineration improvements take on added importance because of environmental concerns and the region's future sludge disposal needs. The MDC is now in the process of upgrading incinerator number three with BACT technology.

1.2.4 *Looking Ahead*

The district's MHFs will be able to provide the region with an environmentally effective and efficient means of processing and disposing of sludge for the useful life of the equipment, conservatively estimated to be 15 years. The district is committed to ensuring that the upgraded incinerators will operate at optimum performance levels and, in doing so, meet or exceed all applicable emissions standards on an ongoing basis. Nonetheless, we will continue to evaluate alternative methods of sludge disposal or reuse as part of our commitment to the environment in general and the neighborhood in which the HWPCF is located.

2.0 REGION OF PEEL LAKEVIEW INCINERATOR

The Region of Peel, with approximately one million inhabitants in 2001, was facing predicted growth of 1.6 million by 2031 because of significant immigration and urbanization. In addition, a consensus had been reached on the pressing need to address odors and reduce truck traffic in an increasingly populated urban community. To address these needs, the team of KMK Consultants, Ltd., and Black & Veatch was commissioned to develop a long-term biosolids management strategy for the region and implement the best option.

The biosolids management strategy study involved compiling potential disposal and recycling approaches. These approaches included beneficial reuse on agricultural lands, co-incineration with municipal solid waste or in an industrial process, incineration with ash disposal, land reclamation and rehabilitation, pelletization with fertilizer distribution, composting, and tree farming. In consultation with the region's steering committee and the public, a selection was made that focused heavily on reducing odors, maximizing reliability, and minimizing capital and operating costs.

Expansion and upgrading of the existing incinerator facility with a fluid bed incinerator at Lakeview Wastewater Treatment Plant (WWTP) was considered the best option for several reasons: (1) a fluid bed incinerator had been successfully implemented 20 years earlier; (2) it minimizes environmental effects of truck traffic; (3) it virtually eliminates odor generation; and (4) it offers a potential reuse of waste heat from the process. See Figure D.2 for a schematic of this fluid bed incinerator.

The Region of Peel; KMK Consultants, Ltd.; Black & Veatch; and a value-engineering consultant proceeded with a hot wind box design in lieu of the existing warm wind box incinerators with waste heat boilers. The existing incinerators used natural gas or oil to assist in the combustion of dewatered biosolids. Some of the waste heat from each incinerator was then captured in a dedicated boiler, converted to steam, and used primarily to feed the thermal conditioning process, a form of wet air oxidation whereby raw sludge is heated to high temperatures and pressures. Despite achieving nearly 40% solids content following vacuum filtration, the existing thermal conditioning process was maintenance-intensive, a significant source of odors, and nearing the end of its useful life.

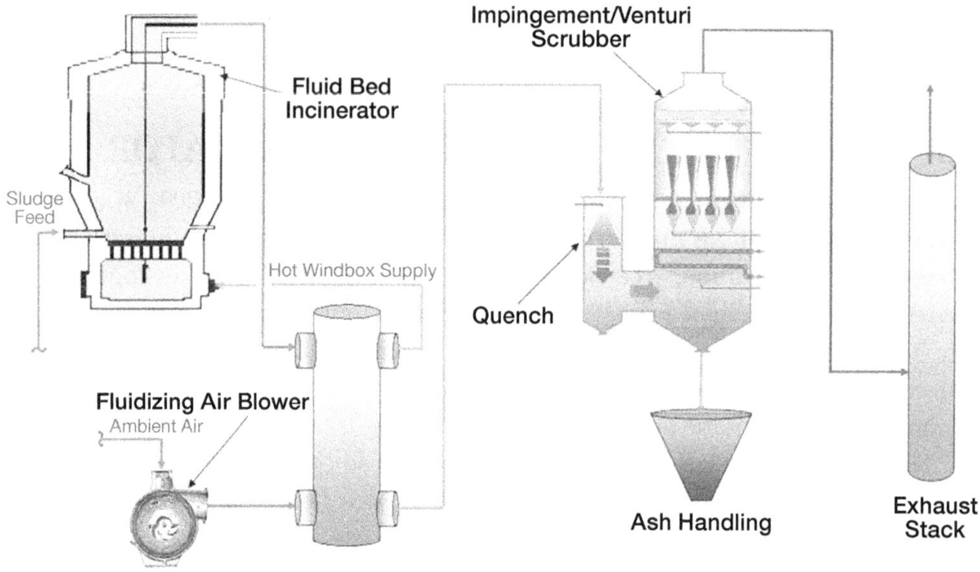

FIGURE D.2 Lakeview WWTP fluid bed incineration process schematic.

The region and its design partners elected to replace both the aging thermal conditioning process and the warm wind box incinerators. The thermal conditioning process was replaced with new high solids dewatering centrifuges. However, with centrifuge discharge solids content of approximately 28%, significant supplemental fuel would be required for combustion if the region was to resort again to warm wind box incineration. Consequently, the region chose to proceed with hot wind box fluid bed incineration to reduce the need for supplemental fuel and simplify the incineration process.

Hot wind box fluid bed incineration uses a heat exchanger that captures heat from the incinerator reactor combustion gases to preheat the incoming fluidizing air in excess of 650°C (1200°F). By introducing centrifuge dewatered biosolids to the fluidized sand bed at these temperatures, the biosolids contain enough heating value to combust autogenously (without auxiliary fuel) and maintain steady-state conditions.

A hot wind box design coupled with a waste heat boiler also was considered. The potential for waste heat recovery was examined for numerous uses, including building heating and cooling, indirect drying, process equipment (fluidizing blower), and power generation (steam turbine). Through extensive evaluation, however, the addition of a waste heat boiler and an induced draft fan to maximize heat recovery was outweighed by several factors. The boiler-fan combination resulted in more complex operations and maintenance requirements; continued reliance on a licensed steam operator; increased training requirements; increased space requirements; increased safety concerns; and involved higher capital costs. The region chose to keep the incinerator design simple and reliable and accept moderate heat recovery with little to no auxiliary fuel costs.

The most recent hot wind box fluid bed incinerator (fluid bed incinerator) addition, installed in a new building, has been operational since December 2005. Currently, the three existing warm wind box incinerators are being replaced with new hot wind box fluid bed incinerators. All incinerators are expected to be operational by 2008.

3.0 METROPOLITAN SANITARY DISTRICT (MSD) OF GREATER CINCINNATI LITTLE MIAMI WASTEWATER TREATMENT PLANT

The MSD selected fluid bed incineration for the Little Miami WWTP to increase solids processing reliability and consistency and reduce maintenance costs (see Figure D.3). The new fluid bed incinerator replaced two MHFs that were originally

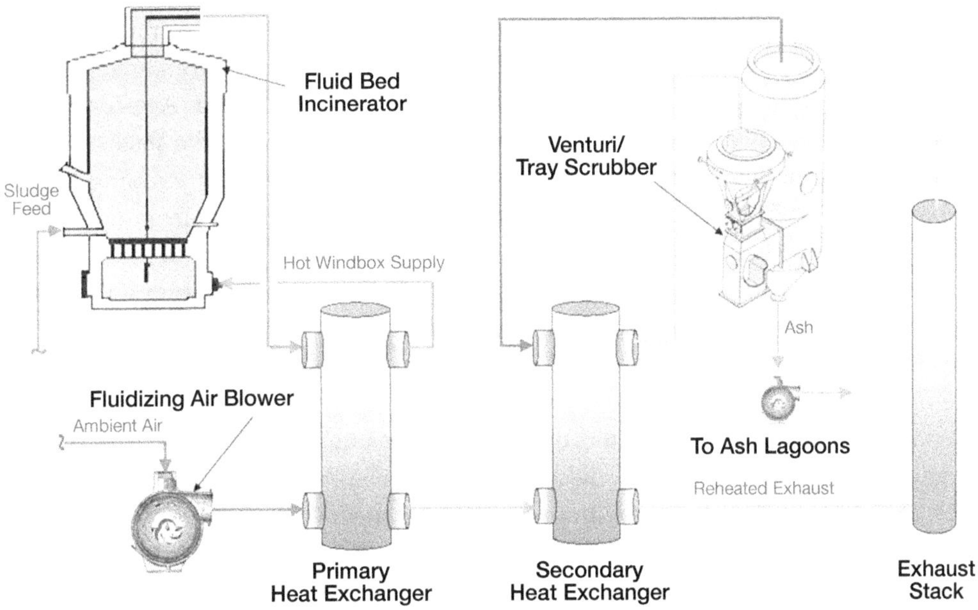

FIGURE D.3 Little Miami WWTP fluid bed incineration process schematic.

designed to burn thermally conditioned biosolids dewatered by vacuum filters. Since that time, thermal conditioning has been removed and the vacuum filters have been replaced by belt filter presses. Because of the resulting changes in biosolids feed and approximately 20 years of operation, the MHFs were in need of upgrades to reliably oxidize the average 27 tonnes (30 tons) per day of dry solids processed and received at the plant. A present worth cost analysis revealed that the most cost-effective solution would be to replace one of the MHFs with a fluid bed incinerator installed in the same location. In addition, modifications would be made to the other MHF to keep it in service during the two-year construction period.

The fluid bed incinerator has been in operation since 2000 and is designed to process up to 65 tonnes (72 tons) of dry biosolids per day. As part of the fluid bed installation, a new dewatered solids receiving facility was installed to receive solids from nearby MSD plants. A combination of the biosolids received and processed at the plant are collected in live bottom bins before incineration. New high-pressure piston pumps transfer and split the biosolids flow into one of four feed points along the perimeter of the fluid bed reactor. This feed arrangement has proven to be effective at maintaining consistent temperatures across the reactor bed.

Following incineration of the biosolids in the fluid bed reactor, exhaust gases are directed through two heat exchangers in series. The first heat exchanger is designed to transfer heat to the fluidizing air before entering the incinerator wind box. Heat transfer is controlled through a bypass damper that allows operators to adjust the amount of fluidizing air that passes through the heat exchanger. A second heat exchanger is designed to transfer heat to the exhaust about to exit the stack, thereby providing plume suppression. Installation of the secondary heat exchanger was important to MSD because they consistently received complaints from neighbors about the plume exiting the stack when it was in operation. In fact, after the fluid bed incinerator system had been commissioned, MSD received a call from a neighbor thanking them for turning off their incinerators.

Because of the success of the fluid bed incinerator installation at Little Miami and the operation and maintenance experience gained over the last six years, MSD is preparing to construct three fluid bed incinerator s at the Mill Creek WWTP, the largest of their plants. Similarly, these fluid bed incinerators will replace six existing MHFs that have been plagued with maintenance issues. The new fluid bed incinerators will be designed to process 87 tonnes (96 tons) per day of dry biosolids each and will be installed in a new building. As at the Little Miami facility, the construction arrangement will allow for the MHFs to remain in operation during the estimated two-year construction period, which was scheduled to begin in spring 2007.

4.0 METROPOLITAN COUNCIL ENVIRONMENTAL SERVICES OF MINNEAPOLIS/ST. PAUL METRO WASTEWATER TREATMENT PLANT

The Metropolitan Wastewater Treatment Plant (MWWTP) serves a population of 1.7 million people and 20 major industries. It is the largest of eight WWTPs operated by the Metropolitan Council Environmental Services (MCES) and treats approximately 850 000 m3/d (200 mgd) on average and processes approximately 180 tonnes (200 tons) of dry solids per day. The new solids management system (SMS) was nearly a decade in the making. Master planning for the metro plant in the mid-1990s identified the need to replace the aging and increasingly inefficient incinerators. After an extensive analysis of a variety of solids processing technologies, the council chose a fluid bed incinerator in 1998. Following further planning and detailed design, three years of construction commenced in fall 2001. The system, constructed in a new building, replaced a solids handling system that included a Zimpro sludge conditioning system, dewatering, and six MHFs. The MCES selected the new SMS to improve operational reliability and reduce costs and odors. The MCES determined that the incineration system would maximize energy recovery and reduce emissions. The existing solids handling system remained in operation during construction and was shut down immediately following commissioning of the SMS.

Commissioning of the SMS began in late 2004 and was completed in early 2005. The SMS is designed to incinerate up to 258 tonnes (285 tons) of dry biosolids per day and produce as much as 5.0 MW of power or 27 200 kg (66 000 lb) per hour of lower pressure steam. As part of the SMS, a new dewatering facility using centrifuges, an incineration system, and an emergency alkaline stabilization system were installed.

The new process consists of three identical fluidized bed combustion trains. Each fluidized bed has the ability to process 95 tonnes (105 tons) per day of wastewater treatment sludge containing 26 to 32% dry solids. Two trains in operation have the capacity to fulfill the plants present needs. The third train is provided for future capacity.

Each fluidized bed combustion train (see Figure D.4) includes a hot wind box reactor that is equipped with ammonia injection to control NOx emissions. The

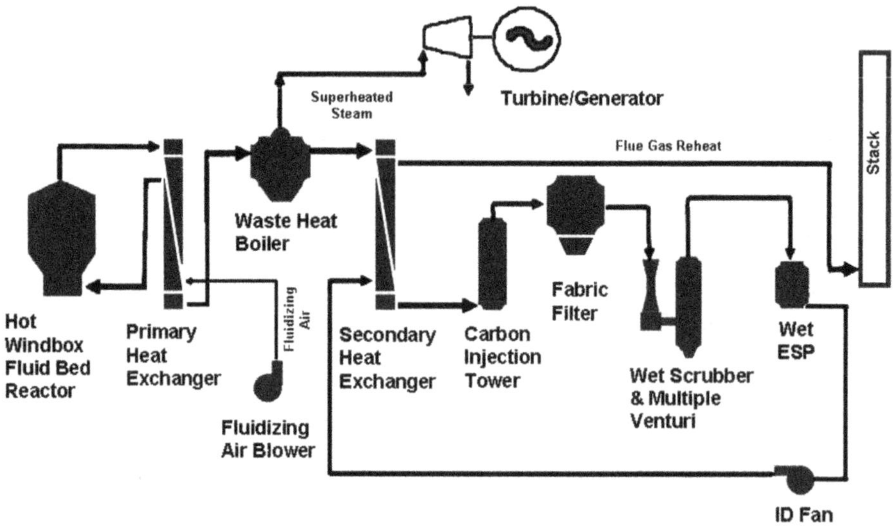

FIGURE D.4 A fluidized bed combustion train (ID = induced draft and ESP = electrostatic precipitator).

exhaust gases exiting the reactor pass through a gas-to-air heat exchanger to preheat the fluidizing air with the reactor exhaust gases. The exhaust gases then pass through a waste heat recovery boiler and economizer where the exhaust gases are cooled and superheated steam is generated. The exhaust gases then pass through a secondary heat exchanger,where they are further cooled and cleaned and reheated to provide plume suppression. The gases then pass through a reaction tower where carbon is injected for mercury removal and a baghouse where particulate matter (including carbon and mercury) is removed. The exhaust gases then pass through a wet scrubber where acid gases are removed, followed by a wet electrostatic precipitator where any remaining fine particulate and mist are removed. The gases finally pass through an induced draft fan, the secondary heat exchanger, and exit the stack. The superheated steam is used to fulfill a number of the MWWTP's energy requirements.

A turbine generator is used to direct the recovered energy for the production of electricity or low-pressure steam for heating and cooling. The new fluid beds are fitted with air distribution systems and process temperature control systems that cause combustion without the need for auxiliary fuel. This is accomplished even with sludge moisture contents as high as 74%.

Emissions from the SMS are much lower than the previous MHF facility, as illustrated in Tables D.1 and D.2.

TABLE D.1 Performance of fluid bed combustion systems.

Performance	Guarantee*	Actual
Throughput	95 tonnes (105 ton) per day	109 to 118 (120 to 130)
Power use	281 kW/tonne (342 hp/ton)	218 to 238 (266 to 290)
Auxiliary fuel	0	0
Caustic consumption	42 L/tonne (10 gal/ ton)	5.8 to 10 (1.4 to 2.4)
Carbon consumption	3 kg/tonne (6 lb/ ton)	1.5 to 2 (3 to 4)
Steam production	5.8 GJ/tonne (5.5 × 106 Btu/ton)	7.9 to 9.5 (7.5 to 9.0)

* tonne(s) are on a dry basis.

5.0 NORTHEAST OHIO REGIONAL SEWER DISTRICT

Faced with potential regulatory changes that could have eliminated incineration as a viable biosolids management option, thereby increasing its management costs, the Northeast Ohio Regional Sewer District (NEORSD) conducted a study to determine how it could most cost-effectively manage its biosolids over the next 25 years. The study's project team investigated various, common biosolids management practices in the United States. These options were compared with one another using standard economic and other criteria. The highest ranked biosolids management alternatives, for a variety of reasons, were landfilling and incineration.

TABLE D.2 Emission rates of multiple-hearth furnaces versus fluid bed reactors.

	Multiple-hearth incinerator 2000 to 2004 (kg/tonne)	Fluid bed reactor 2005 (kg/tonne)	Mass reduction (kg/a)	Reduction (%)	Fluid bed reactor new process % of air permit emission rate limit
Total particulate	0.38	0.0095	25 468	97.50%	< 10%
PM_{10}*	0.415	0.013	28 014	96.90%	< 10%
Lead	0.0005	0.000022	32	96.90%	< 0.1%
Mercury	0.001	0.00001	68	98.80%	< 1%
Carbon monoxide	10.65	0.028	740 159	99.70%	< 10%
Sulfur dioxide	0.175	0.0125	11 323	92.90%	N/A
Nitrous oxides	4.75	0.485	297 191	89.80%	N/A
Hydrochloric acid	0.004	0.00255	100	36.30%	< 10%

* PM_{10} = particulate matter of 10 μm or smaller.

Based on site-specific conditions at the NEORSD's three WWTPs, a detailed evaluation of the potential capital costs and annual O&M costs were developed for the various alternatives. Net present values (NPVs), based on the detailed upfront capital costs and ongoing operations and maintenance (O&M) costs, were developed and used to compare relative cost-effectiveness. Site-specific issues (e.g., permitting requirements, sensitivity to increases in natural gas and transportation costs, staffing and maintenance requirements, biosolids storage requirements, and number of trucks required to transport biosolids) also were analyzed.

The analyses served as the basis for the district's long-term biosolids management plan, which consists of the following:

- Continue incineration of biosolids at the NEORSD's Southerly and Westerly WWTPs, with landfilling as a backup.
- Replace Southerly's four existing MHFs with three new fluid bed incinerators.
- Continue pumping solids from the NEORSD's Easterly WWTP to the Southerly WWTP for processing and management.
- Continue to incinerate biosolids in the Westerly WWTP's two existing MHFs for at least the next 10 years. Reinvestigate potential long-term management alternatives for Westerly WWTP's biosolids in 2012.
- Continue to store ash in the Southerly WWTP's existing ash lagoons and haul ash to a municipal solid waste landfill once every two to three years. Continue hauling Westerly's ash to a municipal solid waste landfill. Investigate potential ways to reduce NEORSD's ash-related landfilling costs, including other potential uses.

Appendix E

Glossary

absolute pressure The sum, at any particular time, of the gauge pressure and the atmospheric pressure.

absolute temperature The temperature of a substance measured on an absolute scale.

absolute zero The temperature (–459.6°F or –273.1°C) at which the molecular motion of a substance theoretically ceases. This is the temperature at which the substance theoretically contains no heat energy.

accm Actual cubic centimeters per minute.

acfm Actual cubic feet per minute.

Actual cubic centimeters per minute (accm) The volume of gas flowing anywhere in a system independent of its density. If the system were moving air at exactly the "standard" condition, then accm would equal standard cubic centimeters per minute.

Actual cubic feet per minute (acfm) The volume of gas flowing anywhere in a system independent of its density. If the system were moving air at exactly the "standard" condition, then acfm would equal standard cubic feet per minute.

adiabatic An adjective pertaining to or designating variations in volume or pressure not accompanied by gain or loss of heat. When a substance undergoes adiabatic expansion, because it does not receive heat from without, the work that it does is at the expense of its internal energy; therefore, its temperature falls. Similarly, when it is adiabatically compressed, its temperature rises.

aerobic bacteria Any bacteria living or occurring in the presence of oxygen.

airborne pollutants Contaminants borne by air that cause harm to human health or the environment.

air-setting refractories Compositions of ground refractory materials that develop a strong bond on drying, including mortars, plastic refractories, ramming mixes, and gunning mixes. They are available in both wet and dry condition, the latter requiring addition of water to develop the necessary consistency.

air-to-fuel ratio The ratio of air supply flow rate to the fuel supply flow rate when both are measured under the same conditions. It is the reciprocal of the fuel-air ratio and may be used interchangeably in qualitative discussions.

anaerobic bacteria Any bacteria that can survive in complete or partial absence of air.

analog Representation of numerical quantities by means of physical characteristics. Hence, an analog input to a computer could be an electrical signal representing flow, temperature, or pressure.

ash Noncombustible mineral matter.

ash fusion temperature The lowest temperature at which the ash can melt and blend into a vitrified substance or clinker; also eutectic point.

atomization The process of breaking a liquid into a multitude of tiny droplets.

atomizing air The part of the air supplied through a burner (typically approximately 10%) that is used to break the oil stream into tiny droplets. The atomizing air is also used for combustion after it has broken up the oil stream.

autogenous combustion The burning of a wet organic material where the moisture content is at such a level that the heat of combustion of the organic material is sufficient to vaporize the water and maintain combustion. No auxiliary fuel is required except for start-up. Also called autothermic combustion.

automation The ability of a control system to react to upstream and downstream variations of a unit process and to maintain a reference standard that defines good performance without human intervention.

available carbon Carbon not combined chemically with oxygen in any way and, therefore, available for combustion.

available heat The gross quantity of heat released within a combustion chamber minus both dry flue gas loss and the moisture loss. It represents the quantity of heat remaining for useful purposes and to balance losses to walls, openings, and conveyors.

available hydrogen Hydrogen not chemically combined with oxygen in any way and, therefore, available for combustion.

bacteria Single-cell microbes that grow in nearly every environment on earth. They are used to study diseases, to produce antibiotics, to ferment foods, to make chemical solvents, and in many other applications.

bioenergy Bioenergy technology in the wastewater process can encompass many diverse technologies that use biological solids to produce energy. Just like biosolids is a subset of residuals, the subset bioenergy would be those processes using biological solids from wastewater treatment processes to produce energy.

biosolids (municipal) Generally used after applicable beneficial recycling criteria have been achieved, i.e., at the outlet of the stabilization process. Common stabilization processes include the following: aerobic digestion; autothermal thermophilic aerobic digestion; anaerobic digestion; composting; alkaline stabilization; thermal drying, including flash, rotary, fluid bed, paddle, hollow-flight, disc, and infrared dryers; thermophilic pozzolanic fixation; acid oxidation/disinfection; and heat treatment/acid digestion. *Biosolids* is intended to be used in reference to municipal/domestic solids. While there are some industrial residuals that can be beneficially recycled, these residuals are generally referred to as *sludge, solids*, etc., as appropriate.

boiler horsepower The equivalent of 9.809 5 kW (33 475 Btu/hr), which is the energy rate needed to evaporate 15.65 kg (34.5 lb) of water at 100°C (212°F) in 1 hour. This is equal to a heat output of 970.2 × 34.5 = 33 471.9 Btu/hr.

bottom ash Non-airborne combustion residue from burning fuel in a furnace. The residue falls to the bottom of the furnace.

breeching A passageway leading from a furnace to its emergency exhaust stack or air emissions control equipment.

British thermal unit (Btu) The mean British thermal unit is the heat required to raise the temperature of 1 lb of water 1°F at a specified temperature (such as 39°F).

burner A device that positions a flame in the desired location by delivering fuel and air to that location in such a manner that continuous ignition is accomplished. Some burners include atomizing, mixing, proportioning, piloting, and flame-monitoring devices.

burner refractory A refractory block with a conical or cylindrical hole through its center that helps to maintain ignition and reduces the probability of flashback or blowoff. It is mounted in such a manner that the flame fires through the hole. Also called burner block, burner tile, combustion tile, combustion block, refractory tile, or refractory block.

calibration The determination, checking, or rectifying of the graduation of an instrument that is providing quantitative measurements.

calorie The mean calorie is a measure of the energy required to raise the temperature of 1 g of water 1°C.

castable refractory A mixture of a heat-resistant aggregate and a heat-resistant hydraulic cement. For use, it is mixed with water and rammed or poured into place. Also called hydraulic setting refractory.

CEMS, CEM/DAS, CEM/DAHS Continuous emission monitoring system, data acquisition system, data acquisition and handling system.

Checksum The total of a group of data items or a segment of data that is used for error checking.

clinkers Ash that is fused together which results from excessive heat or a combination of various compounds within the sludge, causing fusion into large chunks.

closed burner A sealed-in burner that, in most cases, supplies all the air for combustion through the burner itself.

coefficient of transmission The amount of heat transmitted from air to air in one hour per unit area of the wall, floor, roof, or ceiling for a difference in temperature of 1°F between the air on the inside and that on the outside of the wall, floor, roof, or ceiling.

combination burner A burner capable of burning either gas or oil.

combustibles Materials that can be burned.

combustion Burning or rapid oxidation.

combustion products Matter resulting from combustion, such as flue gases, water vapor, and ash.

composite walls Walls made up of a series of materials of various qualities. Used in heating chambers to resist temperature, abrasion, and heat loss.

conductance The amount of heat transmitted from surface to surface in one hour through unit area of a material or construction, whatever its thickness, when the temperature difference is 1°F between the two surfaces.

conduction The transfer of heat through a material by passing it from molecule to molecule.

conductivity A measure of the amount of heat transmitted in one hour through unit area of a homogeneous material of unit thickness for a difference in temperature of 1°F between the two surfaces of the material.

constant relative humidity line Any line on the psychometric chart representing a series of conditions that may be evaluated by one percentage of relative humidity; there are also constant dry-bulb lines, wet-bulb lines, effective temperature lines, vapor pressure lines, and lines showing other physical properties of air mixed with water vapor.

continuous pilot A pilot that burns throughout the entire period that the furnace or oven is in service whether or not the main burner is firing. Also called constant, standby, or standing pilot.

convection The transfer of heat by moving masses of matter. Convection currents are set up in a fluid by mechanical agitation (forced convection) or because of differences in density at different temperatures (natural convection).

course A horizontal layer or row of bricks in a structure.

data highway A shared communications "wire" (copper, fiber) connecting remote parts of a system.

DCS Distributed control system.

DDC Direct digital control.

dewpoint temperature The temperature corresponding to saturation (100% relative humidity) for a given moisture content.

digestion Decomposition of organic waste materials by the action of microbes; the process of wastewater treatment by the decomposition of organic matter.

digital Representation of numerical quantities by discrete levels, or digits conforming to a prescribed scale of notation.

distributed control system (DCS) A computer network using field inputs/outputs to remotely monitor and physically control treatment processes. This is one type of supervisory control and data acquisition (SCADA) system.

draft A difference of pressure that causes a flow of air or gases through a furnace or chimney.

dry-bulb temperature The temperature of the air indicated by any type of thermometer not affected by water vapor content or relative humidity of the air.

dry flue gas Gaseous products of combustion exclusive of water vapor. Separation of the vapor from the flue gas is a theoretical concept that is used in combustion calculations.

dynamic setpoints (As opposed to fixed setpoints in a proportional, integral, derivative system.) Expert automation control strategies can provide a dynamic control system to deal with nonlinear processes where changing conditions do not permit the use of fixed process setpoints. There are several advantages to using a dynamic as opposed to a fixed setpoint system. Many parameters affecting unit process performance can and do change over the course of time, including pH, percent solids, or changing from one digester to another. The control sensors, even with self-cleaning features, may gradually become filmed over, affecting the control in a setpoint based system. To overcome this difficulty, an expert system looks for relative changes, not absolute numbers.

electrostatic precipitator (ESP) A device that collects particulates by placing an electrical charge on them and attracting them onto a collecting electrode.

enthalpy Total heat content above an arbitrary set of conditions chosen as the base or zero point.

entropy A measure of the unavailable energy in a closed thermodynamic system so related to the state of the system that a change in the measure varies with a change in the ratio of the increment of heat taken in to the absolute temperature at which it is absorbed. For practical purposes it is evaluated by dividing the heat content of a unit weight of a substance by its absolute temperature. It is useful in examining changes during a heat cycle. Entropy is constant during a reversible adiabatic change of state.

equilibrium The condition that exists when the walls of a furnace have absorbed all the heat they can hold at a specific furnace temperature so that any further flow of heat to the walls results in an equal amount of heat being transferred to the outside.

ESP Electrostatic precipitator.

eutectic point Temperature at which fractions of the ash soften or melt.

excess air The air remaining after a fuel has been completely burned or that air supplied in addition to the quantity required for stoichiometric combustion.

exit temperature The temperature of combustion gases as they leave a furnace.

exothermic reaction A chemical reaction that liberates heat such as the burning of a fuel.

extraction test procedure A series of laboratory operations and analyses designed to determine whether, under severe conditions, a solid waste, stabilized waste, or landfill material can yield a hazardous leachate.

feedback control Measurement of a process output variable to make an adjustment to an input variable. The simplest form of this is an on-off control. The sensor responds to a variable dropping below a setpoint by turning on a feed device.

fertilizer Any one of several natural and synthetic materials, including manure and nitrogen, phosphorus, and potassium compounds, spread or worked into the soil to increase its fertility.

field device, field element Items located remotely from the central control system.

fireclay brick A refractory brick manufactured substantially or entirely from fireclay.

firing rate The rate at which air, fuel, or a fuel-air mixture is supplied to a burner or furnace. It may be expressed in volume, weight, or heat units supplied per unit time.

flame blowoff The phenomenon that occurs when a flame moves away from a burner. When the fuel-air mixture leaves the burner at a velocity greater than the velocity with which the flame front progresses into the mixture, the flame blows off and is often extinguished.

flame-monitoring device A device for flame surveillance: UV detector, flame rod, flicker detector, infrared detector, photocell, thermopile, or bimetal wrap switch. Also called flame sensor, flame scanner; formerly called flame safety device.

flammability limits The maximum and minimum percentages of a fuel in a fuel-air mixture that will burn. Sometimes called limits of inflammability.

flash point The lowest temperature at which evaporation of a substance produces sufficient vapor to form an ignitable mixture with air near the surface of the liquid.

flocculation The process of forming aggregated or compound masses of particles, such as a cloud or a precipitate.

flue gas All gases that leave a furnace, recuperator, or regenerator by way of the flue, including gaseous products of combustion, water vapor, excess oxygen, and nitrogen.

flue gas analysis A statement of the quantities of the various components of a sample of flue gas, typically expressed in percentages by volume.

flue gas loss The sensible heat carried away by the dry flue gas plus the sensible and latent heats carried away by the water vapor in the flue gas. Also called stack loss.

flue gas desulfurization The operation of removing sulfur oxides from exhaust gas streams of a combustion process.

fluid bed incinerator The reactor is a vertical steel shell, lined with refractory and composed of four sections:

1. Wind box—lower section is the area below the refractory arch distributor.
2. Refractory arch—contains alloy tuyeres or nozzles that allow hot air to be distributed homogeneously throughout the bed.

3. Bed area or combustion zone—section immediately above the distributor that is filled with sand. Air from the distributor causes the bed of sand to fluidize and to mix with the solids.
4. Freeboard or disengagement zone—section above the bed that completes combustion of any volatile hydrocarbons escaping from the bed.

fly ash Airborne combustion residue from burning fuel.

forced draft Difference in pressures that blows air into a furnace, typically produced by a fan located in the inlet air passage to the furnace.

fossil fuel A hydrocarbon fuel, such as petroleum, derived from living matter of a previous geologic time.

friable Easily reduced to a granular or powdery condition.

fugitive emissions Emissions other than those from stacks or vents.

fumes Particles of solid matter resulting from such chemical processes as combustion, explosion, and distillation, ranging from 0.1 to 1.0 μm in size.

furnace An enclosed space in which heat is intentionally released by combustion, electrical devices, or nuclear reaction.

furnace pressure The gauge pressure that exists within a furnace combustion chamber. The furnace pressure is said to be positive if greater than atmospheric pressure; negative if less than atmospheric pressure, and neutral if equal to atmospheric pressure.

greenhouse gases Gaseous components of the atmosphere that contribute to the "greenhouse effect." When sunlight reaches the earth's surface, some is absorbed and warms the earth. Because the earth is much cooler than the sun, it radiates energy at much longer wavelengths than the sun; some of these longer wavelengths are absorbed by greenhouse gases in the atmosphere before they are lost to space. The absorption of this longwave radiant energy warms the atmosphere.

grout A suspension of mortar material in water; of such consistency that when it is poured on horizontal courses of brick masonry, it will flow into vertical open joints.

hardware Physical equipment used for measuring and controlling functions.

HAZOP Hazardous operation analysis.

heat A form of energy generated by the transformation of some other form of energy, as by combustion, chemical action, or friction. According to the molecular theory, heat consists of the kinetic and potential energy of the molecules of a substance. The addition of heat energy to a body increases the temperature or the kinetic energy of motion of its molecules (sensible heat) or increases their potential energy of position but does not increase the temperature, as when melting or boiling occurs (latent heat).

heat capacity The measure of the heat energy required to increase the temperature of a unit quantity of a substance by a certain temperature interval.

heat content The sum total of latent and sensible heats stored in a substance minus that contained in an arbitrary set of conditions chosen as the base or zero point.

heat exchanger Any device for transferring heat from one fluid to another without allowing the fluids to mix.

heat of combustion The heat released by combustion of a unit quantity of a fuel.

heat transfer Flow of heat by conduction, convection, or radiation. Often used to mean heat transfer rate.

heating surface The exterior surface of a heating unit. Two primary types: (1) Extended heating surface (or extended surface), which has air on both sides and is heated by conduction from the prime surface. (2) Prime surface, which has a heating surface with the heating medium on one side and air (or extended surface) on the other.

heavy metals Metallic elements having a high density. They can be precipitated by hydrogen sulfide in acid solution, for example, lead, silver, gold, bismuth, and copper. They can be present in industrial, municipal, and urban runoff.

high fire Input rate to a burner or combustion chamber is at or near its maximum.

high heating value Total heat obtained from combustion of a specified amount of fuel and its stoichiometrically correct amount of air, both being at 16°C (60°F) when combustion starts and the combustion products being cooled to 16°C (60°F) before the heat release is measured. Also called gross heating value.

high-pressure gas Gas at pressures greater than 14 kPa (2 psi).

HMI Human-machine interface.

horsepower A unit to indicate the time rate of doing work equal in the United States to 746 W and nearly equivalent to the English gravitational unit of the same name that equals 550 ft-lb / s or 33 000 ft-lb / m.

humidity Water vapor mixed with dry air in the atmosphere. Absolute humidity refers to the weight of water vapor per unit volume of space occupied. Specific humidity refers to the weight of water vapor carried by unit weight of dry air. Relative humidity is a ratio, typically expressed in percent, used to indicate the degree of saturation existing in any given space resulting from the water vapor present in that space. Relative humidity is either the ratio of the actual partial pressure of the water vapor in the air to the saturation pressure at the dry-bulb temperature or the ratio of the actual density of the vapor to the density of saturated vapor at the dry-bulb temperature. The presence of air or other gases in the same space at the same time has nothing to do with the relative humidity of the space.

ignitable mixture A mixture that, when ignited, is capable of the initiation and propagation of flame away from the source of ignition.

ignition The act of starting or initiating combustion.

incineration An engineered process using controlled flame combustion to thermally degrade waste materials while producing heat, dry inorganic ash, and gaseous emissions.

incomplete combustion Combustion in which fuel is only partially burned and is capable of being further burned under proper conditions.

inputs/outputs (I/Os) Signal inputs and outputs, either analog or digital, to a supervisory control and data acquisition or distributed control system.

induced air Air that flows into a furnace through openings because the furnace pressure is less than the atmospheric pressure. Also, air brought into a furnace by entrainment in a high-velocity system.

induced draft Gas flow caused by a furnace exit pressure less than the furnace pressure. It may be produced by natural or artificial means.

induced draft fan A fan or blower that produces a negative pressure in the combustion chamber either by taking its suction from the combustion chamber or by induced draft.

inorganic material Material derived from nonorganic, or nonliving, sources.

insulating firebrick A refractory brick characterized by low thermal conductivity and low heat capacity.

interlock An electrical, pneumatic, or mechanical connection between elements of a control system that verifies conditions satisfactory to a proper operating sequence that also commands a shutdown of the system when a dangerous or unwanted condition develops.

intermittent pilot A pilot that burns during "lighting-off" and the entire period that the main burner is firing and is shut off with the main burner.

interrupted pilot A pilot that burns during the flame-establishing period or trial-for-ignition period and is cut off (interrupted) at the end of this period. Also called ignition pilot.

I/O Inputs/outputs.

landfill A large, outdoor area for waste disposal; landfills where waste is exposed to the atmosphere (open dumps) are now illegal; in sanitary landfills, waste is layered and covered with soil.

latent heat The heat energy required to cause a change of state at constant temperature, as in the melting of ice or the vaporization of water.

laws of thermodynamics The first law states that the total energy of an isolated system remains constant and cannot be increased or diminished by any physical process. The second law states that no change in a system of bodies that takes place of itself can increase the available energy of a system.

lean mixture A mixture of fuel and air in a premix burner system in which an excess of air is supplied in relation to the amount needed for complete combustion.

lintel A horizontal supporting member spanning a wall opening.

loop drawing A drawing showing the relationship of sensors and read-outs.

low fire The input rate to a burner or combustion chamber is at or near the minimum.

low heating value The high heating value minus the latent heat of vaporization of water vapor formed by combustion of hydrogen in the fuel. For a fuel with no hydrogen, high and low heating values are the same. Also called net heating value.

manometer An instrument for measuring pressures; essentially a U-tube partially filled with a liquid, typically water, mercury, or a light oil, so the amount of displacement of the liquid indicates the pressure being exerted on the instrument.

manual automation A concept that puts decision-making into the hands of the operators, with modern control technology providing them with all of the pertinent data in a user-friendly format. This technology provides the operator with a "feel" for the role that the variables perform in a control algorithm. When the operator becomes familiar with the effect of changing these variables, they can confidently select the variables appropriate for their facility, and can put the unit process into full automation mode when they choose.

manual reset safety shutoff valve A fuel shutoff valve that automatically closes by spring action when its hold-open mechanism is electrically or pneumatically tripped by any connected interlock sensing a dangerous condition. It must be reopened manually after the dangerous condition is rectified and the hold-open mechanism is reenergized. Also called manual reset valve or M-R valve.

MCC Motor control center.

mechanical equivalent of heat The mechanical energy necessary to produce 1 Btu of heat energy.

MHF Multiple-hearth furnace.

MMI Man–machine interface.

Mol = mole = pound mol, the molecular weight of a substance expressed in pounds, e.g., 32 lb O_2 constitute 1 lb mol oxygen (or 32 g O_2 constitute 1 g mol). For perfect gases, a pound mol occupies 379 cu ft at standard temperature and pressure; a gram mol, 22.414 L at standard temperature and pressure.

monolithic lining A furnace lining without joints, formed of material that is rammed, cast, gunned, or sintered into place.

mortar (refractory) A finely ground refractory material that becomes plastic when mixed with water and is suitable for use in laying refractory brick.

MR&R Monitoring, reporting, and recordkeeping.

Multiple-hearth furnace (MHF) A vertical, cylindrical, refractory-lined steel shell furnace. It contains 6 to 12 horizontal hearths and a rotating center shaft with rabble arms. The solids enter the top hearth and flow downward while combustion airflows from the bottom to the top. The rabble arms are shaped to sweep the solids in a spiral motion, alternating in direction from the outside in, to the inside out, between hearths.

MUX/DeMUX Multiplexer-demultiplexer.

natural convection Transfer of heat by currents set up in fluids by differences of density resulting from differences in temperature. Also called free convection.

natural draft A difference in pressure resulting from the tendency of hot gases to rise up a chimney, creating a partial vacuum in the furnace.

neutral atmosphere An atmospheric condition in firing a furnace or kiln that is neither oxidizing nor reducing.

nine-inch equivalent A brick volume equal to that of a standard 9 in × 4.5 in × 2.5 in straight brick; the unit of measurement of brick quantities in the refractories industry.

nonautogenous combustion Combustion that requires continuous addition of auxiliary fuel. Also called subautogenous combustion.

nonlinear variables Many of the physical components and processes within dewatering and thickening systems have a non-linear relationship between each other. For example, chemical dose-rate and solids capture. Historically, in cases where there is a known setpoint at which a process will operate satisfactorily, proportional, integral, derivative (PID) controllers have been used successfully to manage many such operations.

nutrient An element or compound, such as nitrogen, phosphorus, or potassium, that is necessary for plant growth.

organic material Material derived from organic, or living, things; also, relating to or containing carbon compounds.

orsat analyzer An absorption apparatus used to determine the percentages (by volume, dry basis) of CO_2, O_2, and CO in flue gases.

overall coefficient of heat transfer, *U* The coefficient relating the heat transferred from one point to another to the temperature difference between the two points and the cross-sectional area of heat transfer, including the combined effects of several resistances in series as in composite walls with surface and film resistances.

oxidizing atmosphere A furnace atmosphere with an oversupply of oxygen that tends to oxidize materials placed in it.

oxygen sensor A device for measuring the oxygen content of a gas.

PCE Pyrometric cone equivalent.

PCS Plant control system, process control system.

percent air The actual amount of air supplied to a combustion process, expressed as a percentage of the amount theoretically required for complete combustion.

percent excess air The amount of air supplied in excess of that required for complete combustion, expressed as a percentage of the amount required for complete combustion. For example, 20% excess air is 120%.

perfect combustion Combining of chemically correct proportions of fuel and air in combustion so that both the fuel and oxygen are totally consumed.

photocell flame detector A device that generates or rectifies an electric current while exposed to the light from a flame. Failure of the current or lack of rectification may be used to close an automatic fuel shutoff valve.

PIC Products of incomplete combustion.

P&ID Piping and instrument diagram.

PID Proportional, integral, and derivative control algorithm.

pilot A small flame used to light a burner. Also known as continuous pilot, interrupted pilot, and intermittent pilot.

plastic refractory A blend of ground refractory materials in plastic form, suitable for ramming into place to form monolithic linings.

PLC Programmable logic controller.

postpurge An acceptable method for scavenging the combustion chamber, boiler passes, and breeching to remove all combustible gases after flame failure controls have sensed pilot and main burner shutdown and fuel shutoff valves are closed.

power The rate of performing work, expressed in units of horsepower or watts.

preheated air Air heated before its use for combustion. Frequently the heating is done by hot flue gases.

preignition purge An acceptable method for scavenging the combustion chamber, boiler passes, and breeching to remove all combustible gases before the ignition system can be energized. Also called prepurge.

primary treatment The first process in wastewater treatment that removes settled or floating solids.

products of incomplete combustion (PICs) Products of incomplete oxidation of C and H.

programmable logic controller (PLC) An industrial computer that monitors and controls processes. A PLC is designed to exchange inputs/outputs with other instruments and mechanical equipment. A PLC may also contain logic for automating a process.

proportional integral derivative Controllers that use proportional control to provide response to current levels of the control variable; integral control provides response to the history of the control variable; derivative control anticipates future levels of the control variable. The cruise control on an automobile is an example of a PID.

purging The eliminating of an undesirable substance from a pipe, piping system, or furnace by flushing it out with another substance.

purple peeper A device that energizes an electronic current when it "sees" the small amount of ultraviolet radiation that is present in all industrial burner flames. Also called UV flame detector.

pyrolysis Chemical decomposition of a material by heat in the absence of oxygen.

pyrometer An instrument for measuring high temperature.

pyrometric cone One of a series of pyramidal-shaped pieces consisting of mineral mixtures used for measuring time-temperature effect. Each cone is numbered and is of a definite mineral composition so that when heated under standard conditions, it bends at a definite temperature, the pyrometric cone equivalent.

pyrometric cone equivalent (PCE) The number of the standard pyrometric cone whose tip would touch the supporting plaque simultaneously with a cone of the refractory material being investigated when tested in accordance with the method of test for PCE of refractory materials (ASTM Designation C24).

radiation Heat transfer in which the heat travels by electromagnetic waves rather than by conduction or convection.

ramming mix A ground refractory material mixed with water and rammed into place for patching shapes or forming monolithic furnace linings.

reducing atmosphere A furnace atmosphere that tends to remove oxygen from substances placed in the furnace. It may be produced by supplying inadequate air to the burners, intentionally making combustion incomplete.

refractories Highly heat-resistant materials used to line furnaces and kilns.

rich mixture A mixture of fuel and air in a premix burner system in which an excess of fuel is supplied in relation to the amount needed for complete combustion.

rise of an arch The vertical distance between the spring line and the highest point of the undersurface of an arch.

RTU, MTU Remote terminal unit, master terminal unit.

sampling systems Often referred to as sample filtration systems, or sample conditioning systems, sampling systems are designed to prepare a sample for an online instrument before taking a reading by its sensor. The primary objective of a sampling system is to obtain a sample from the process in such a form that it is representative of the material in the process. In addition, sampling systems provide transport and conditioning.

saturated air Air containing as much water vapor as it can hold without any condensing out; in saturated air, partial pressure of the water vapor is equal to vapor pressure at the existing temperature.

SCADA Supervisory control and data acquisition.

sccm Standard cubic centimeters per minute.

scfm standard cubic feet per minute.

scrubbing The removal of impurities from a gas stream by spraying of a fluid.

secondary treatment The wastewater process where bacteria are used to digest organic matter in the wastewater.

sensible heat Heat given up or absorbed by gases between two temperatures that are often widely different.

sensors An element or device that receives information in the form of one quantity, and converts it into information in the form of another quantity. An example is a device to measure the level of suspended solids in a liquid.

skewback The course of brick, having an inclined face, from which an arch is sprung.

sludge Sludge is generally used before applicable beneficial recycling criteria have been achieved, which typically occurs at the outlet of the stabilization process. It should be used in tandem with a specific process descriptor, e.g., primary sludge, waste activated sludge, secondary sludge, etc.

slurry A pumpable mixture of solids and fluid.

smoke Carbon or soot particles smaller than 1.0 μm in size that result from the incomplete combustion of carbonaceous materials.

soak (soaking) To hold the load in a kiln or furnace at one temperature for a time to allow equalization of temperatures throughout the load.

software A set of programs, procedures, or rules that runs in a computer or programmable logic controller.

soldier course A course of brick set on end.

solids Solids, residuals, or another appropriate term should be used for general description, e.g., solids handling, plant solids, etc. In addition, sludge should not be used as a general descriptor, e.g., dewatering and not sludge dewatering.

soot A black substance consisting of small particles of carbon or heavy hydrocarbons that appears in smoke as a result of incomplete combustion.

spalling of refractories The loss of fragments (spalls) from the face of a refractory structure through cracking and rupture with exposure of inner portions of the original refractory mass.

specific gravity The ratio of the weight of a body to the weight of an equal volume of water at some standard temperature, typically 4°C (39.2°F).

specific heat The quantity of heat required to raise the temperature of a substance one degree temperature rise compared with that required for the same weight of water. 1 Btu/lb°F = 1 cal/g°C.

spanner tile A piece of refractory that bridges an opening to provide support above a burner tile.

specific volume The volume of the unit weight of a substance.

spring line or spring of refractory arch The line of contact between the inside surface of an arch and the skewback; typically used to specify the height of an arched refractory chamber.

sprung arch A curved structure supported only by abutments at the side or ends.

SQC/SPC Statistical quality control / statistical process control.

standard air As defined by the American Society of Heating, Refrigerating, and Air-Conditioning Engineers codes, standard air is air weighing 11.996 kg / m^3 (0.748 8 lb / cu ft), which is air at 20°C (68°F) dry-bulb and 50% relative humidity with a barometric pressure of 101.040 kPa (29.92 in Hg). Most engineering tables and formulae involving the weight of air are based on air weighing 12.002 kg / m^3 (0.074 92 lb / cu ft), which is dry air at 21°C (70°F) dry-bulb with a barometric pressure of 101.043 kPa (29.921 in Hg). The error involved in disregarding the difference between the above two weights is slight and, in most instances, may be neglected.

Standard cubic centimeters per minute (sccm) A volumetric flow rate corrected to standard-density conditions.

Standard cubic feet per minute (scfm) A volumetric flow rate corrected to standard-density conditions; scfm is volumetric flow rate at a "standardized" pressure, temperature, and relative humidity. The "standard" ambient conditions are defined by 14.7 psia, some temperature (e.g., 68°F) depending on the "standard" used, and some relative humidity (e.g., 36%, 0%) depending on the "standard" used.

starved air combustion Substoichiometric combustion in which 40 to 80% of the ideal combustion air volume is supplied to the furnace.

static pressure The compressive pressure existing in a fluid. It is a measure of the potential energy of the fluid.

steam turbine A device for converting energy of high-pressure steam (produced in a boiler) into mechanical power which can then be used to generate electricity.

stoichiometric ratio The chemically correct ratio of fuel to air, that is, a mixture capable of perfect combustion with no unused fuel or air.

stretcher A brick laid flat with its length parallel to the face of the wall.

supervisory control system A system providing oversight or higher control over simple functions.

supervisory control and data acquisition (SCADA) A network of field inputs/outputs, programmable logic controllers, and computers used to remotely monitor and physically control treatment processes. Typically expanded to include functions of data collection storage, archiving, display trending, and reporting.

surface conductance The amount of heat transmitted by radiation, conduction, and convection to or from a surface to the air or liquid surrounding it in one hour per unit area of the surface for a difference in temperature of 1°F between the surface and the surrounding air or liquid.

system integrator A company which amasses hardware, software, firmware, and equipment to function as one system.

telemetry Receiving information or data from a distant location.

theoretical air The chemically correct amount of air required for complete combustion of a given quantity of a specified fuel. Also known as stoichiometric air, on-ratio air, correct air, or ideal air.

therm Symbol used in the gas industry representing 100 000 Btu.

thermal conductivity, *k* The quantity of heat that flows in one second across a slab of unit area and unit thickness when the temperatures of the faces of the slab differ by 1°F. Also called coefficient of thermal conduction, specific thermal conductivity.

thermal resistance The reciprocal of conductance.

thermal resistivity The reciprocal of conductivity.

thermal shock A sudden temperature change.

ton A unit of mass. In the United States, it typically refers to the short ton, which is 2000 lb. In the United Kingdom, a ton typically refers to the long ton and is 2240 lb.

transducer A sensor, such as pressure, temperature. A field element.

transmitter A device that converts a process signal from a primary sensor such as pressure, temperature, or flow to an analog value.

turbulence A state of being highly agitated. In turbulent fluid flow, the velocity of a given particle changes constantly both in magnitude and direction.

ultimate analysis A statement of the quantities of the various elements of which a substance is composed, typically expressed in percentages by weight.

unit process A self-contained process, which may be an element of a larger process. For example, a thickening unit process is an element of the larger wastewater process.

volumetric analysis A statement of the various components of a substance (typically applied to gases only), expressed in percentages by volume.

wall loss The heat lost from a furnace or tank to or through its walls.

weight flow rate The quantity (measured in units of weight) of a fluid flowing per unit of time.

wet-bulb temperature The lowest temperature that a water-wetted body will attain when exposed to an air current. This is the temperature of adiabatic saturation.

Appendix F

Tables and Conversions

TABLE F.1 Properties of saturated air at different temperatures (range 32°F to 211°F).

Temperature (°F)[a]	Humidity (lb H_2O/lb dry air)[b]	Volume (cu ft/lb dry air)[c]	Enthalpy[e] (Btu/lb dry air)[d]
32	0.003 79	12.46	11.76
33	0.003 94	12.49	12.17
34	0.004 11	12.52	12.58
35	0.004 27	12.55	13.01
36	0.004 45	12.58	13.44
37	0.004 63	12.61	13.87
38	0.004 82	12.64	14.32
39	0.005 01	12.67	14.77
40	0.005 21	12.69	15.23
41	0.005 42	12.72	15.70
42	0.005 64	12.76	16.17
43	0.005 85	12.78	16.66
44	0.006 09	12.82	17.15
45	0.006 33	12.85	17.65
46	0.006 58	12.88	18.16
47	0.006 84	12.91	18.68
48	0.007 10	12.94	19.21
49	0.007 37	12.97	19.75
50	0.007 66	13.00	20.30
51	0.007 95	13.03	20.86

(continued on next page)

TABLE F.1 Properties of saturated air at different temperatures (range 32°F to 211°F) *(continued).*

Temperature (°F)[a]	Humidity ($lb\ H_2O$/lb dry air)[b]	Volume (cu ft/lb dry air)[c]	Enthalpy[e] (Btu/lb dry air)[d]
52	0.008 26	13.06	21.44
53	0.008 57	13.10	22.02
54	0.008 89	13.13	22.62
55	0.009 23	13.16	23.22
56	0.009 58	13.19	23.84
57	0.009 93	13.23	24.48
58	0.010 30	13.26	25.12
59	0.010 69	13.29	25.78
60	0.011 08	13.33	26.46
61	0.011 49	13.36	27.15
62	0.011 91	13.40	27.85
63	0.012 35	13.43	28.57
64	0.012 80	13.47	29.31
65	0.013 26	13.50	30.06
66	0.013 74	13.54	30.83
67	0.014 24	13.58	31.62
68	0.014 75	13.61	32.42
69	0.015 28	13.65	33.25
70	0.015 82	13.69	34.09
71	0.016 39	13.72	34.95
72	0.016 97	13.76	35.83
73	0.017 57	13.80	36.74
74	0.018 2	13.84	37.66
75	0.018 8	13.88	38.61
76	0.019 5	13.92	39.57
77	0.020 2	13.96	40.57
78	0.020 9	14.00	41.58
79	0.021 6	14.05	42.62
80	0.022 3	14.09	43.69

Temperature (°F)[a]	Humidity (lb H_2O/lb dry air)[b]	Volume (cu ft/lb dry air)[c]	Enthalpy[e] (Btu/lb dry air)[d]
81	0.023 1	14.13	44.78
82	0.023 9	14.17	45.90
83	0.024 7	14.22	47.04
84	0.025 6	14.26	48.22
85	0.026 4	14.21	49.43
86	0.027 3	14.35	50.66
87	0.028 2	14.40	51.93
88	0.029 2	14.45	53.23
89	0.030 2	14.50	54.56
90	0.031 2	14.55	55.93
91	0.032 2	14.60	57.33
92	0.033 3	14.65	58.78
93	0.034 4	14.70	60.25
94	0.035 6	14.75	61.77
95	0.036 7	14.80	63.32
96	0.037 9	14.86	64.92
97	0.039 2	14.91	66.55
98	0.040 5	14.97	68.23
99	0.041 8	15.02	69.96
100	0.043 2	15.08	71.73
101	0.044 6	15.14	73.55
102	0.046 1	15.20	75.42
103	0.047 6	15.26	77.34
104	0.049 1	15.32	79.31
105	0.050 7	15.39	81.34
106	0.052 3	15.45	83.42
107	0.054 0	15.52	85.56
108	0.055 8	15.59	87.76
109	0.057 6	15.65	90.03

(continued on next page)

TABLE F.1 Properties of saturated air at different temperatures (range 32°F to 211°F) *(continued)*.

Temperature (°F)[a]	Humidity (lb H_2O/lb dry air)[b]	Volume (cu ft/lb dry air)[c]	Enthalpy[e] (Btu/lb dry air)[d]
110	0.059 4	15.72	92.34
111	0.061 4	15.80	94.72
112	0.063 3	15.87	98.18
113	0.065 4	15.94	99.71
114	0.067 5	16.02	102.31
115	0.069 6	16.10	104.98
116	0.071 8	16.18	107.73
117	0.074 2	16.26	110.55
118	0.076 5	16.34	113.46
119	0.079 0	16.43	116.46
120	0.081 5	16.52	119.54
121	0.084 1	16.61	123.72
122	0.086 8	16.70	125.98
123	0.089 6	16.79	129.35
124	0.092 4	16.83	132.8
125	0.095 4	16.98	136.4
126	0.098 4	17.09	140.1
127	0.101 6	17.19	143.9
128	0.104 8	17.29	147.8
129	0.108 2	17.40	151.8
130	0.111 6	17.52	155.9
131	0.115 2	17.63	160.3
132	0.118 9	17.75	164.7
133	0.122 7	17.87	169.3
134	0.126 7	17.99	174.0
135	0.130 8	18.12	178.9
136	0.135 0	18.25	183.9
137	0.139 3	18.39	189.0
138	0.143 9	18.53	194.4
139	0.148 5	18.67	199.9

Temperature (°F)[a]	Humidity (lb H_2O/lb dry air)[b]	Volume (cu ft/lb dry air)[c]	Enthalpy[e] (Btu/lb dry air)[d]
140	0.153 4	18.82	205.7
141	0.158 4	18.97	211.6
142	0.163 6	19.13	217.7
143	0.168 9	19.29	224.1
144	0.174 5	19.46	230.6
145	0.180 3	19.63	237.4
146	0.186 2	19.81	244.4
147	0.192 4	19.99	251.7
148	0.198 9	20.18	259.3
149	0.205 5	20.38	267.1
150	0.212 5	20.58	275.3
151	0.219 7	20.79	283.6
152	0.227 1	21.01	292.4
153	0.234 9	21.23	301.5
154	0.243 0	21.47	310.9
155	0.251 4	21.71	320.8
156	0.260 2	21.96	331.0
157	0.269 3	22.22	341.7
158	0.278 8	22.49	352.7
159	0.288 7	22.77	364.2
160	0.299 0	23.07	376.3
161	0.309 8	23.37	388.8
162	0.321 1	23.69	402.0
163	0.332 9	24.02	415.7
164	0.345 2	24.37	429.9
165	0.358 1	24.73	445.0
166	0.371 6	25.11	460.7
167	0.385 8	25.51	477.2
168	0.400 7	25.92	494.4
169	0.416 3	26.36	512.4

(continued on next page)

TABLE F.1 Properties of saturated air at different temperatures (range 32°F to 211°F) *(continued)*.

Temperature (°F)[a]	Humidity ($lb\ H_2O$/lb dry air)[b]	Volume (cu ft/lb dry air)[c]	Enthalpy[e] (Btu/lb dry air)[d]
170	0.432 7	26.81	531.5
171	0.450 0	27.29	551.5
172	0.468 2	27.79	572.7
173	0.487 5	28.33	594.9
174	0.507 8	28.89	618.3
175	0.529 2	29.48	643.2
176	0.551 9	30.10	669.4
177	0.576 0	30.76	697.3
178	0.601 6	31.46	726.9
179	0.628 8	32.21	758.3
180	0.657 8	33.00	791.8
181	0.688 7	33.84	827.4
182	0.721 8	34.74	865.7
183	0.757 2	35.71	906.5
184	0.795 3	36.74	950.5
185	0.836 3	37.85	997.7
186	0.880 5	39.05	1049
187	0.928 3	40.35	1104
188	0.980 2	41.76	1164
189	1.037	43.29	1229
190	1.099	44.96	1301
191	1.166	46.79	1378
192	1.241	48.81	1464
193	1.324	51.04	1559
194	1.416	53.52	1666
195	1.519	56.29	1784
196	1.635	59.42	1918
197	1.767	62.96	2069
198	1.917	67.01	2243

Temperature (°F)[a]	Humidity (lb H_2O/lb dry air)[b]	Volume (cu ft/lb dry air)[c]	Enthalpy[e] (Btu/lb dry air)[d]
199	2.091	71.68	2443
200	2.295	77.14	2677
201	2.532	83.43	2948
202	2.820	91.2	3279
203	3.170	100.5	3684
204	3.61	112.4	4189
205	4.18	127.6	4849
206	4.94	148.0	5710
207	6.00	176.6	6941
208	7.59	218.8	8760
209	10.25	290.4	11 840
210	15.54	432.9	17 931
211	30.45	831.9	35 068

[a] (°F − 32) × 0.555 = °C.

[b] lb/lb × 1000 = g/kg.

[c] cu ft/lb × 0.062 43 = m^3/kg.

[d] Btu/lb × 2.326 = kJ/kg.

[e] Enthalpy basis: 0°F.

TABLE F.2 Heat content of various gases at different temperatures (Btu / lb)[a] (courtesy of Fluidized Combustion Systems, Inc., Oakbrook, Illinois).

	Combustion gas													
Temperature (°F)[b]	Air	Sludge organics	No. 2 fuel oil	No. 6 fuel oil	Natural gas	N_2	O_2	CO_2	H_2O	SO_2	CO	CH_4	H_2	HCl
70	0	0	0	0	0	0	0	0	0	0	0	0	0	0
100	7.3	69.7	85.2	75.4	134.3	7.5	6.6	6.3	1067.2	4.6	7.5	16.8	103.0	5.7
150	19.4	82.1	97.9	88.0	147.4	19.9	17.7	16.8	1088.1	12.3	20.0	44.8	274.8	15.3
180	26.6	89.6	105.4	95.5	155.2	27.4	24.4	23.1	1100.1	16.9	27.5	61.6	377.8	21.0
200	31.5	95.0	111.1	101.0	161.5	32.4	28.8	27.3	1116.4	20.0	32.5	72.8	446.5	24.8
300	55.8	120.2	131.9	126.3	187.7	57.4	51.4	49.9	1154.8	36.3	57.5	134.9	792.5	44.0
400	80.2	146.1	162.8	152.4	215.1	82.5	74.1	72.8	1201.9	52.9	82.7	198.5	1138.9	63.2
450	92.6	159.3	176.2	165.6	228.9	95.1	85.8	85.1	1225.5	61.7	95.5	234.3	1312.3	72.8
500	105.0	172.5	189.6	178.9	242.9	107.8	97.5	97.5	1249.1	70.5	108.2	270.0	1485.7	82.5
600	129.9	199.2	216.6	205.7	271.0	133.3	121.1	122.5	1296.8	88.3	134.0	343.8	1832.9	101.8
700	155.4	226.4	244.1	233.0	299.5	159.1	145.2	148.5	1345.2	106.8	160.1	423.9	2180.9	121.2
800	181.2	254.0	272.1	260.7	328.5	185.3	169.6	175.0	1394.3	125.6	186.6	508.0	2529.6	140.8
900	207.2	282.1	300.4	288.8	357.9	211.8	194.4	202.1	1444.3	144.7	213.3	596.5	2878.9	160.6
1000	233.5	310.5	329.2	317.4	387.7	238.6	219.5	229.9	1495.1	164.1	240.4	689.3	3229.8	180.5
1100	259.9	339.2	358.2	346.2	417.8	265.6	244.8	258.0	1546.8	183.7	267.8	784.1	3581.4	200.4
1200	286.9	368.5	387.8	375.5	448.5	293.0	270.5	286.8	1599.5	203.7	295.6	886.2	3935.8	220.8
1300	313.9	397.9	417.5	405.0	479.3	320.5	296.2	315.9	1653.0	223.8	323.5	989.6	4291.0	241.3
1400	341.4	427.9	447.8	435.0	510.7	348.5	322.4	345.6	1707.5	244.2	352.0	1098.0	4650.6	262.2
1500	369.0	458.2	478.5	465.3	542.6	376.8	348.7	375.5	1763.4	264.7	380.6	1208.1	5009.5	183.2
1600	397.0	488.8	509.4	496.2	574.7	405.4	375.3	405.8	1819.3	285.5	409.6	1322.7	5366.6	304.2
1700	425.2	519.7	540.7	527.1	607.2	434.1	401.9	436.3	1877.2	306.4	438.6	1439.5	5732.9	325.7
1800	453.7	550.6	571.9	558.1	639.6	462.8	428.5	467.0	1934.9	327.2	467.8	1558.6	6109.1	347.2
1900	482.4	582.0	603.7	589.6	672.7	492.0	455.5	498.2	1993.9	348.3	497.3	1680.3	6484.7	369.0
2000	511.1	613.7	635.8	621.4	706.1	521.4	482.6	529.7	2053.6	369.4	526.9	1803.2	6860.2	390.9
2100	540.0	645.6	668.0	653.4	739.6	551.0	509.8	561.2	2114.0	390.6	556.7	1929.4	7238.7	413.1
2200	568.9	677.4	700.2	685.3	773.2	580.5	537.0	592.7	2174.8	411.8	586.7	2056.5	7618.0	435.3
2300	598.3	709.7	732.9	717.7	807.3	610.4	564.4	624.6	2237.1	432.9	616.8	2187.2	8002.9	457.8
2400	627.5	742.0	654.6	750.0	841.3	640.3	591.8	656.4	2299.3	454.1	646.9	2318.9	8387.5	480.4
2500	656.6	774.3	798.2	782.4	875.3	670.0	619.0	689.0	2361.0	475.1	676.9	2453.3	8771.8	503.3

[a] Btu / lb × 2.326 = kJ / kg.
[b] (°F − 32) × 0.555 = °C.

TABLE F.3 Densities of exhaust gas components at standard conditions (courtesy of Degremont Technologies – Infilco).

Gas component	Gas density (kg/m³)[a]	Gas density (lb/cu ft)[b]
Air	1.2910	0.0748
H_2O	0.8037	0.0466
N_2	1.2498	0.0724
O_2	1.4276	0.0827
CO_2	1.9634	0.1138
SO_2	2.8580	0.1656
HCl	1.6266	0.0943
HBr	3.6097	0.2092
CO	1.2496	0.0724
CH_4	0.7157	0.0415
C_2H_6	1.3415	0.0777
C_3H_8	1.9673	0.1140
C_6H_{14}	3.8447	0.2228
NO_2	2.0524	0.1189

[a] At 0°C, 1.013 bar.
[b] At 70°F, 14.69 psig.

TABLE F.4 Conversion of main air pollutant emission concentrations (courtesy of Degremont Technologies – Infilco).

Air pollutants	Multiply	Abbreviation	by	To obtain
Particulates	grain per dry standard cubic foot	gr/dscf	2.465	mg/Nm³
	pound per dry ton sludge	lb/dry ton	0.5	kg/dry tonne
Carbon monoxide (CO)	part per million by volume	ppmv	1.250	mg/Nm³
Total organic as methane (CH_4)	part per million by volume	ppmv	0.716	mg/Nm³
Sulfur dioxide, (SO_2)	part per million by volume	ppmv	2.858	mg/Nm³
Hydrochloric acid (HCl)	part per million by volume	ppmv	1.627	mg/Nm³
Nitrogen dioxide as NO_2	part per million by volume	ppmv	2.052	mg/Nm³

Index

D

E

www.ingramcontent.com/pod-product-compliance
Lightning Source LLC
Chambersburg PA
CBHW080703050126
37695CB00002B/4
9781572783898